A BRIEF HISTORY OF SCIENCE

As seen through the development of scientific instruments

By the same author

The Law for Everyman
(Collins, 1963)

Man and his Kind
(Darton, Longman & Todd, 1973; Praeger, 1973)

The Phenomenon of Money
(Routledge & Kegan Paul, 1980)

The Death of an Emperor
(Constable, 1989; OUP Paperback, 1990)

The Anthropology of Numbers
(Cambridge University Press, 1991)

The Japanese Numbers Game
(Routledge, 1992)

Solar Eclipse
(Constable, 1999)

A BRIEF HISTORY OF SCIENCE

As seen through the development of scientific instruments

Thomas Crump

CARROLL & GRAF PUBLISHERS
New York

Carroll & Graf Publishers
An imprint of Avalon Publishing Group, Inc.
161 William Street
New York
NY 10038 2607
www.carrollandgraf.com

First published in the UK by Constable,
an imprint of Constable & Robinson Ltd 2001

First Carroll & Graf edition 2001

ISBN 0–7867–0907-3

Printed and bound in the EU

Library of Congress Cataloging-in-Publication Data is available on file.

Contents

Illustrations

In the text

Plates

Preface

IN WRITING this book I have had to make up for much lost time. In about 1960 the history of science, as a topic of general interest, lost its appeal to authors. In 1957 Professor Herbert Butterfield, a historian rather than a scientist, had published *The Origins of Modern Science 1300–1800*,[1] which in England was almost certainly the last book of its kind. One reason is given in the introduction to A. R. and M. B. Hall's *Brief History of Science*[2] (the most recent American book, but still nearly forty years old), where the authors note that 'if this book were planned according to the volume of scientific discovery in different periods, everything before 1800 would have to be summarized on the first page'. So much for Butterfield, who stopped at 1800. Now, forty years later, the first page would have to be reduced to the first paragraph. In any case, if one is to stop at 1800, few later studies can equal William Whewell's immensely influential *History and Philosophy of the Inductive Sciences*, finally published in the complete edition in 1860.

In the twentieth century the landmark study was W. C. Dampier's monumental *History of Science*, which ran to four editions after first being published in 1929. This allowed the author to take into account the quantum revolution of the 1920s. Working in Cambridge, Dampier also had the advantage of knowing many of the leading actors. With the last edition in 1948 (by which time J. J. Thomson, Lord Rutherford and many others whom Dampier had known had died) he had become a stranger in the new world of science. He could not otherwise have written in his new introduction that 'some new work, especially in England and America, done to solve definite war problems, has led incidentally to an increase in scientific knowledge' – an incredible understatement.

What all this involves for an author writing at the beginning of the twenty-first century can be seen by comparing the summary introduction to particle physics, presented by Dampier, with Gerard 't Hooft's *In Search of the Ultimate Building Blocks*,[3] a popular study of the same subject, from the same publisher, but two generations on in time. The world of elementary particles that 't Hooft presents to the *general* reader is complex in a way that Dampier could hardly have conceived of.

The difference between the two reflects more than an incidental increase in scientific knowledge. The transformation in scale, both of the international scientific establishment, and of the equipment and apparatus at its disposal, is fundamental. Dampier, if he wished to observe the state of the art in contemporary physics, needed only a five-minute walk from his college to the old (and original) Cavendish Laboratory, where sooner or later he could be sure of meeting almost all the leading men in the field. 't Hooft's world of science is defined by CERN (just outside Geneva) or the Fermilab (just outside Chicago), and if he is in Cambridge he will go to the new Cavendish Laboratory – a whole scientific campus on the edge of town (where the museum still contains such treasures from the Old Cavendish as the apparatus James Chadwick used to discover the neutron in 1932).

't Hooft's book is exemplary for today's presentation of science to the public: a study of some particular aspect of science, whether molecular biology or quantum mechanics, now defines the realm of popular science. Many such studies, often written by leading men in the field, such as 't Hooft, have been useful sources for the present book. There is, however, a problem: they can only demand from readers the most limited competence in mathematics, the fundamental tool of almost all science – as witness the fact that the computer is today indispensable for any scientist. But as 't Hooft notes in the introduction to his book, 'to really appreciate the rock-solid logic of the laws of physics, one cannot actually avoid math.' None the less he does do so, but reluctantly: in this I follow him, more or less. In some places, however, such as in my treatment of Niels Bohr (1885–1962) (a great scientist by any standard), I stray from the narrow path. In doing so, the mathematics introduced is much less forbidding than it appears at first sight. It often involves little more than counting in a systematic way. At the same time it is not only mathematics that the general reader finds daunting: the curious reader need only look up 't Hooft's Table 6,[4] which I forbear to include in my text. Chemical formulae, which present much the same problem as the mathematics of science, are impossible to avoid: the Appendix is designed to make them intelligible.

The very complexity of the world of 't Hooft (and other Nobel prizewinners of the last half century) explains why, if authors still write general histories or art, philosophy, warfare, or whatever, science has become simply too daunting. There has been just too much of it in the last fifty years. No-one can come to terms with the whole of science, so it is hardly surprising that no general history of science, up to the beginning of the third millennium, is to be found in publishers' catalogues. At the same time, the history of science, as a university study, has been served by a number of texts, such as the two books by Abraham Pais listed in the Bibliography, which are only intelligible to specialists.

In the twenty-first century it is perhaps foolhardy even to try writing a general history of science. The task would be impossible without guidelines, particularly in relation to the last 200 years. To start with, there is the problem that science is so much a part of everyday life that we take it for granted. While we cannot conceive of a world without science, we have no difficulty in accepting that science, as we know it today, is the product of very recent historical times. The word itself is evocative, and what it evokes, almost immediately, are institutions of the present day – like CERN,[5] where, in a 26 kilometre circular tunnel, deep underground, elementary particles, with energies measured in billions of electron-volts, collide with each other in experiments in particle physics. Such institutions belong to 'Big Science',[6] which only took off in World War II. My final chapter is an attempt to show what this has meant for the world of science in the last fifty years.

Although the instrumentation of big science, such as Ernest Lawrence's cyclotron – the original particle accelerator – began to appear in the 1930s, this book takes 2 December 1942 as the day when big science came into its own. On that day, at 4 p.m., Enrico Fermi's atomic pile, painstakingly constructed in a doubles squash court at the University of Chicago, went critical for the first time. This landmark event in the history of science was witnessed by a remarkable number of the world's top physicists, who were working together with one common purpose – to be ahead of Nazi Germany in developing a nuclear bomb. (In the event, the bomb that came as a result of that seminal day in Chicago destroyed a Japanese city, Nagasaki, on 9 August 1945 – one of the most destructive incidents in history of science.[7])

The emergence of big, or post-modern,[8] science was to bring to an end a period of what I choose simply to call modern science (which, essentially, was what Dampier was writing about). This I take to begin with the publication, in 1543, of Nicolaus Copernicus' (1473–1543) *De Revolutionibus Orbium Coelestium*, a book stating for the first time a fundamentally correct astronomy of the solar system.

The new Copernican astronomy would become a *cause célèbre* when Galileo, using the newly invented telescope to observe the night sky, discovered the essential truth in it, which established a new canon for scientific thought and practice. The debate started by Galileo would be fought out in public, and until well into the twentieth century leading scientists expected to share their results, not only with professional colleagues, but with educated people generally. This was certainly the approach of Ernest Rutherford (1871–1937), who discovered the atomic nucleus in 1910.

Although this book's main focus is on modern science, as it evolved during the period of nearly 400 years separating the publication of Copernicus' *De Revolutionibus* and Fermi's atomic pile going critical in Chicago, the first chapter looks at science as it was before Copernicus, while post-modern big science is the subject of the final chapter. The hard core, Chapters 2 to 9, is mainly devoted to the 400-year period, 1542–1942, from Copernicus to Fermi.

Science, whatever it is, is defined by its practitioners, the institutions that support them, the instruments they use, the phenomena they observe, and above all their mind-cast. In a broad sense this statement has always been true, so long as there have been any people around who thought and acted scientifically. This may be a monstrous tautology, but everything depends on the meaning of 'science' as an abstract noun, and the other forms of speech derived from it. In the seventeenth century, when Molière presented his comedy *Le bourgeois gentilhomme*, the lead character, M. Jourdain, in his attempt to become an educated gentleman, conceived of the world of learning as defined by 'Science, Logic, Meteorology and Philosophy'; this was not far from the truth as it was seen in his day (although in the end he came to focus his studies on language).

When it comes to a definition, the first problem is to distinguish between science and knowledge – difficult in a language such as Dutch, where *wetenschap* has both meanings (as *scientia* does in Latin). The prefix *natuur* brings us closer, equating science with knowledge of nature. But then *natuurkunde*[9] means physics, and the English word is in turn derived from the Greek *physis*, which means no more than nature. Obviously there are both broad and narrow definitions of nature, and getting to terms with the meaning of science involves finding a path through a semantic jungle in which it is all too easy to go round in circles.

From dictionary definitions in various languages, I have derived the following definition to fit the scope of this book: science is the aggregate of systematised and methodical knowledge concerning nature, developed by speculation, observation and experiment, so leading to objective laws governing phenomena and their explanation. The process is one of trial and error, so that the 'objective laws' are not necessarily correct. The historical process consists very largely of established laws being replaced by new ones: the case of Ptolemaic and Copernican astronomy (discussed in Chapter 2) is exemplary.[10]

It is the explanations, in any field, which constitute the leitmotiv of any history of science – noting, once again, that they may well be mistaken. The reference to speculation is critical. Without speculation there would have been no science, but the problem has often been not too little speculation, but too much. The result is scientific overkill: historically there have been simply too many explanations, given the rudimentary methods available for the systematic study of phenomena. Science, even though an aggregate of knowledge, can also refer specifically to one branch of it, say physics or geology. We see this reflected in words, such as geology, biology, physiology, with their common ending derived from the Greek *logos*. Although generally translated as 'word' it means something much more fundamental, as shown by the opening verse of St. John's Gospel: 'In the beginning was the Word . . . and the Word was God.'

Indeed, until the seventeenth century, theology was accepted as ruling all other sciences: this, essentially, was the problem that faced Galileo. Before leaving the field of definition, it is useful to look at the Japanese *kagaku* in the same light. The word means science, in the most general sense, but it embraces all the different branches, from *sûgaku* (mathematics), *butsurigaku* (physics), to include such subjects as *tetsurigaku* (philosophy). The concept of 'gaku' is close to learning, particularly in an institutionalised context. In Japanese, *gaku*, in any of its branches, is pre-eminently the subject matter of education.

This is a good focus for any introduction to the history of science, so that we talk of the school of Aristotle, or Aquinas – it is not for nothing that Aquinas, together with others of the same tradition, are known as 'scholastics'. If we no longer regard such people as scientists, this reflects no more than a change in our mind-cast – in a process of cultural change over the past four or five hundred years. Finally the Japanese definition emphasises not only system or method, but also the complete generality of the target group: science is essentially popular, and the loss of this perception, in the Western world, has been characteristic of the dominance of big science in the last half century.

The cultural change reflects a shift in the foundations of science from intuition to reason. This reflects the restrictions on the horizons of *Homo sapiens sapiens* through the greater part of the 100,000-odd years of the existence of this species – to which all of us belong. To start with, the focus will be on so-called preliterate populations, that is, human groups without written language. This means not only everyone who lived before writing first emerged with cuneiform something over 5000 years ago, but all those still around today who live in communities without the benefit of the written word. There are still hundreds if not thousands of languages spoken in the world without any written texts (save possibly missionary translations of scripture), and the science of those who speak them is circumscribed in a way comparable to that of prehistoric populations.

For the beginnings of science, in prehistory, the net can be cast quite widely, as Chapter 1 shows. The key to science, at its earliest stage, is fire, which humankind has always exploited for its own ends. Without fire it would have been nowhere; with fire the range of habitat and diet extended to the point that no later than 10,000 years ago this one species was at home in the greater part of the world as we know it today. At this stage, the hearth, the source of heat for both warmth and cooking, was at the heart of every home – anywhere where there was human habitation.

The open hearth, with its fire often kept alight, night and day, regardless of the season, provided constant evidence of ongoing reaction, the most fundamental of all scientific phenomena, now still being explored, as Chapter 6 shows, at the furthest reaches of the universe. This was part of human life everywhere, long before there was any domestication of plants and animals, or any all-purpose means of recording information. (The purpose of cave paintings or megaliths, whatever it was, must have been quite specific.)

The scope of the book begins to narrow down at the dawn of civilisation. As humankind began to make pottery, work with metals and glass, and trade over long distances – particularly across the seas – skills became specific to certain recognised classes, while, at the same time, certain populations became dominant in their own part of the world. This was greatly helped by the invention of writing and the systematic measurement of time and distance, a process almost always accompanied by the use of money. Against this background can be seen the origins of the nation state and the class society, both critical to the development of science as we understand it today.

The result, for this book, is to focus on one tradition, which is that of Western civilisation, with its origins in the Middle East some five thousand years ago. I accept that other civilisations, notably that of China, have an ancient scientific tradition, with achievements, at certain periods of history, more impressive than anything in the Western world. I accept also that key scientific resources, such as Arabic numerals (which in fact originated in India[11]), and indeed paper to write them on, were not indigenous in the West. Even so, the tradition known to history by which individuals systematically developed and recorded knowledge and understanding of the world and the cosmos is uniquely Western: its origins were in Greece, some six centuries before the Christian era.

This is the point reached at the end of Chapter 1. What, then, was the inheritance from the ancient world? On the positive side are men who were true scientists, in a modern sense: of these Archimedes (*c.* 287–212 BC), still known for his principle governing the weight of liquid displaced by an immersed body, is exemplary. The great monuments of the ancient world, including engineering works such as aqueducts, and technology such as the sail and the wheel, are also to its credit. The debit side, however, proved to be critical for the development of science. Galen's medicine, Ptolemy's astronomy, and above all Aristotle's physics, all put science on the wrong track – for hundreds, if not thousands, of years.[12] Worse still, when the Western Church developed its own powerful intellectual tradition, it chose to incorporate all the mistaken ideas inherited from antiquity. St. Thomas Aquinas was as disastrous for the development of scientific ideas as Aristotle, if not more so.

Chapter 2 relates how some 500 years ago, science began to be developed by a community of scholars in which the contributions of individual members were decisive for the whole process. The invention of printing and the discovery of the New World had transformed the realm of knowledge and vastly extended the horizons of scholars. The need to govern and administer vast empires led to the emergence of a class of literate administrators independent of the Church, although its language, Latin, would continue to be used in scholarship – even, in some cases, until the nineteenth century.[13] Arabic numerals, paper and printing, new accounting systems, mathematical tables, increasing uniformity in weights and measures, were all reforms favourable to scientific progress.

Finally, the new science triumphed largely because of the instruments available to it, and, of these, two, the telescope and the microscope, were decisive. Although both revealed new worlds, that of the cosmos, opened up by the telescope, proved to be far more open to exploration and understanding by the great thinkers of the day. The microscope, although a seventeenth-century invention, did not really come into its own until the nineteenth century, when it became a key instrument in fields such as medicine and geology. On the other hand, the telescope, from the day of its very first use by Galileo to look at the night sky, was indispensable to the advance of astronomy, which made continuous progress right up to the present day.

The bias of this book is towards the exact sciences, following the principle once stated by Ernest Rutherford, that 'all science is either physics or stamp-collecting'.[14] This would have been acceptable to Copernicus, the key figure who started, somewhat reluctantly, the revolution described in Chapter 2, and even more so to Isaac Newton, who brought it to the point where its results would become definitive in the practice of science. Significantly, where Copernicus' ideas developed within the community of the Catholic Church, Newton's were formulated in an unmistakably secular world, whose triumph over the Church had been sealed by Galileo a generation before Newton was born. In this new world, science was broadly conceived of as 'natural philosophy', although both optics and mathematics were recognised as autonomous. This division was reflected in the instruments used for science, which were classified as 'philosophical', 'optical' and 'mathematical'.

Alessandro Volta's invention of the electric battery in 1800 can be taken as a landmark in the historical process by which science came to be divided up into the different branches which now define it. The Voltaic pile (as it was originally known) immediately opened up vast areas for research, particularly in chemistry. Electrolysis, described in Chapter 6, made possible the separation of familiar compounds into their basic elements, a process which continued throughout the nineteenth century. At the same time, such basic terms as 'science' and 'physics' came into everyday use. They were introduced by the philosopher of science William Whewell (1794–1866), who also coined such specialist terms as 'anode', 'cathode' and 'ion'. His *History and Philosophy of the Inductive Sciences* (1837–60) also went a long way towards establishing the distinction between physics and stamp-collecting.

If, in this book, the emphasis is on physics rather than stamp-collecting, there must be some explanation of what is involved in these two approaches. It is a question of where the starting point is. The physicist, by inclination, wants to see facts fit theory: he works in terms of hypotheses likely, according to the present state of the art, to be confirmed by experiment and observation. He has – or should have – the integrity to accept that this is not always so, in which case theory may have to be modified, sometimes radically. Instances occur throughout this book.

Ideally, there is a continuous interaction between theoretical development and empirical results: the pace of the process can vary enormously. Scientific discovery is at its most exciting when the pace is rapid, so that new results have continuously to be taken into account. Chapter 7, in describing what it was like to work with Rutherford in Manchester a hundred-odd years ago, provides a classic example. Apparatus must become ever more sophisticated and innovative, accuracy in measurement and counting is at a premium, which explains why I treat this aspect as fundamental. (It has also an obvious appeal to the non-specialist, as one can see by visiting the Science Museum in London.)

The stamp-collector works with a vast range of instances, and his most important instrument is his notebook. Carolus Linnaeus (1707–78) is exemplary, and significantly he started off studying medicine and then switched to botany. His introduction of specific and generic names for plants and animals led to the hierarchical system of classification that is still used. He also turned his hand to geology, where the vast range of different minerals and crystals could be reduced to order in the same way. This, in evolutionary terms, is a more basic approach: Chapter 1, describing how Indians in southern Mexico classify hundreds of different plants according to the purposes for which they can be used, provides an instance in a preliterate society, where the mind-cast of the physicist would be quite incomprehensible. Such people could understand what Linnaeus was getting at; the science of his contemporary, the physicist, Henry Cavendish – described in Chapter 8 – would leave them baffled. Classification is characteristic of the earth and life sciences. The former focus on the earth and the way the weather, the tides, and violent events such as earthquakes and volcanic eruptions change its form and composition; the latter focus on the living organisms that exploit the earth's resources. Here in particular, the notebook, long before the end of the twentieth century, had been supplemented by more sophisticated apparatus, notably the electronic computer, with actual observation depending of such advanced techniques as X-ray crystallography (described in Chapter 6) and electron microscopy. Even so, the complexity of the recently decoded human genome is of an order comparable to that confronting, say, particle physicists.

In any case, earth and life sciences, mineralogy or botany or whatever, occur only incidentally in my book. The focus is on astronomy, physics, chemistry and the instruments used to explore and develop these branches of science. (This is also true of Dampier's book.) I concentrate, wherever possible, on the basic apparatus rather than the specific instance. Even at a place like CERN (which has produced eight Nobel prizewinners) there is a governing principle: the need to accelerate fundamental particles to the highest possible energy level. This then defines a vast range of possible experiments, which goes to justify the scale on which CERN operates.

The transformation characteristic of post-modern science would not have been possible without the electronic computer: in writing this book I long considered what I should say about computers. The underlying theory, as developed notably by Alan Turing (1912–54), is remarkable and fundamental to any understanding of mathematics and mathematical processes. The same is true of solid-state physics, that branch of the subject largely defined by the theory of the electrical semiconductors indispensable to modern computers. In the end I decided not to open this particular can of worms. The whole subject is endlessly documented – just look in any bookshop – and if allowed a place in this book would add, at popular level, little that would be new to my readers. Even so, vast banks of computers, with their human operators rapt in concentration, were common to all the different research institutions visited in the course of gathering material for this book. In Los Alamos I saw the world's largest computer (which also appears in the film *Jurassic Park*), but to judge from a report presented to the National Astronomy Meeting at Cambridge, on 5 April 2001, Britain is not far behind. A computer at the Astrophysical Fluids Facility at the University of Leicester, with 128 processors working in parallel, confirmed a result predicted by theory that heavy elements, such as gold and platinum, originally occurred not only as a result of the explosion of supernovae (as related in Chapter 8) but also as the apocalypse of binary neutron stars merging to form a black hole – a phenomenon well known to today's astronomers.

Finally, therefore, I must express my thanks to all those who made these visits possible and guided my uncertain steps in an unfamiliar but none the less remarkable world. Starting in England, some fifty years ago, as an undergraduate at Cambridge, I came to know four remarkable men, Prof. A. S. Besicovitch FRS, Sir Hermann Bondi FRS, Prof. O. R. Frisch FRS and Sir George Thomson FRS, all of whom talked with me in ways useful for this book. At Cambridge also I would like to thank the staff of the University Library, the Cavendish Museum and the Whipple Museum of the History of Science for the help given to me while writing this book.

Moving half a century forward to London, I have been greatly helped by Prof. William Wakeham of Imperial College of Science, Technology and Medicine and his colleagues, Prof. David Caplan, Prof. Marin van Heel, Dr. Rob Iliffe, Prof. Gareth Jones, Prof. Tom Kibble, Dr. David Klug, Dr. Steven Curry and Prof. Gerard Turner. I would also like to commend the Science Museum, next door to Imperial College, where many of the scientific instruments described in my text are to be seen. At the National Physical Laboratory, Dr. A. Hartland and Dr. D. Henderson introduced me to the science of metrology, showing me clocks that were accurate to one ten-millionth of a second and instruments for measuring electric currents so small that electrons could be counted. At the

Royal Institution, Dr. Frank James showed me the unique collection of scientific instruments, particularly those of Michael Faraday – the greatest of all popularisers of science.

Outside London and Cambridge, I enjoyed a remarkable visit to the Joint European Torus (JET), the EU version of the Russian tokamak, at the Culham National Laboratory, just south of Oxford. Before leaving England, I must thank one of my oldest friends in the world of science, Prof. Leon Mestel FRS of the University of Sussex, who taught me most of what I know about astronomy.

I have also received immeasurable help from American scientists. At Los Alamos, Stirling Colgate, who as a boy at the Ranch School had seen Ernest Lawrence and Leslie Groves first come to inspect the premises as a possible scientific base for developing the atom bomb, was able to recount, from his own experience, the whole history of the National Laboratories (whose contribution to science is related in Chapter 10). What Colgate could not tell about the history of Los Alamos was filled in by his contemporary, Louis Rosen – a major figure in particle physics. Their colleagues, Charles Bowman, Dave Forslund, Joyce Goldstone, Joe Martz, James Mercer-Smith, Brian Newman, Jim Phillips, John Reynders, John Richter and Hywel White, then showed me what Los Alamos was working on at the approach to the new millennium.

At the Lawrence Livermore National Laboratory, just outside San Francisco, Mortimer Mendelsohn (whom I had met earlier in Hiroshima) and Keith Thomassen showed me the cutting edge of fusion physics. At the University of Hawaii, Gareth Wynne-Williams told me of the wide range of astronomical research based on the remarkable complex of telescopes at the summit of Mauna Kea, which I later visited.

East of the Mississippi I was able to visit the Oak Ridge National Laboratory in Tennessee, and in Huntsville, Alabama, the Marshall Space Flight Center (famous for Wernher von Braun's rocket research, described in Chapter 3). In both these centres I was able to see the material side to today's travel in space. At the Goddard Space Flight Center, just outside Washington, Fred Espenak showed me the state of the art in planetary research.

In Japan, Dr. Satori Ubuki, of the Research Institute for Nuclear Medicine and Biology in Hiroshima, showed me the other side to the nuclear equation, leaving me overwhelmed by the destructive potential of modern science.

Back in continental Europe (where I have now lived for thirty years) Neil Calder and Gerard Bobbink showed me round two of the giant underground work stations of CERN, Delphi and L3, and led me on a short walk through the 26 kilometre long tunnel housing the Large Electron Positron (LEP) collider (which is now being replaced by the Large Hadron Collider). In the summer of 2001 Dr. David Ward, also of CERN, brought me up to date with the latest developments. In Brussels, Frédéric Clette and his colleagues at the Royal Observatory introduced me to the remarkable world of solar physics.

In the Netherlands, I have been helped by any number of visits to the Teylers Museum in Haarlem and the Boerhaave Museum in Leiden. Hans Balink conducted me round Euratom's High Flux Reactor at Petten, a major European source for research isotopes. Last, but not least, Gerard 't Hooft of the University of Utrecht – in a half-hour telephone conversation – answered any number of questions about the state of the art in particle physics, a field in which his contribution earned him a Nobel prize in 1999.

Even with so much help, and my own checking and double-checking, there are bound to be mistakes. For these I can only apologise: the field was just too large to get everything right. I am comforted by the fact that the history of science is itself largely a chapter of errors. But then, as G. K. Chesterton once said, 'A man who has never made a mistake, has never made anything.'

In writing this book, I have often wondered how the wives of the greatest scientists were able to put up with them, noting also that it could not have been all that easy to be married to Marie Curie (but then Pierre did share in her work). It is just as well that some of the most difficult characters, notably Isaac Newton and Henry Cavendish, never married. I am, therefore, endlessly grateful to my own wife, Carolien, for her patience and good humour when I have been lost in the world of science. Writing this book has been something of an ego-trip, as anyone visiting our home in Amsterdam could observe. While I often preferred to hold my head in the skies, Carolien, helped on occasion by our children, Maarten and Laurien, saw to it that I kept my feet on the ground.

1

From the mastery of fire to science in antiquity

> We cannot too carefully recognise that science started with the organisation of ordinary experiences.[1]
>
> A. N. Whitehead

Science and the human mind

THE EARLY history of science relates to the general study of preliterate thought, which is pre-eminently the domain of the anthropologist, as is reflected by the title of Claude Lévi-Strauss's classic *La pensée sauvage*.[2] At this stage, the limitations of human physiology are critical. Whatever the sum of objective knowledge, it must be subject to what can be perceived with the five senses. The starting point must always be what humankind can see, hear, touch, taste or smell, and in practice the world both of the individual and the culture to which he or she belongs is defined, overwhelmingly, by sight and sound. The one adjunct which distinguishes *Homo sapiens sapiens* from all other species is the power of speech. Whatever achievements have been noted or instilled by science in members of other species, we may take it that speech, in any form useful for science, is 'uniquely human'.[3] In the long run, the power of speech overcomes all the limitations on the range of what can be perceived by the senses. What is more, as the Russian psychologist L. S. Vygotsky (1896–1934) showed in his classic *Thought and Language*,[4] the basis of all human thought (except that of very young children) is linguistic. Even the cleverest laboratory primates, benefiting from years of intensive private tuition given by human instructors, hardly reach the stage at which children's thought begins to develop the adult forms demonstrated by Vygotsky. These pampered primates give a new twist to the meaning of the term 'educationally subnormal'.

For humankind, what is remembered is just as important as what is perceived. What the child first has to remember is the language spoken in its immediate circle. The human predisposition for language is debatable, but the publication in 1957 of Noam Chomsky's seminal book, *Syntactic Structures*,[5] ensured its place as an active field for research. In any case, whatever obstacles a child faces in learning its mother tongue, it will be proficient at a very young age – four or five years old.[6] That is, the process of communicating the contents of a local culture to a child can start at a very early stage, and this is what defines the world as it seen by any small local population. This is the starting point for any investigation into its potential for science.

Both language and topography confine the populations we are looking at to a very restricted domain. We grow up knowing of the existence of a wide range of habitats, from the snow of the Arctic to tropical rain forests, across desert and mountain, knowing at the same time that the vast oceans are no barrier to visiting any of them. The preliterate population, on the other hand, with no access to the media or any of the mains services we take for granted, knows little about what the world is like on the other side of the mountain range or across the ocean on its doorstep. Outside their own familiar territory, those belonging to any traditional local culture will sooner or later encounter other people, speaking incomprehensible languages and with unfamiliar customs, so it is better to stay at home and like Candide cultivate one's own garden. This was how things were for the entirety of the world's population until some 7000-odd years ago, and still are for remarkably many people in today's developing countries. What do these people then do about science, especially without any of the means for keeping the records that are indispensable to science as we know it?

The natural habitat is a preoccupation of any small-scale society, particularly in the absence of any but the smallest human settlements. Whatever the beginnings of science, they are to be found in contexts where observers had nature on their doorstep. The problem then is that the natural scene simply conveys too much information: this is where the differencc between seeing and perceiving is so important. To understand what this involves requires a digression into the physiology of perception. This will concentrate on sight, because of all the senses it is the only one that is absolutely indispensable in coming to terms with the world around us. (We will later look also at the part played by hearing.)

An individual's field of vision can be defined in two stages: first, it consists of the whole of that part of the environment which transmits light, generally by reflection from a recognised source such as the sun, to his eyes; second, it consists of what he consciously perceives, which is that part of the whole to which his attention, consciously or subconsciously, is directed. What David Hyndman has to tell of the Wopkaimin of New Guinea (who number only 700) is true of almost any population:

> Their behaviour is highly affected by that portion of the environment they actually perceive. They cannot absorb and retain the visually infinite amount of environmental information that impinges on them daily.
>
> Their culture acts as a perceptual filter screening out most information in a very selective manner . . . Through mental mapping they acquire a sense of place by acquiring and storing essential information about their everyday spatial environment and using it to decide where to go, how to get there, and what to do with it.[7]

The retina can be taken to be the interface between the two stages mentioned above. On one side is the impact of light focused by the lens of the eye; on the other side are the neural signals transmitted to the visual cortex, the part of the brain concerned with sight. The retina is a complex of a very large number of rods and cones, sensitive to light: in humans the cones, which are receptive to bright light, divide into three categories, each sensitive to light at different wavelengths. This gives us colour vision.

The inherent nature of the retina imposes two severe limitations on the power of observation. First, the fact that the number of rods and cones is finite places a critical lower limit on the size of any part of the field of vision capable of providing a stimulus in such a way that a signal is then transmitted to the brain. This defines, irrevocably the limit to the power of resolution, such as is tested by an optician's eye-chart. Quite simply, objects that are too small cannot be observed by the naked eye. Without some way of breaking through this barrier, a whole universe of micro-phenomena is closed to human knowledge. This is why the invention of the microscope around the beginning of the seventeenth century is so critical in the history of science (as Chapter 2 will explain). Second, the fact that the neural signals take the form of discrete pulses means that any phenomenon to be visible must last for a finite time – measured in milliseconds. Phenomena that are too transitory, such as the trajectory of a bullet, will simply not be observed, at least not directly. Recording such phenomena had to await the invention of photography in the 1830s and other more sophisticated technology in the years since then.

The stage now reached in the argument is that, outside the world's literate cultures, the range of observations essential to science is severely constrained by the limitations both of the habitat of any local population, which may be cultural, natural or geophysical, and of human physiology, particularly as it relates to vision. This statement is true both historically and anthropologically. Moreover, without writing, the accumulation, development and, above all, diffusion of knowledge become extremely difficult.

None the less the natural world still provides the raw material for many different branches of science – geology, zoology, botany, meteorology, and so on – and in its own way a local culture will incorporate them. Each one will have a distinctive content and range of application. Zoology and botany illustrate this point in different ways.

Zoology, based on the observation of the lives of animals, is largely based on intermittent phenomena that make systematic observation very difficult. Except for domestic animals, a culture without zoos (a recent historical development[8]) will have to be content with chance observation of certain facets of the life of any fauna. It is surprising not only how much a culture can attribute to an animal that is seldom observed but also how often such attributions are false. The anthropological record contains any number of instances.

Corpus Christi College, Oxford, has a statue of a pelican pecking at its breast to produce blood to feed its young. This evokes the Christian sacrifice, in which the wine offered in the mass represents the blood of Christ and the sacrifice of Jesus upon the cross. The symbolism, and its relation to the body of Christ, hardly requires any exegesis. The behaviour of the pelican implicit in this medieval iconography is totally false: this is not how pelicans, or any other species of fauna, feed their young. No matter: the symbolism is much more important than the science in any primitive Christian culture. The self-sacrifice of the pelican feeding its young was recognised throughout medieval Europe.

Not only are the presumed habits of the pelican characteristic of prehistoric zoology, but the same is true of their symbolic use. Even in our own popular culture many an animal is the basis for particular attributes – as reflected in such adjectives as foxy, feline, bovine – and just think what it means to call someone a 'rat' or a 'shark'. This is not very helpful to an objective science of zoology.

When it comes to plants, the position is rather different. Plants are rooted in the ground, and have limited defences against humans who interfere with them. (Stinging nettles and poison ivy are the exception rather than the rule.) Local plants can be studied at leisure, and their changes in the course of the year are well known. (As I sit at my computer I can see the maple tree in my garden just coming into leaf.) To a degree they can be subject matter for a cult: just go to Japan in the spring and observe the almost ecstatic popular reaction to cherry trees in blossom.

The plant world is more than just a spectacle. It is a resource to be exploited for food, fuel and material for making almost anything – clothes, paints and dyes, tools, houses, containers, and medicine. Scientific knowledge, according to any of the definitions at the beginning of this chapter, is implicit in such exploitation of the environment. Once again, local knowledge is often mistaken, a point well illustrated by the question of edible fungi in Europe. This case is interesting because modern biochemistry has identified a general principle for determining whether or not a particular species is toxic.[9] Its application, however, requires a laboratory test, so in practice the fungi acceptable for human consumption are determined according to the local culture. The result is that the fungi accepted as edible vary widely across Europe, although the actual species vary comparatively little. Russian housewives cook mushrooms which British housewives would look at with horror, even though reliable, scientifically based guides to edible fungi have been available for well over a hundred years.[10]

The nature of fungi as pathogens is just one instance of a general concern of traditional botany. While relatively few plant species are suited for human consumption, others may be recognised either for being toxic or with the power to cure sickness. This is characteristic of cultures in which knowledge is based on oral, rather than literate, tradition. The distinction is critical, for 'it takes only a moderate degree of literacy to make a tremendous difference in thought processes'.[11]

What then is the essential difference between memory and written records as the foundation of scientific knowledge. We must return to basics and look at the distinction between sight and sound as human faculties. This is neatly stated by Ong:[12]

> Sight isolates, sound incorporates. Whereas sight situates the observer outside what he views, sound pours into the hearer. Vision comes to a human being from one direction at a time... When I hear, however, I gather sound simultaneously from every direction at once; I am at the centre of my auditory world, which envelops me, establishing me as a kind of core of sensation and existence... By contrast with vision, the dissecting sense, sound is thus a unifying sense. A typical visual ideal is clarity and distinctness, a taking apart... The auditory ideal, by contrast, is harmony, putting together.

In a world without writing it is the mind's disposition to unify knowledge transmitted orally that determines its content. Analytically, such knowledge at any time consists of what is stored in the memory of certain individuals, which means that it is represented only in one, necessarily isolated, spoken language. This is far from the universality of modern science, for which a common language is essential. (The point has been made by Gerard 't Hooft: 'Today, to the regret of some, all science happens to be in English.'[13])

In the absence of any means of recording knowledge, individual memory is constrained to retain countless different instances of any category occurring in nature. Village research in southern Mexico showed how a typical inhabitant could identify from memory hundreds of different plants,[14] while back in the United States more than a hundred professional botanists had to be consulted to find the equivalent Latin names.[15] The reason for this is to be found in the character of the written record. Botany, as a science based on writing, has an unlimited capacity for rearranging its own material, classifying plants in different ways, so that there is always the possibility of some new ordering leading to an original scientific insight.

Nothing need ever be lost, but this gain comes at the cost of ordering and accessing a steadily increasing corpus of material. The range of such processes was greatly increased by the invention of printing (which enabled identical copies of written material to be stored in any number of different places), but this gain was as nothing compared with that following the invention of the electronic computer in the middle years of the twentieth century. What is more, the invention of the microscope some 400 years ago (discussed in Chapter 2) marked the beginning of a process in which the instrumentality at the disposal of botanists, and the range of results that it made possible, steadily increased.

In spite of the extraordinary detail in which natural phenomena are recorded in preliterate cultures, there is still a remarkable disposition to state general principles based on a process of induction common to human thought at any stage in cultural development. A particle physicist, woken by rain early on a summer morning, and looking forward to a day at the beach, could well remark, 'rain before seven, fine before eleven', stating a rule he would find difficult to defend in argument with a colleague who was a meteorologist. If such a rule were acceptable to meteorology (which is doubtful in this new millennium), it would be stated somewhat more circumspectly:

> There is a significant positive correlation between precipitation in the early hours of the morning and fine weather before midday.

This statement (ignoring the principle of never using two words where one will do) would then be supported by calendar records showing that in, say, only 0.5% of recorded cases did rain occurring before 7 a.m. continue until after 11 a.m. The case, so stated, seems extremely unlikely, but if it were true the scientist would look for a theoretical explanation (and a modern meteorologist would probably start talking about isobars and weather fronts). Here the modern scientist is not alone: the process goes back to the first appearance of *Homo sapiens sapiens* with power of speech.

The Mexican village, once again, is exemplary. In the early morning the valleys are filled with mists. The limestone hills are full of caves, which, in popular belief, are the source of the mists. Get up early and you can see the mist coming out of the side of the hills. The local culture adds a supernatural dimension, but still observation consistently confirms the principle. Moreover, since the caves occur throughout the region, which consists of one range of hills after another, there is no way that a local inhabitant would ever consider the possibility of morning mists occurring in a region with neither hills nor caves.

We have here an attribute of human thinking present most probably throughout the entire history, recorded and unrecorded, of *Homo sapiens sapiens*. This is the disposition to overexplain. Mexican Indians do not actually need to know what causes the morning mists. Nor does Catholic doctrine require a geocentric universe; the last four centuries have shown that it can get on very well without one. Seven centuries ago, St. Thomas Aquinas would never have accepted this. For him philosophy (which *a priori* meant the Catholic version) was a theory of everything, or it was nothing. In spite of their vast achievements, which have pushed back the frontiers of the universe to almost unimaginable distances, today's scientists are humble in a way that their ancestors in past millennia were not. Scientific method today would achieve nothing without the rigorous control of its observations by means of instruments of almost unbelievable accuracy. Four hundred years ago, observation still went no further than what the five senses could transmit to the related parts of the brain.

The limitations were most critical when it came to sight: objects too small, too far away or simply too fleeting, were outside the realm of scientific investigation. The same, however, was true of temperatures or pressures that were too high or too low, or of sounds of frequencies that did not resonate in the human ear. This was only half the story. Whatever our human faculties, their usefulness in science is restricted by the local environment, with its limited range of phenomena. The scientific breakthrough began less than 10,000 years ago among populations where the life of some members, at least, brought them into contact with the world outside their own frontiers, and, when these and traders invented writing, the way to scientific knowledge was open. What this involved before the era of modern instruments (which we take to open with the invention of the microscope) is examined later in this chapter. First, however, we must look at the instrumentality developed by humankind before this time. This turns on one critical factor in the life of any human population at any time – fire.

Fire

Fire has always been part of human life: when *Homo sapiens sapiens* first walked the earth some 100,000-odd years ago, he not only knew fire, but worked with it, reckless of the risks he ran. The earliest hominids' use of fire may have been for guarding against animal predators, which may be the sum of their legacy to *Homo sapiens sapiens*. By the time that writing first appeared this use of fire had been considerably extended. What then did all this involve?

Fire is essentially an epiphenomenon which takes many different forms: in everyday life these extend from the flash of an explosion to the smouldering of tobacco. The fire characteristic of the stars can be seen in the sun, still our main source of heat and light, but the process of combustion within the sun only began to be understood in the twentieth century.

This chapter, dealing with times long before the development of modern solar physics, is focused on fire confined to our own planet. Even so, the true nature of what we know as fire only began to be discovered in the late eighteenth century, by the French chemist, Antoine Lavoisier (whom we shall meet again in Chapter 6).

Fire, following Lavoisier, is the result of a chemical reaction between oxygen and some other chemical, generally an organic compound, that is, one based on carbon. Since oxygen is the main reactive component of the earth's atmosphere and organic matter (which includes all vegetation) is widely distributed both on and under the earth's surface, the basic raw materials for fire occur together in countless different contexts. When what is known technically as combustion occurs, oxygen and a carbon compound (or sometimes just carbon, as in charcoal) combine in a reaction which generates heat, and is generally accompanied by the phenomenon known as incandescence, that is, the particles of matter involved in it emit photons at different frequencies. Those within the visible spectrum we experience as light, those outside it, as heat. Until late in the nineteenth century, incandescence, produced as a result of combustion, was an essential part of any useful process that produced light or heat artificially.

Fire, because of the nature of the reaction that produces it, is essentially destructive. Natural fires have occurred since long before the time of our first primate ancestors, and from the very first appearance of humankind the control of fire has been an essential part of existence. All human populations have had some mastery of fire: not one other species has ever shared it. As Charles Darwin noted, 'the discovery of fire, possibly the greatest ever made by man, excepting language, dates from before the dawn of history.'[16]

In what way, then, has humankind's mastery of fire developed in the course of time? Scientifically, there was much to learn from the fires spontaneously generated in nature. The way that new growth appeared after the original vegetation went up in flames or the reaction of local fauna were phenomena containing many lessons, including particularly the advantages of cooking. Even more fundamental was the chance to preserve and exploit fire, by selecting and transporting combustible material occurring in nature. Once this lesson was learnt, a fire, whatever its natural origins, could be kept alive indefinitely, and the open hearth as a human institution was born. If there is one characteristic of all human habitations, it is that there will be a fire burning at the centre, even in the warmest climates. The need to keep the home fire burning must always have been a major factor in human migration.

The one essential principle in the use of fire is its capacity to transform: '*ignis mutat res*'. And transformation is at the heart of science. This explains the unequalled importance of fire, for as the Dutch scientist Herman Boerhaave stated in 1720:

> If you make a mistake in your exposition in the Nature of Fire, your error will spread to all the branches of physics, and this is because, in all natural production, Fire . . . is always the chief agent.[17]

What then are the problems to be solved in exploiting fire to best advantage? Three are fundamental, although in varying degrees, according to circumstance. First is the problem of ignition. For all that fire constantly occurs in nature, it is not always there when needed. What technique will then cause combustion? Friction between two dry inflammable surfaces will sooner or later raise the point of contact to a temperature at which combustion will occur. This is the principle of the match, which, in its present form, the safety match, was described by the sociologist Herbert Spencer as 'the greatest boon to mankind in the nineteenth century'.[18] The invention of its forerunner in the eighteenth century followed the discovery in 1669 of phosphorus by the German chemist Hennig Brand.[19] This, the first new chemical element to be discovered since ancient times, is highly inflammable (and, be it noted, it is not organic). The smallest amount of friction, applied to the end of a wooden stick coated in phosphorus will cause it to burst into flame. As with so many inventions, the underlying principle (which will be considered in Chapter 6) was not discovered until much later – in the nineteenth century.

On the other hand, the use of friction to produce fire goes back to prehistoric times and can still be witnessed in remote undeveloped corners of the world. One method is to twist a stick, pointed like a pencil, in a cavity, filled with wood shavings, in another piece of wood. The point will become very hot (as can be confirmed today by touching the head of a drill immediately after use). After about half a minute, the shavings will begin to smoulder and can be blown into flame, on the same principle as bellows applied to the glowing embers in a hearth. The essential property of fire, to propagate itself amid combustible material, can then be used to create a blaze sufficient for any domestic, or other, purpose. The process is laborious, so it is not surprising if domestic fires are kept burning day and night, nor that where new fire is needed it is made by taking embers from another fire nearby. Particularly in a cold damp climate, this is a great trouble-saver.

The second problem is the choice of fuel. As the use of fire progressed to meet many different applications, which are looked at later in this section, the need for greater heat became ever more compelling. Almost any vegetative material will burn, but wood is the most efficient producer of heat, and some types of wood are better than others.

The technological breakthrough came with the original production of charcoal. This, an impure form of carbon, is produced by heating organic matter, generally wood, in a way that deprives it of oxygen, thereby preventing combustion. The process is elaborate and laborious,[20] but effective in driving out the volatile elements in the raw material so as to convert it into the desired product – a fuel of greatly enhanced efficiency. Only with the use of fossil fuels, starting with coal, was greater efficiency achieved, and coal, together with its by-products, only came into its own in historical times.

The third problem with fire is to extend its range of uses, on the principle of *ignis mutat res*, into a realm in which an open hearth cannot meet the demands of the technology. With an open fire, the application of the principle is largely confined to cooking by roasting. In the early evolution of humankind this represented a considerable breakthrough, for it presented the option, denied to any other living species, of consuming food in two essentially different forms. It is not for nothing that the first volume of Claude Lévi-Strauss's monumental *Introduction to a Science of Mythology* is entitled 'The Raw and the Cooked'.[21] None the less, however succulent meat, roots and tubers – the characteristic diet of early humankind – may become as a result of roasting (as any barbecue gourmet knows), the transformations achieved are not sufficiently different to arouse scientific curiosity. The limitations of the process, such as the impossibility of applying it to water, are palpable.

The ceramic breakthrough

At least in the prehistory of science, there was no significant development in the use of fire until the invention of pottery. This occurred some time in the fifth millennium BC in a 'nuclear zone' now divided between Iran, Syria and Turkey, where some two millennia earlier agriculture, carried out by settled communities, first came into existence.

The link between agriculture and pottery is not fortuitous. Early agriculture focused on two cereals, barley and wheat. Barley, by the process of malting, becomes beer, with yeast as a by-product. A mixture of flour and water, kneaded into dough to which yeast is added, will become bread, provided that it is baked in a sufficiently hot oven. The earliest ovens were little more than holes dug into the ground, but stone and eventually brick were used for purpose-built structures. Brick compounded of various earths was originally dried in the sun, but the process invited extension to products made in different shapes and fired at greater heat. This required more careful selection of raw material, but the main problem was to make the oven in which it was baked sufficiently hot. Charcoal was adequate as a fuel; the problem was to ensure a sufficient draught of air – to increase the rate of combustion and produce higher temperatures. With the invention of bellows, the potter's kiln and, later, the smith's forge came into their own.

It is impossible to exaggerate the importance of pottery and of the processes by which it is made. Its first use was in cooking: pottery is useful or indispensable in the preparation and preservation of most human foodstuffs. It is particularly useful in the preparation of grain by boiling as opposed to baking. The materials for making it are almost universal.

The jar is the basic product used for storage or as a cauldron for boiling. There had been earlier vessels made of wood or stone, or consisting simply of gourds, but their range of usefulness came nowhere near to that of the jar. Scientifically, the process of firing is significant for the one-way chemical transformation of the material subject to it. The soft malleable clay becomes a product that is hard and brittle. Firing pottery is irreversible, but the end-product has so many useful properties that there is no doubt about the economic value of the process. The final product is rigid and durable, can withstand a high level of heat, and, subject to evaporation, can contain liquids indefinitely. Chemically it is highly unreactive, so processes such as oxidation are no threat to it (which explains why shards, the broken fragments of ancient pottery, are so important to archaeologists).

Originating from the art of pottery, ceramics has always belonged to technology rather than science, although in the twentieth century, with its constant need to improve heat insulation, a great deal of science has been applied to developing ceramic-based heat shields for spacecraft, as well as any number of other hi-tech products. In the history of science, ceramics is important, above all, for establishing boiling, together with its extension, melting, as part of the instrumentality of science, although (in spite of being noted by Aristotle[22]) it was a long time before this came to be recognised.

The one salient weakness of pottery is that it is so easily breakable but in practice this is counteracted by the cheapness of production. None the less, alternative materials to kiln-fired clay have a definite value of their own. This leads us to the realm of metals, where the fire-based technology extended that developed for pottery.

The technology of metals

In the periodic table of the elements (Appendix A) about half of those listed are metals, and of these the twenty-nine 'transition elements' include iron, copper, zinc, silver, platinum, gold and mercury, all known in antiquity, together with the 'poor metals', tin and lead. In modern chemistry the distinction between metals and non-metals is fundamental, but it only became significant in the nineteenth century. Before then, all the metals listed above had been known since the first millennium BC, and some for up to 3000 years earlier. Even so the discovery and use of metal came after the invention of pottery, and close to the time when language was first reduced to writing. This coincidence marks the dawn of civilisation as we know it.

How then did the revolution that started with what we now call the Bronze Age come about? Metals occur very rarely in a pure and recognisable form: in nature they are a component of many different forms of rock, known as ores, in which the basic element is compounded with oxygen, silicon, carbon or sulphur, all extremely common non-metallic elements.[23] Although each metal has its distinctive ores (which are often shared with other metals), no single ore suggests by its appearance the characteristics of the metals it contains. Within certain late prehistoric cultures, whose geographical area contained both copper and lead ores, a distinction was certainly made between the two: the recognition of ores was an essential first step in discovering and then extracting their metallic content.

How did this breakthrough occur? There can be only one answer to this question. The ore, recognised for its distinctive and often attractive appearance, was put into a furnace and subjected to the sort of heat that until then had been applied only to firing pottery. At a certain stage, the fragments of ore begin to break up and a molten mass separates, leaving a solid residue. The furnace can be designed with a conduit allowing the liquid mass to flow to the outside; an alternative is to have the process take place in a crucible, from which the liquid mass can be poured. In either case, the resulting product little resembles the ore from which it was extracted: it has become quite recognisably something new – a metal.

Metal, in modern science, is above all a good conductor of electricity. Until Volta invented his electrochemical battery in 1800, there was no means of producing an electric current to be conducted anywhere, so we must look elsewhere for the important properties of metal in the ancient world. With the notable exception of mercury, which is already liquid at standard temperatures, metals, when heated, will melt and become liquid, and in that state they can be poured into moulds to create a great variety of objects, both practical and decorative. What is possible varies from one metal to another, the critical factor being the temperature at which any one metal melts.

In contrast to firing in ceramics, casting metal is a reversible process: the objects made can always be recast into something else. This explains why metals are often traded in some standard form, like the 750 ounce gold ingot. The crucible also allows different metals to be mixed in the molten state, so producing alloys with valuable new properties.[24] This process developed at a very early stage, so that copper was alloyed with tin to make bronze or with zinc to make brass. When at a much later stage it became possible to work with molten iron, which has a much higher melting point (1535°C) than copper (1083°C), zinc (420°C) or tin (232°C), steel alloys were produced, and these were to become indispensable in the modern industrial world. On the other hand, the ability to produce alloys had disastrous consequences in the history of science, for it gave rise to the false idea that if the right combination could be found, common base metals could be transformed into rare precious metals. This was the basis of alchemy.

Working with metals led to the discovery that the process of heating was useful at levels much below the melting point. Hot metals could become malleable, so that they could be hammered into different shapes, or ductile, so that they could be stretched out to make wire. The possibilities seemed endless, as appears from the objects discovered by archaeologists in different parts of the world and coming from different periods of time.

The manufacture of metal objects revealed two further useful properties: a metal surface could be polished to a shine, making it useful as a mirror; and metals cast in the right form were sonorous, making them useful for bells. Metals also share two properties fundamental in modern chemistry. First, they tend to be reactive. One consequence of this is that they are subject to corrosion by processes such as oxidation, which with iron produces rust, or with copper, verdigris. A great deal of practical metallurgy has been devoted to counteracting this process, to which rare and precious metals, such as silver and gold, are largely immune – one reason why they are so valued. Second, metals, together with many other elements and compounds in their solid state, are crystalline, so that they have a cellular structure in which each identical cell has the form of a parallelepiped (defined on page 191) in the geometry of three dimensions. This form can most easily be conceived of by starting with a solid block with six rectangular faces and made of some elastic material, say India rubber. All possible crystalline forms can then be obtained by squashing it in different directions, but always in such a way that not only opposite faces but also the sides of each face remain parallel. Given this restriction, only seven forms – each with their own name – are possible. The simplest form is cubic, with all faces perfect squares, the most complex, triclinic, in which no adjacent faces, nor the angles between them, are equal. A moment's thought will show that any uniform substance, will, if crystalline, contain only one form of crystal. (With aggregates, needless to say, any number of different forms are possible.) The fact that solid material, whether metallic or not, is largely crystalline, is a key factor in modern chemistry, relevant and useful, as will be shown in many different contexts in the rest of this book.

Some ninety-odd chemical elements occur in nature,[25] of which the great majority have only become known in the last 200 years. Only ten elements were known to antiquity: listed according to their abundance in the earth's crust, these were iron, sulphur, carbon, zinc, copper, tin, lead, mercury, silver and gold. On a broad definition these are all metals except carbon and sulphur. The majority of metals were only discovered in the nineteenth century. This is true of aluminium, the third most common element in the earth's crust and atmosphere, which because of the difficulty of separating it from its ores, was not discovered until 1825.

The discovery and development of glass

Humankind's constant and sometimes innovative use of fire led some 5000 years ago to the production of one of the most remarkable and useful of all chemical compounds – glass. Its basic component is quartz, the most common of all minerals, which is itself a simple compound of silicon,[26] the most abundant element after oxygen in the earth's crust. Quartz is crystalline and, with its trigonal form, occurs in the natural form of distinctive six-sided prisms, colourless and transparent: this is known simply as rock-crystal. There are also coloured forms, and the rarest and most beautiful of these are valued as precious stones, such as amethyst.

Quartz is the main component of sand, an aggregate of small particles derived from the weathering of quartz-bearing rocks (of which there are many different types). Because it is a compound of silicon, sand can be found in every part of the world, and so sooner or later the kiln, developed for potters, was certain to be used for its effect on aggregates containing sand. At the earliest stage this was no doubt something of a hit-or-miss process, but potters were to discover that their products could be given a special finish if, before firing, a surface based on some composition of sand was applied to them. This was the origin of glazing, a manufacturing process that has continued in use for some five millennia.[27] There is hardly an end to what can be achieved with different forms, each depending upon a specific choice of the materials compounded with sand to make the original glaze.

Glass manufacture carried this process a stage further, by subjecting a compound of sand to the heat of the kiln, without it being applied to pottery. It emerged that when the compound included appropriate quantities of lime[28] and soda,[29] both extremely common in the earth's crust, the result was a molten mass, which when rapidly cooled produced a uniform translucent material. Like pottery glazes, this material could take many different forms. For a very long time, these were mainly ornamental, which is not surprising given that the technology could produce only small pieces. (Although glass is essentially a manufactured product, volcanic processes can also produce small fragments, and these have even been discovered on the moon.)

Chemically, the distinctive property of glass is that it is non-crystalline. According to the *Oxford Concise Science Dictionary*, this means that 'the atoms are random and have no long-range ordered pattern. Glasses are often regarded as supercooled liquids. Characteristically they have no definite melting point, but soften over a range of temperatures.' Randomness and supercooling would have meant nothing to those who first made and worked with glass, but they are essential to the optical properties of glass, such as were discovered by Isaac Newton and others from the seventeenth century onwards – a history told in Chapter 2.

The absence of a definite melting point is, however, the key to the usefulness of glass as the right material for almost any sort of container or vessel, or tube connecting them. The glass-blower, taking advantage of the fact that glass is soft and elastic over a wide temperature range, can produce an astonishing range of artefacts, both practical and ornamental, and, with modern industrial processes, often extremely cheap. Glass, like ceramics, has the advantage of being highly unreactive, so it is suited for containers and conduits for all kinds of material. Its poor conductivity of heat, added to its transparency, makes it ideal for windows, although modern plate-glass is a comparatively recent invention.

Writing and the scientific record

Science as we know it is inconceivable without writing. Although a 'system of graphic symbols' must underlie any definition, writing only became useful as a scientific tool when it became capable of conveying 'any and all thought', and this relates it inescapably to spoken language. The relation is in every case specific. With the above definition, there is no possibility of an 'all-embracing universal system of writing which can be used for any spoken language'.[30]

It follows that every spoken utterance corresponds to a written text, and vice versa, and every such correspondence is embedded in one language.[31] In practice, texts have always been produced without corresponding to any actual utterance; indeed, this is the general case. The result is that a writer may go a long way in disregarding the canons of speech when producing a text. There is no alternative to doing so, nor is anything important lost in the process. Everyday speech is characterised by variations in volume, pitch, pace and coherence, which writing does not record. The basic capacity of writing to record in permanent form what otherwise is ephemeral far outweighs these shortcomings (which in a written text would be no more than distractions). At the same time, writing has its own modulations, such as punctuation and paragraphing, which are absent in speech.

Notwithstanding the essential correspondence between writing and speech, the domain of the written word is quite different from that of the spoken word. If it were otherwise, writing as an institution would add little to the resources of any given culture. In fact it transforms them: this is the simple lesson of history. When it comes to content (which in the end is all that counts), writing and speech deal with essentially different topics: this is true, even though the whole process of classroom education is based on the opposite premise. So be it, but then education, if it succeeds in its objects, is always self-effacing. When it comes to the appropriate form for writing, the orthographic system has, in the history of the last 3000 years, outpaced every known alternative. In such a system, the symbols that constitute the written language relate only to the pronunciation, and not the meaning, of spoken words.

This a principle of extreme economy, because the number of phonemes – that is, distinctive sound units – occurring in spoken languages varies between 15 and 60 (English has 44). There is no objection to a phoneme (which may be either a vowel or a consonant) being represented by more than one character, so that, for instance, the English word 'cheese', although containing only three phonemes, is still written with twice as many letters. The opposite can also be true, so that 'ox', written with two letters, also contains three phonemes.

The most user-friendly orthographic system is alphabetical writing – with characters representing both vowels and consonants – which first appeared with Greek some time around the beginning of the first millennium BC. This was the final point in a stage of development which started with written systems that were not orthographic: the earliest forms of written Greek, although orthographic, used a syllabary, not an alphabet.[32] It is no accident that Greek, after some 3000 years, is still written with the same letters, nor that the letters, developed for Latin from the Greek, are today used by any number of languages in a world in which Latin is no longer a spoken language. This is no problem, since there is no essential relationship between the form of a letter, and the sound or sounds it represents in different languages.

In practice, written as opposed to spoken language can only flourish where it is supported by an institutional system founded and maintained for the purposes of teaching it. Spoken language you learn at your mother's knee; written language you learn at school. One result of this is that the people involved as teachers have a recognised special status. This extends beyond just giving lessons to becoming self-appointed guardians of that part of the culture recorded and preserved in writing.

There is no reason why the institution should be open to all: everyone has a mother, but not necessarily a teacher. More often than not, the opposite is the case. Schools, and the written language taught there, are for a minority, and belonging to it will be counted as a privilege. Just who is admitted is something about which any given society makes it own rules.

A number of consequences then follow. For one thing, the investment needed for maintaining a literate subculture requires a relatively large population. But then, within this population, not all adult members need be literate, so that literacy tends first of all to be the prerogative of men rather than women, at the same time being the preserve of a particular class whose position in society depends upon their exploiting their power to record and communicate in writing. Historically this has meant that writing has been the hallmark of religious specialists, who as often as not are also responsible for education. The consequences of this, more often than not, have been unfavourable for science.

At this stage in the argument, writing, because of its unbreakable tie to the spoken word, remains confined to one linguistic domain. Those who first wrote Greek also spoke Greek. So be it, but then, in spite of the unbreakable link, there is no reason why people who speak Greek should not write Latin, as happened in Byzantium in the days of its empire. The result, historically, is that written cultures tend to coalesce about certain dominant literate traditions. This process has a pronounced snowball effect, so that in today's world the percentage of written material in English far exceeds that of native English speakers.

Particularly in the last thousand years, this process has been helped by the use of all kinds of written notations that supplement language. Since these are freed from the need to 'convey any and all thought', they need not be tied to any spoken language at all (except for the purpose of teaching them). One obvious example of such a notation is that used for music.

When it comes to the history of science, numbers are much more important than music, and in this case one system, that of place-value, can be adapted to any language domain.[33] The prime example of such a system is that of our so-called Arabic numerals (which actually originated in India). Surprisingly, ancient Greece, having developed the world's first alphabet, got nowhere in developing a workable numerical system: instead letters were assigned to numbers in a way that made even elementary arithmetic extremely laborious and opaque. In the event, Arabic numerals only came into use in the West in medieval times.

Measurement – essential to all science

Measurement is fundamental even in the most rudimentary science, or so one would think. None the less, its critical importance has often been ignored. The reason is that the process is inimical to free intuitive thought. Too many ideas fail to survive the test of accurate measurement. But first, we had better have a definition.

Measurement is the means by which numbers can be assigned to different things so as to be able to compare them on the basis of some property common to all of them.[34] The process implies some unit, which by being counted, defines the measure. Until the early eighteenth century, only three such abstract properties were recognised: these, the dimensions of time, length and weight, had been recognised for thousands of years. The fact that no means had evolved to measure other properties, such as heat, does not mean that they had some sort of subordinate status. If anything, the opposite was the case. Heat was always a key factor in both physics and medicine, and with the absence of any way of measuring it acquired any number of spurious attributes. When Daniel Fahrenheit, at the beginning of the eighteenth century, invented the thermometer, and a scale according to which it could be calibrated, this was a breakthrough in the history of science. The process, once started, continued with any number of new units of measurement, often relating to previously unknown properties, such as electrical resistance,[35] which only became important in the nineteenth century.

The three fundamental dimensions recognised in the ancient world can be extended both by combining them in different ways, and, in the case of length, raising it to the second and third powers, to produce area and volume. Length combined with time gives speed and acceleration, mass combined with volume gives density, and all three dimensions combine in different ways to give force, momentum and energy, although such combinations only became significant with the physics of Isaac Newton.

When it comes to establishing units of measure, in any dimension, two important practical questions arise as to use and context: one relates to scale, the other to the choice of a standard. Both can be illustrated by reference to time. As to scale, what period is one interested in – how long it takes to boil an egg, to walk to the nearest town, to bring in the harvest, to see one's ewes come to lambing, or one's son come of age? These questions are clearly chosen to relate not so much to the present as to the past, but this is the perspective of the present chapter. In each case, the scale is based on a different unit, respectively the minute, the hour, the day, the week or the month, and the year. The series is open at both ends, so at one we end continue with seconds, and at the other with centuries, but here we are moving to the fringes of everyday life.

The time scale differs from that of length or mass, in that the cosmic order itself defines three fundamental units, the day, the month and the year, which according to today's cosmology (which accepts the Copernican revolution as establishing a heliocentric planetary system) are the periods, respectively, of the earth's rotation, the moon's orbit around the earth,[36] and the earth's orbit of the sun. These periods so dominate everyday life, at almost any point in time or space within the human compass, that they have been used for measuring time since prehistoric days.[37]

None the less, the day, the month and the year have proved to be extremely problematic, and that for two related reasons. First, their duration is not constant, but subject to small, continuous, secular variation, so that there are more days in summer than in winter. Second, no precise numerical relationship links them to each other. To an observer in ancient times, the month, measured from one new moon to the next, has 29 or 30 days, and the year, measured from the spring equinox, has something over 365 days. A calendar based on twelve lunar months is about eleven days short of the solar year, which means that if solar and lunar calendars are to be coordinated, an extra intercalary month must be added about once in every three years.[38]

At the same time, adopting a 365-day year means that the cosmic events that define it – the solstices and the equinoxes – occur steadily later in the calendar.[39] This anomaly was first corrected by Julius Caesar's introduction of the leap year, but, even so, by the sixteenth century the calendar had gained ten days. This was corrected in 1582 by Pope Gregory XIII's revised calendar, in which the century years, unless divisible by 400, were deemed not to be leap years. England waited until 1752 before adopting the Gregorian calendar, and, by this time, eleven days (in September of that year) had to be dropped in order to come into line. Even so, the Gregorian calendar is not quite accurate, but since it is out by only one part in a million, it will not need to be corrected for several thousand years.

Having sorted out, more or less, the day, the month and the year as basic units for measuring time, the question arises as to how to measure time for periods falling within the day. Here the cosmic order is not very helpful. The position of the sun can be traced by observation, which becomes more accurate by relying on the shadow of a gnomon or pointer on a flat surface. This can then be calibrated in standard units – leaving aside the question as to how one comes by the standard in the first place. The problem is that the area swept by the shadow of the gnomon varies considerably according both to the seasons and to the latitude of its location. Whatever the basic units of measurement, these factors make it difficult for them to be defined by any apparatus based on a gnomon, such as a sundial.[40] (The modern world would perhaps have solved this problem by adopting, by convention, a fixed location on one specific day of the year, say the vernal equinox. This was how Greenwich came to be established as the base for reckoning longitude, but the late nineteenth century was a far cry from the ancient world.)

Let it be: the ancient world never faced up to the problem of establishing a constant unit of time. It was content with the practice, first adopted by the Egyptians, of dividing both day and night into equal periods of twelve hours. These could be counted off according to the position of the sun in the sky during the day, and of the stars during the night. Instruments for measuring time without relying on celestial observation were also developed: the most successful of these was the clepsydra, or water-clock, which originated in ancient Mesopotamia.[41] None the less it was only with Christiaan Huygens' invention of the pendulum clock at the end of the seventeenth century that science gained the means of measuring time accurately enough for its own purposes. And in the 300 years since then accuracy has improved by a factor measured in millions: the related science, which is beyond anything that Huygens could have conceived of, belongs to Chapter 10.

Leaving time for the two other dimensions, length and mass, we find that measurement in antiquity developed to meet special cases rather than to provide universal standards.[42] So, also, the units applicable to a given context were decided by scale: even today we know, almost intuitively, that the domain of miles is quite different from that of inches, and that of ounces quite different to that of tons. There was also considerable local variation: until well into the nineteenth century the pound in Germany had twelve ounces, in Britain, sixteen. At the same time, the process of establishing a standard arithmetical relationship between different units was very slow to get under way. We know (at least if we take the trouble to work it out) that there are 63,360 inches in a mile: although both units originated in ancient Rome,[43] with a cognitive link between them, there was no way of establishing the arithmetical relationship. The result of all this for science is that it had to get by with metrical systems valid only in special contexts, often defined by professions, such as those of the apothecary, coiner or architect.

To complete the picture, it is worth noting how often one dimension was translated into another. This is still part of everyday language, with expressions such as 'ten-minute walk' or a 'two-day journey'. The preference for units of time is significant, for they alone could claim to be universal. At a later stage, money achieved something of the same property, but this required a common domain arising as a result of political and economic factors. As the Roman Empire came to its end, such a domain did develop in the West, with its basic unit, the pound (*libra*), being the standard both for money and for weight. The relationship between the two was defined by equating a pound of silver to the same amount in money: in commerce this was represented by the denarius, a pennyweight of silver, of which there were 240 in a pound.[44] The monetary system that then resulted became standard in Western Europe under Charlemagne at the beginning of the ninth century: it survived in the United Kingdom until 1971.

Confusion in metrology was a major obstacle to the development of science, and standardisation was long in coming. It is not for nothing that the introduction, some 200 years ago, of the metric system (described in Chapter 3), now universally adopted by scientists, coincided with the need to establish any number of new units, above all in electricity and magnetism. It is just worth noting that its decimal basis had been adopted in Chinese metrology some 2000 years earlier.[45]

Science and the ancient world

The time has come to ask what were the resources, useful to science, of the literate cultures that had spread across the Middle East and the Mediterranean littoral by the year 1000 BC? And what, in the end, did science make of these resources?

Today's inheritance from the ancient world is testimony to remarkable achievements in astronomy, architecture, pottery, ornamentation, metal-working and writing – to name only some of the more important cultural domains. None of this would have been possible without a considerable practical understanding of the way the material world was constituted. So much must be clear from the earlier sections of this chapter. At the same time, there was continuous speculation about underlying causes, leading to established perceptions that then governed the practical application of the state of the art.

Medicine, defined by its focus on human pathology, reflects, at almost any time or place, a compulsive dedication to a proactive approach to curing sickness. This is no more than what patients demand, and what practitioners are all too ready to offer. Even in the twentieth century, the result can be either primary health care as we see it in the modern industrial state, or treatment according to traditional folk medicine. In surprisingly many different contexts, not all in developing countries, medical practice can vary between these two extremes.[46] When it comes to the ancient world, the only medicine was folk medicine, and treatment was hardly effective. And what goes for medicine is true, if in varying degrees, also for other ancient crafts and professions. However remarkable the products, the underlying science was primitive and destined to remain so for hundreds if not thousands of years.

If we take Athens as the ideal of the ancient world, we see a civilisation which we cannot help admiring. There, more than anywhere else in the ancient world, some, among the most gifted citizens, tried to transcend the limitations of the Mediterranean culture they had inherited and to think scientifically. Their achievements continue to nourish modern scholarship, but even so the enterprise failed, as it was bound to from the very beginning. The reason for this is that they lacked the resources to achieve a breakthrough.

The point has already been made, but once again the Greeks of antiquity, just as all their contemporaries in any corner of the earth, only had the five senses with which to explore the world around or the heavens above them. They had no technology to extend the range of observation, nor any outside source of power, beyond wind and water, when these could be harnessed for practical use. For transport over land they had little beyond beasts of burden. At the same time the Greeks inherited all the technology related to pottery, metals and glass described earlier in this chapter, but they could not carry it any further.

Their kilns and furnaces were no hotter, they discovered no new metal or alloy, and such glass as they had was too flawed to be useful for optical experiments. Greek triumphs were in the realm of the mind. This was just the trouble, and how much trouble it meant for the advance of science is the final leitmotif of this section. And for the first time we can focus on individual achievement.

The story begins with Thales (624–545 BC), traditionally the founder of Greek philosophy: he came from Miletus, now part of Turkey, but in his day an important Greek settlement. Thales belongs to legend, rather than history, but in the writings of Herodotus and Pliny, the legends passed as history, so that Thales is recorded – contrary to the astronomical record – as having predicted a solar eclipse which put an end to a war between the Medes and Lydians.[47] On the other hand he did propound a cosmology identifying water as the original substance and the basis of the universe, thereby initiating a trend towards fundamental thinking about matter which has lasted to the present day.

The next great name is that of Pythagoras. Little is known about his actual life save that he lived in the sixth century BC, coming, most probably, a generation later than Thales. Even so, the well-known theorem about right-angled triangles is ascribed to him, and the first proof that the square root of 2 was irrational imputed to his followers. It is significant that Pythagoras was a mathematician, because across the whole spectrum of Greek science, only the work of the mathematicians retains any validity: this point will come up again in the discussion of Euclid. In astronomy, followers of Pythagoras were the first to recognise that the earth is a sphere, although the actual proof came only when Aristotle pointed out that the earth's shadow on the moon's surface during a lunar eclipse was circular.

We now come to the fourth century BC, at the end of which Athens had become established as a centre of learning with a reputation unequalled in history. When it comes to science, this has proved to be regrettable. Returning to the beginning of the century, the man to look at is Democritus (*c.* 460–*c.* 370 BC), a prolific author of works in many different fields – ethics, physics, mathematics, music, you name it – which survive only in fragments. According to Democritus, the world 'consists of an infinite number of minute particles, whose different characteristics and combinations account for the different properties, and qualities of everything in the world, animate as well as inanimate'.[48]

Democritus is regarded as the founder of atomic theory, and it is not for nothing that when, in 1911, Ernest Rutherford discovered the atomic nucleus, an admirer described him as the greatest atomic physicist since Democritus. Until at least the end of the eighteenth century, there was hardly anyone worth counting in the whole of the 2000 years that separated them. The fault must be reckoned to Aristotle, together with Plato, the best known of all Athenian thinkers. The great misfortune for Democritus, who was no Athenian, is that his works are known to posterity mainly through references to them by Aristotle, and Aristotle had come to the conclusion that Democritus was mistaken.

As a scientist, Aristotle (384–322 BC) is unequalled, both for his influence on posterity and for his capacity to be mistaken about fundamentals. Both aspects require further explanation.

In the Athens of Aristotle, politics, science and philosophy were dominated by a number of outstanding men, whose names are still familiar to us. Above all, the leading intellects of the day would be known from the disciples they attracted. The greatest of them established their own institutional base, of which the Academy of Plato (428–348 BC) and the Lyceum of Aristotle were the most important (although the original inspiration may have come from Socrates (469–399 BC), known to us only through Plato's *Dialogues*). The institutional base encouraged the great teachers to record their scholarship in writing, and the most prolific of these was Aristotle.

First, however, we must look at Plato. Our examination can be cursory, since Plato's dismissal of 'the objects of this world as mere imperfect shadows of an eternal world of ideal objects or "forms" existing outside of space and time'[49] disqualifies him as a scientist. He could still be a mathematician, for the essence of mathematics was purely abstract: this made it perfect for training the mind in pursuit of forms, and allowed his followers to see in mathematics 'the key to the essential nature of God, the soul, and the world soul which was the universe'.[50]

In the Athenian world of learning, Aristotle, much more than Plato, wrote about almost every possible subject – logic, metaphysics, ethics, politics, rhetoric, poetry, biology, physics and psychology. Since the greater part of his work still survives, we can study his methods and aims in detail. As to the former, Aristotle, to put it bluntly, was a reason-freak. Human intelligence, if correctly applied to the interpretation of the cosmos, would be able to explain how it was constituted and everything that happened within it.

Despite the astonishing range of Aristotle's interests, all his scientific thinking was dominated by his belief in fundamental principles, such as are stated in his *Physics*.[51] Of these, change, and nature as the source of change, were the most important, but Aristotle (also the author of *Metaphysics*) went further, and insisted that all the changes observable in the universe must have had a first cause. This must be God.

Aristotle's God embodies the principle of reason, as opposed to observation and experiment, which Aristotle saw as irrelevant to such general topics as matter, space, time and motion, although not to natural history. Aristotle built upon many of his predecessors' theories, such as that of Eudoxus (*c.* 400–*c.* 347 BC), according to which the heavenly bodies belonged to a system of homocentric spheres, whose centre was the earth, or that of Empedocles (493–433 BC) propounding four elements – fire, air, water, earth – as the constituents of all matter. In doing so, he established new models, which were to prove exceptionally durable.

In astronomy, Aristotle's theory of spheres started with that in which the moon orbited the earth. This defined a sublunary realm subject to Empedocles' four elements: everything outside was the exclusive realm of a fifth element, ether. This quintessential realm was finite and bounded by the outermost of the celestial spheres, that containing, on its surface, all the stars, that is, all heavenly bodies except the five planets[52] (Mercury, Venus, Mars, Jupiter and Saturn), the sun and the moon, each of which, in Aristotle's model, had its own set of spheres. In this ethereal realm, defined by its homocentric spheres, there was no essential change: the heavenly bodies continued forever in set courses, so that each sphere revolved at its own constant rate. In every case, the period of rotation could be derived, at least in theory, from the observations of astronomers.

In practice this could have been none too easy since 55 spheres were necessary to account for the observed paths of the planets. Since, for Aristotle, the basic movement of the heavens is their rotation every night about the unmoving earth at the centre of all spheres, the outermost sphere is the driving force for all the others. That this can happen without any change of place is made to depend upon an argument that 'the world as a whole is not a place'.[53] (As to the outermost sphere Hipparchus (*c.* 180–125 BC) would soon note that every year there was a small, but consistent, change in the stars observable at any given time. This, the so-called precession of the equinoxes, established a new period of some 26,000 years.)

The whole concept defies our imagination: the spheres fit together like Russian dolls; they are uniform, consisting only of ether, and the interface between any two is free of any friction. Every single sphere contains the mechanism for driving the one immediately inside it, but the only visible evidence of the system (other than the stars of the outermost sphere) are the sun and the planets, which *a priori* can only consist of ether, which is otherwise invisible. This is theory-building with a vengeance, but intelligent people continued to accept it long after Aristotle's day.

Aristotle described the sublunary realm in much greater detail. Here change is unremitting, and, when anything is changing, the underlying force must continue until the process is complete. In particular, continuous force is needed to sustain any movement. This principle is on the face of it counterintuitive, so that, for instance, Aristotle had to find some continuing force to explain why a ball thrown at a wall bounces off it – an instance of his principle that the primary kind of change is movement, that is, change of place.[54] Here we come to a fundamental distinction between self-movers (basically animate beings) and objects that can only move in response to an external force. This forces Aristotle to concede that 'it is not the wall, but the thrower of the ball who causes it to move' – even, so it would seem, after he had let go of it. Not necessarily – even though, according to Aristotle, the ball belongs to a wide category of homogeneous objects, which, by their nature, are incapable of being self-movers. But then 'each of them does contain within itself a source of movement; it is source which enables them to be affected, however, rather than to cause movement or to act'.[55]

Today this seems to be the most monstrous casuistry, but reasoning of this kind is pervasive in Aristotle's *Physics*. Since many of the principles contained in this text were generally accepted for more than 2000 years, we must continue to look at them. One of the most fundamental, already mentioned on page 27, is that of the four elementary constituents of all matter in the sublunary sphere: fire, air, water and earth. Nature and change are the keys to understanding them. Each element has a natural place, and a natural motion to that place; earth goes down, and fire goes up, so that the natural place for earth is at the centre, for fire at the periphery, with water covering the earth with air above it – once again a model of concentric spheres. Change also means that one element can turn into another, so that, for example, water becomes air on boiling – a process in which fire plays an essential part. That the world as we know it only roughly corresponds to the model (so that outside the oceans earth is in direct contact with air) is the result of the endless change inherent in the nature of the sublunary sphere.

Although Aristotle explained how the four elements are made of a common matter, which is earth when it is cold and dry, water, when cold and wet, air when hot and wet, and fire when hot and dry, this theory is much less important in the *Physics* than that relating to change.[56] Hot, cold, dry and wet are then the four primary opposites. Any actual material is then compounded out of the four elements, in proportions measured according to these properties. Place has no existence apart from bodies, and time has no existence apart from changes. Changes are of three kinds: generation, by which the subject of the change comes into being; destruction, when it ceases to be; and variation, when some attribute of it is changed. Because changes are not substances, they have no properties and *a fortiori* no properties that can vary. Change, being one of the fundamental concepts in natural science, cannot be defined in terms of anything more fundamental and need never be proved.[57]

When it comes to change of place by a solid object falling to the ground, speed increases with weight as it seeks its natural place at the centre.[58] This principle should be contrasted with that already stated in relation to a ball bouncing of a wall: that change of place requires continuing force. Here the ball's propensity to seek its natural place is counteracted by the force applied to it by an agent, that is, the person who throws it.

One could go on indefinitely, but the question still remains as to why almost all the principles stated above continued to be accepted for more than 2000 years, when they are so fundamentally mistaken. Some, it is true, were modified by scientists who followed Aristotle (often centuries later), but none the less the Aristotelian mind-cast must have had some inherent appeal to human reason for it to remain dominant for so long.

Before considering this question further, it is important to look at two fields of interest to Aristotle but in which the standard model was established centuries after his death. These are astronomy and medicine, and in each case the accepted canon was established by one dominant figure: in astronomy this was Ptolemy[59] (*c.* 90–168); in medicine, Galen[60] (*c.* 130– *c.* 201).

Both Ptolemy and Galen were essentially encyclopedists: Ptolemy's essential knowledge of astronomy derived from Hipparchus, and Galen's, of medicine, from Hippocrates.[61] Even so, Ptolemy and Galen established the canon that would remain definitive of their respective sciences for well over a thousand years. Although Ptolemy's model of the heavens was not the same as Aristotle's, both were based on geometry rather than physics. The essential physical base, that the heavens revolved around a fixed unmoving earth, was taken for granted.

From this starting point, the geometrical problem was to establish a model that would accord most closely with what the astronomers had observed of the motions of the heavenly bodies.

The model adopted by Ptolemy replaced Aristotle's system of homocentric spheres, with one based on two circular systems in the plane of the ecliptic, which in antiquity was defined by the orbits of the sun and the planets around the earth. (The moon was also included, although its orbit is inclined to the ecliptic at an angle of about 5°.) In Ptolemy's system, the orbits of the moon and all five planets were each conceived of in terms of two circles, the deferent and the epicycle (see Figure 1.1). At any given time, each planet was at some point on its epicycle, while the centre of the epicycle was, in turn, at some point on the deferent circle, whose centre was the earth.

The sun was a special case: it had no epicycle, but a simple circular orbit round the earth. It was essential to the model that this orbit and all the deferent circles were 'homocentric' – that is, with the earth as their common centre. The essential configuration is shown in Figure 1.1. Now, astronomers had long observed that each of the three outer planets passes, with complete regularity, through a phase when the direction of its normal path through the stars is reversed. Ptolemy's model was acceptable because it accommodated this retrograde motion of the three outer planets, Mars, Jupiter and Saturn. It also accommodated the two inner planets, Mercury and

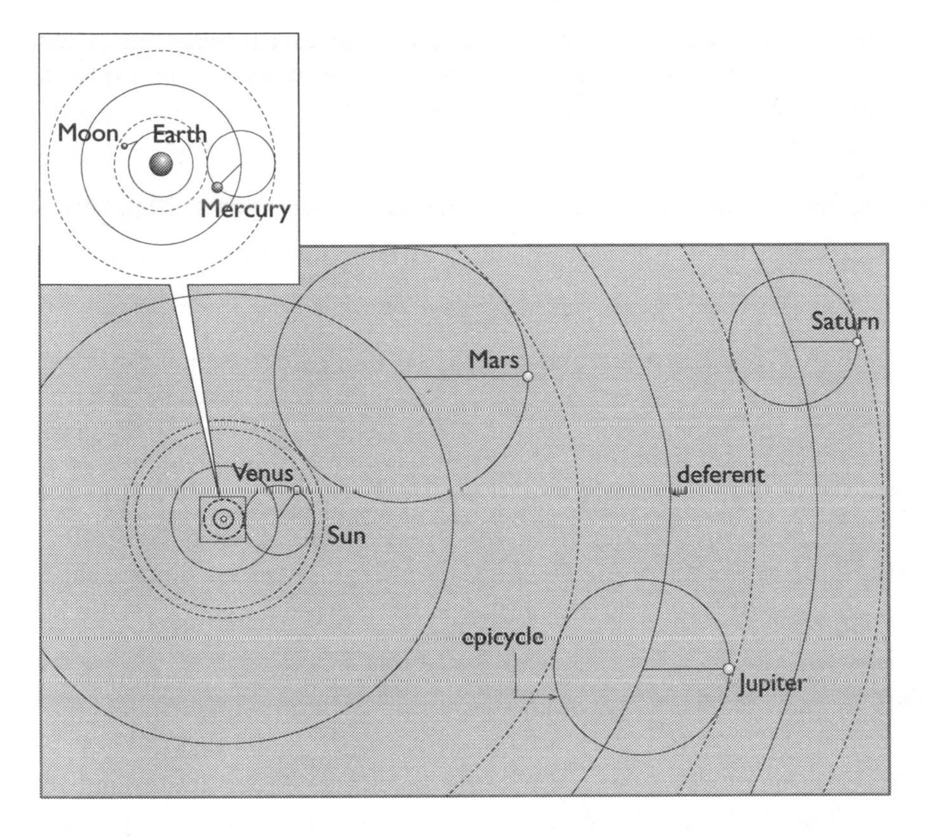

Figure 1.1 The Ptolemaic universe.

Venus, but much less satisfactorily. This basic model must now be refined in a number of ways. To begin with, each epicycle defines a circular band, with an inner and outer diameter, defined respectively by the points at which the planet is closest to or furthest from the earth. The bands are taken to be contiguous, so that, for instance, the outer diameter of that of Mars is the same as the inner diameter of that of Jupiter. The innermost band is that of the moon, so that its inner diameter also defines the boundary of Aristotle's sublunary sphere. The following bands are those of Mercury and Venus: this is necessary to account for the fact that both planets are observed either to set in the evening sky or rise in the morning sky, so that in the middle hours of the night they are never visible at any time of year. The sun must come next, with the bands of Mars, Jupiter and Saturn outside its orbit. This order then accounts for the fact that the three outer planets can be observed at any time of night. The model makes a further key distinction between the two inner and the three outer planets. The centres of the epicycles of Mercury and Venus are always on the line joining the earth to the sun. For Mars, Jupiter and Saturn, this line is parallel to the line joining each one of them to the centre of its own epicycle. This is apparent from Figure 1.1.

With respect to the rate of revolution, the model allows one degree of freedom to each planet. For Mercury and Venus, this is the rate at which each revolves around the centre of its own epicycle; for Mars, Jupiter and Saturn, this is the rate at which the centre of the epicycle revolves round the earth. If nothing else, Aristotle's principle of change, as it applies outside the sublunary sphere, requires the rates to be constant. Each planet may also be taken to have one degree of freedom in the diameter of its epicycle (which is also the width of its band in the ecliptic). Because the bands are contiguous, the diameters of the epicycles must then determine those of the deferent circles. (The argument can start equally well with the latter, but the conclusion is the same.)

The Ptolemaic model succeeded because, within the two degrees of freedom enjoyed by each planet, numerical values in terms of time and distance respectively could always be assigned in such a way that its path in the heavens accorded with past observation. The model was then also predictive, a property always important to scientists. It was, however, not quite perfect, since it failed to account for the fact, noted by Hipparchus, that the seasons were of different length. Hipparchus had himself dealt with this by allowing the earth to be off-centre in the sun's circular orbit, so that the actual centre was just a point in space.[62] Ptolemy extended the principle to deal with observed discrepancies in the periods of the planets, by establishing, in the plane of the ecliptic, an 'equant point', from which the observed motion of the planet was uniform over time. In the planet's circular orbit, this was placed just off centre, at a distance equal to that of the earth, on the other side of the centre, on the same diameter.[63] This allowed an additional degree of freedom, in the choice of the actual distance separating the equant point from the earth. This is the astronomy to be found in Ptolemy's encyclopedic *Mathematical Compilation*, known to history as the *Almagest*, the name under which it was transmitted, in Arabic translation, in the Islamic world of late antiquity.

When the scholars of medieval Europe began to take up astronomy, they worked with Latin translations of the *Almagest*. In terms of modern physics, the great weakness of the Ptolemaic model is that it was essentially mechanical, in the same sense that today's working models of the solar system in the form of an orrery are mechanical. The whole lot could be made out of Lego. The only trouble is that Ptolemy was working with virtual Lego, which means that something had to replace all those plastic gear-wheels in an actual model. Ptolemy never worked this one out, but as a good Aristotelian he hardly needed to. It was sufficient that the model accorded with observation, but even on this score Ptolemy glossed over well-founded objections. The first, and most obvious, was that following its epicyclic path, the moon would be anything between 33 and 64 Earth radii away from the earth. Its apparent size should then vary in the same proportion, but any such variation would easily be seen with the naked eye. It never was. What is more, the model assumed that the period of both Mercury and Venus was exactly a year. This would also be contradicted by observation, but this was much more difficult given that the two inner planets can never be followed through the middle hours of the night, when they are always below the horizon.

Hellenistic science's legacy to the modern world is completed by Galen's (*c.* 130–*c.* 201) medicine. His work largely consisted of commentaries on Hippocrates (*c.* 460–377 BC), although there are occasional problems of attribution. In particular, Galen saw Hippocrates as the originator of the four elements and the four primary opposites, which were as fundamental to his physiology as they were to Aristotle's physics two generations later. The two were separated by more than five centuries, but in this period all those working in medicine professed their allegiance to Hippocrates, and 'he came to stand for whatever any given writer felt to be most valuable' and scholars often assumed that he 'must have been the author of those treatises that they most admired'.[64]

In modern times Hippocrates has survived better than Galen, largely because the Hippocratic oath is as fundamental in medical ethics as it was when first stated more than 2000 years ago. Scientifically, Greek medicine was as much riddled with error as the physics of Aristotle, or the astronomy of Ptolemy. Because the further history of medicine falls outside the scope of my book, we must leave it with Galen and Hippocrates, noting that their influence extended almost to our own day.

Before leaving the sciences of antiquity, whether physics, astronomy, medicine, or whatever, it is as well to note how they were dominated by principles founded on human reasoning, but which developed into a canon. Today the Bible is recognised, among both Jews and Christians, as being canonical, that is, authoritative in a way that cannot be questioned. Historically both the Old and New Testaments are compilations of scripture dating from the first two centuries AD, and the development both of Judaism and Christianity has been based on commentary and interpretation: the idea that at some stage the canon should be rejected is almost unthinkable – quite simply, it is infallible.

This is the way to fundamentalism, but it is not the way of true science, and in the end Greek science failed because its adherents gave it the attributes of revealed religion. The key figure is Aristotle, whose perception of science, in contrast to that of the Hellenistic astronomers, was cosmological rather than mechanical.[65] He, and his followers over the course of nearly two millennia, claimed too much and knew too little. Today the closest approximation to their mind-cast is to be found in the pronouncements of the Roman Curia, reasoning on such matters as birth control. As Bertrand Russell has noted, 'everything... that Aristotle said on scientific subjects, proved an obstacle to progress'.[66] Before science could begin to discover its true nature, Aristotle had to be dethroned. This proved to be surprisingly difficult, because in the last two or three centuries of late medieval Europe, Aristotelian doctrine had some tenacious and very able defenders. Who they were, what they achieved, and how they were finally defeated are themes of the following chapter.

2

The rebirth of science: Copernicus to Newton

Islam and the medieval legacy

THE LEARNING of Islam bridged the gap between science in the Hellenistic world and that in the medieval world. Islam, established in the seventh century, conquered almost the entire Hellenistic world, and within two centuries established centres of learning from Baghdad in the east to Cordoba and Toledo in the west. T. S. Kuhn has summarised the Islamic contribution to the history of science:

> Muslim scholars first reconstituted ancient science by translating Syriac versions of original Greek texts into Arabic; then they added contributions of their own. In mathematics, chemistry and optics they made original and fundamental advances. To astronomy they contributed both new observations and new techniques for the compilation of planetary position. Yet the Moslems were seldom radical innovators in scientific theory . . . Therefore . . . Islamic civilization is important primarily because it preserved and proliferated the records of ancient Greek science for later European scholars. Christendom received ancient learning first from the Arabs.[1]

Beginning in the tenth century, Latin translations from the Arabic began to appear in Europe, and by the twelfth century scholars were gathering together, first quite informally, in different centres, notably Bologna and Paris, to listen to a master expound new Latin versions of ancient texts. These were the first universities, soon to be followed by new foundations, right across Europe, from Salamanca to Heidelberg, from Oxford to Kraków.

In the early days, scientific and philosophical questions were open to debate, and Aristotle had to compete with Plato for disciples. Then, in the thirteenth century, the Count of Bollstädt (known to history as Albertus Magnus) came on the scene. As a teacher he became known as the 'Doctor Universalis'. His universal knowledge focused quite deliberately on Aristotle,[2] but extended to all that was known in his day of the natural sciences, mathematics and philosophy. (As a somewhat sceptical alchemist, he was the first to describe the element arsenic.)

At this time the world of learning had made little progress in more than a thousand years. We see this in Hereford Cathedral's famous and contemporary *Mappa Mundi*, which, with Jerusalem at the centre of the world, is based on 'geographical knowledge unrevised for 1000 years'. Its range is encyclopedic and its sources are the Bible and the ancient classics, but although it came 'at the end of a phase in fashion in learning, it was in many ways a precursor of times to come'.[3]

This not a bad description of the work of Albertus Magnus. This noble saint had as a pupil another, who was to establish Aristotle at the centre of Christian belief. This was Thomas Aquinas (1225–74), who in his monumental *Summa Theologica* took Aristotle from science into theology, resolving once and for all the conflict with the Platonists. The works of Aquinas and his school may have been too much for the general public, but almost within a generation the gap had been bridged by Dante Alighieri (1265–1321), whose *Divine Comedy* is set in 'a literal Aristotelian universe adapted to the epicycles of Hipparchus and the God of the Holy Church'.[4] In a day when politics was closely tied to the Church, the legacy of Aquinas placed, on the face of it, a severe restriction on scientific thought. In fact Aquinas himself was not the great reactionary that some, such as notably Bertrand Russell, would have him be.[5] He was one of the first to stress the importance of sense perception and the experimental foundation of human knowledge, and from his day, science, in increasing measure, would develop in this spirit.

The Renaissance transformation

A number of different factors helped the translation from medieval to Renaissance science. The success of the *Liber Abaci* (1202) (Book of Calculation) of the Italian mathematician, Leonardo

Fibonacci (*c*.1170–*c*. 1250), in popularising the decimal place-value system of Arabic numerals introduced a system of notation which cannot be improved upon[6] – which explains why it is still in everyday use. This event made good one of the major defects in the written language inherited from antiquity: both Roman and Greek numerals were almost useless for calculation, which was only possible with the help of instruments like the abacus. To begin with, Arabic numerals were mainly used for accounting, particularly after the introduction of double-entry bookkeeping in the fourteenth century, but Fibonacci in his *Liber Quadratorum* (1225) (Book of Squares) also showed their usefulness in pure mathematics.

New mathematical tools appeared also in the fifteenth century, when two German astronomers, Georg van Purbach (1423–61) and Johannes Müller (1436–76), introduced trigonometrical methods and produced the first trigonometrical and astronomical tables. Müller's ephemerides were constantly used by Christopher Columbus, whose discovery of the Americas in 1492 opened up a whole new world to science. The scale of the world had to be drastically revised, and its scope enlarged to include all kinds of natural species and human cultures, previously unknown.

Above all, in the ferment of the late fifteenth century, Johannes Gutenberg's introduction of printing with movable type, some time around 1450,[7] was to revolutionise the transmission of learning. At the same time the Church worried increasingly about the place of Easter in the cycle of the seasons. It occurred too late in the year, so that Julius Caesar's year (which had been standard for one and a half millennia) was clearly too long. But by how much? The need to answer this question gave the Church a practical interest in astronomy.

By the sixteenth century, an international scientific community, modern in spirit, had the means of sharing its results with unprecedented efficiency. The quality of workmanship in its traditional instruments constantly improved, with considerable gains in accuracy. This went hand in hand with new standards of measurement, such as Henry VII's standard yard of 1497.[8] At the same time, scientific research began to attract valuable new patronage, not from within the Church, but from princes. This set the stage for the final act of premodern science, which unfolded in the course of the sixteenth century. The remaining part of this chapter focuses on the scientists who then paved the way for the breakthrough into the modern age.

Copernicus, Tycho and Kepler

Three sixteenth-century astronomers ensured that the days of science, as it was inherited from antiquity, were numbered. Between them, Nicolaus Copernicus, Tycho Brahe and Johannes Kepler unwittingly laid the foundations for the scientific revolution of the seventeenth century. The process started with Copernicus (1473–1543), whose life's work appeared in a book, *De Revolutionibus Orbium Coelestium*, published in the year he died. This book, both 'ancient and modern, conservative and radical',[9] faced in two directions.

Copernicus' aim – to revive the Ptolemaic tradition of mathematical astronomy – was conservative. Ptolemy still offered no definitive solution to the problem of the planets. Copernicus aimed, therefore, to establish a better geometrical model of the their motion. By his day, endless revisions, both Islamic and Christian, to Ptolemy's system, as presented in the *Almagest*, had given rise to any number of different variants, all Ptolemaic in their dependence on deferents and epicycles, but none of them adequate for computing planetary positions which would accord with the increasingly accurate observations of the sixteenth century. The problem was compounded by the long-term accumulation of 'bad data which placed the planets and stars in positions that they had never occupied'.[10]

Copernicus, although a professed monk, was influenced as much by the Neoplatonic as by the Aristotelian tradition in learning. (This was permissible in his day, and he made no secret of this preference.) His first priority, therefore, was to reform the mathematics of astronomy. His basic intuition was simply that a new and alternative model could be developed on the principle that the earth revolved round the sun. On the other hand he remained true to Aristotle in his insistence that 'only a uniform circular motion, or a combination of such motions could account for the regular recurrence of all celestial phenomena at fixed intervals of time'.[11] Copernicus, a devout son of the Church (and the nephew of a bishop whose approval was critically important), was extremely circumspect in deploying his arguments. This he did in two stages: first, he defended the position of a number of Greek astronomers from a time centuries before Ptolemy – all of whom made the earth rotate in the midst of the universe – and then, citing another (the Pythagorean Philolaos, a possible colleague of Plato), he argued that 'since the Planets are seen at varying distances from the Earth, the center of the Earth is surely not the center of their circles' and to clinch the matter claimed 'that the Earth, besides rotating, wanders with several motions and is indeed a Planet'.

This immediately accommodated the retrograde motions of the planets, without any need for Ptolemy's epicycles. On the other hand, Copernicus was always true to Aristotle, and *De Revolutionibus* remains 'classical in every respect that Copernicus can make seem compatible with the motion of the earth . . . The Copernican revolution as we know it is scarcely to be found in the *De Revolutionibus*'.[12]

Today we are so conditioned to the Copernican system that we are oblivious to the fact that in Copernicus' own day, it created almost as many problems as it solved. These were largely the result of two fundamental principles, accepted by Copernicus – following almost all his predecessors – both of which we now know to be mistaken. The first of these is that the outer sphere of the stars was taken to be finite, with every single star in it at the same distance from the centre. The second was that the only proper motion for a heavenly body was in either a straight line or a circle.

The consequence of the first principle is that from a planet, with its own orbit round the sun, the part of the heavens visible above the horizon must always be less than a half. So also, the direction of any star must vary according to the planet's position at any time. This is the phenomenon known as parallax. Now, in both cases, the scope for observing the discrepancy is subject to the accuracy of the best astronomical observations. The state-of-the art degree of accuracy, at any given time, determines a lower limit to the diameter of the sphere of the stars; that is, the greater the degree of accuracy, the greater this diameter must be, if no discrepancies are to be observed. The Aristotelian tradition included, however, a fixed distance of the stars from the earth, which, *a priori*, could not be exceeded. Copernican astronomy contained, therefore, the possibility of contradicting a fundamental Aristotelian principle, once observations became sufficiently accurate. Fortunately perhaps for Copernicus, this point was never reached in his lifetime.

Copernicus dealt with the second principle, the need for circular motion, simply by using Ptolemy's deferents and epicycles to correct the observed discrepancies in the orbits of the planets. In contrast to Ptolemy, however, this was only a minor adjustment,[13] but the expedient made Copernicus' system just as complex as its Ptolemaic predecessors, and when it came to observation, it proved to be no more accurate. Significantly the Church consulted Copernicus about calendar reform, but he advised delay, given that neither observations nor theory were sufficient for a 'truly adequate calendar'.[14] Ptolemy was still ahead of the game, or so one would think.

Tycho Brahe was born under a lucky star, in 1546, three years after the death of Copernicus. His was a noble family, in Denmark, and he was brought up by an uncle who encouraged him, from the age of sixteen, to study astronomy and mathematics, first at the University of Copenhagen and then at several German universities. As the first great scientist to grow up under a Protestant rather than a Catholic regime, he did not have to worry about the discipline of Rome; even so, his scientific freedom was still limited by principles Protestants believed to be established by the Bible.

His first concern was about the appalling inaccuracy of the astronomical record (which had been such a trial to Copernicus). This was to strike him at a very early stage, for when he was sixteen he observed how the conjunction of Jupiter and Saturn predicted for the year 1563 was two days out, even according to the much improved tables based on Copernican astronomy.

In 1572, Tycho, still enjoying the life of a wandering scholar, shared with the rest of humankind the chance to observe a brilliant new heavenly body in the constellation Cassiopeia. On a drawing of Cassiopeia he recorded this with the letter I and the designation '*nova stella*'. His surmise that it was a star, born from his failure to observe any apparent movement relative to stars around it, contradicted the basic Aristotelian principle that there could be no change in the outer heavenly sphere comprising the stars.

Tycho published a tract about the 'nova'[15] – the name by which such stars are still known – but it attracted little notice. None the less, his connections in court circles brought him to the notice of King Frederik II of Denmark, who, in 1576, bestowed upon him the lordship of the island of Hven (in the Sund between Denmark and Sweden), together with an endowment sufficient to build an observatory of unprecedented grandeur. In fact he built two, for when the first, Uraniborg (Castle of the Heavens), became too small, he built a second, Stjerneborg (Castle of the Stars): these were equipped with new instruments made on site. Their unprecedented accuracy, to one minute of arc, was at the limit of what the naked eye could observe.

The greatest of Tycho's instruments was the great mural

quadrant, fixed to an inside wall, aligned precisely in the north–south direction of the meridian. This was a quarter-circle, made of brass, with a radius of more than six feet. Looked at as a clock, it covered the hours from 3 to 6. A small eyepiece could slide along the outer rim, which was calibrated in 90 degrees of arc, with each degree divided into 60 minutes.[16] The eyepiece was directed to a hole, high up, at the centre of the quadrant-circle, in a wall at right angles to it. This provided a view of the night sky through which any star could be observed as it crossed the meridian.

The observer, generally Tycho himself, knew in advance the approximate time for any particular star, waited until it crossed the meridian, as observed through the eyepiece, and then read off its altitude from the scale on the outer rim. An assistant then recorded both time and altitude. Tycho tended to mistrust the time reading, reasonably enough given the inaccuracy of clocks in the era before the pendulum. This did not matter in the long run, since the revolution of the heavens, related to the unprecedentedly accurate observation of the altitude of stars, itself measured time better than any clock.

All this instrumentation came to good use in 1577, when a brilliant comet excited enormous popular interest. Tycho's observations, far more accurate than those of any other astronomer, located its orbit in Aristotle's invisible spheres of the planets. The discovery was revolutionary, since planets as a matter of Aristotelian principle had to orbit within the sublunary sphere, which meant that they belonged to meteorology, not astronomy. This convinced Tycho that the spheres themselves did not exist.

With hindsight, we know that Tycho should have been converted to Copernican astronomy. There were, however, two stumbling-blocks: first, the gravitational problems of a rotating earth; and, second, the problem of stellar parallax if the earth revolved round the sun. As to the latter, Tycho's observations were so accurate that for there to be no observable parallax, the stars would have to be more than 700 times further away than Saturn, the most distant planet then known. That this is in fact the case was not to be demonstrated until more than 200 years after Tycho's death.

If Ptolemaic astronomy could not stand the test of Tycho's new discoveries, Copernicus' alternative still went too far. A compromise, known to history as the Tychonic system, was the answer. In this, the earth remained fixed, with the moon and sun orbiting round it. The orbits of the planets, however, had the sun as their centre. The radii of the different orbits could be chosen to accord with the difference between Mercury and Venus, the two inner, and Mars, Jupiter and Saturn, the three outer, planets. This system hardly had time to gain any currency, for within a generation it would be game, set and match to Copernicus.

Tycho's life ended in a reverse of fortune with profound historical consequences. King Frederik II died in 1588, but since his successor, his son Kristian IV, was an infant, the court was governed by regents. These were all family friends of Tycho, but in 1597 the new king had his own way and ended the royal patronage of the greatest astronomer of the day.

The answer, as so often in Renaissance Europe, was to find another prince: the Emperor, Rudolf II, offered Tycho a home at his court. Tycho, leaving all his instruments behind, moved to Prague, where he planned to devote his time to that common occupation of scientists, writing up the results of his research. He did, however, have a young assistant, Johannes Kepler, and that was to change everything.

Where Tycho, born a child of fortune, was gregarious and wise in the ways of the world, Johannes Kepler came from a more humble home, with a quarrelsome father and a mother suspected of being a witch. Beset by religious hang-ups, which would later influence his thinking as a scientist, he studied for the Lutheran ministry at the University of Tübingen. As was normal at that time, the course included astronomy, and the professor, Michael Maslin (1530–1631), taught the Copernican hypothesis. Kepler proved to be such a gifted student that when, in his third year, the professor at Graz died, he was proposed by Maslin to succeed to the chair.

At Graz, Kepler, accepting Copernicus' six planets, felt constrained to fit their orbits into a divine plan based on concentric spheres, containing, like Russian dolls, the five regular solids – tetrahedron, cube, octahedron, dodecahedron and icosahedron. This may have been beautiful geometry, but otherwise it made little sense. Critically, however, the sun was at the centre, as was made clear by Kepler's first book, *Mysterium Cosmographicum*, published in 1596.

Tycho, still in Hven, was sent a copy, and as a result Kepler was immediately invited to visit. Kepler found Hven too far, but four years later, in February 1600, when Tycho had moved to Prague, a visit became a practical proposition, particularly since the Counter-Reformation was making life difficult for Protestants in Graz. After a preliminary three-month visit, devoted to studying the eccentric orbit of Mars, Kepler came back for good in October 1600.

A year later, Tycho died, and Kepler was appointed his successor. He continued to study Mars, and, working with observations recorded by Tycho, he found that the hallowed circular orbit led to errors of up 8 minutes of arc – quite unacceptable when Tycho's instruments were accurate to within 1 minute. Finally, after trying any number of possible geometrical models, Kepler found that an elliptical orbit for Mars accorded perfectly with Tycho's records. This, the simplest and most elegant solution, was based on geometry known since antiquity.[17] It is now known as Kepler's first law. The stumbling-block was that an ellipse has two foci, with one defining the location of the sun and the other being simply void. This asymmetry did not fit in well with Kepler's search for perfection.

Kepler's achievement was immense, since, in working out the elliptical orbit for Mars, he had to reckon with the fact that all the observations were from another orbiting planet, the earth. This made the mathematics extremely complicated, particularly when working with the rudimentary notation of the time. Not surprisingly, Kepler had been working on the problem for nearly ten years when his book *Astronomia Nova* was published in 1609.

This was just the beginning. Having discovered the elliptical orbit, Kepler went on to research the speed at which a planet travelled round it. The result was his second law: the line joining the planet to the sun sweeps out equal areas in equal times. As illustrated (three times) in Figure 2.1, if the planet takes the same

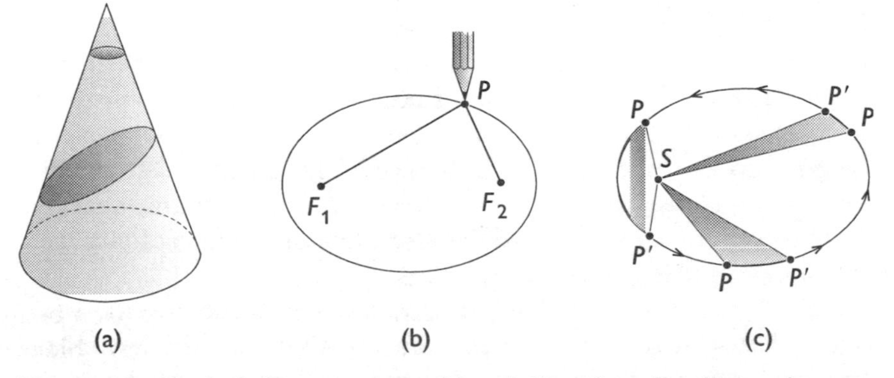

Figure 2.1 Kepler's first two laws: (a) and (b) define the ellipse, the geometric curve in which all planets must move, according to Kepler's first law; (c) illustrates Kepler's second law.

time to traverse the distance PP', then all three shaded areas must be equal. The third law then related the mean distance of the planet from the sun to the period of a complete orbit: the cube of the distance varies as the square of the time. Since the length of the period can be derived from observation (for which Tycho's records were more than sufficient), this third law enables the relative distances of the planets from the sun to be calculated.

Finally, Kepler, at the behest of the Emperor Rudolf, continued Tycho's work in compiling new planetary tables. This was a gigantic task, only completed in 1627, three years before Kepler's death. The new Rudolphine Tables were more than thirty times as accurate as those available to Copernicus. They fitted Kepler's three laws to a degree of accuracy which allowed little room for disputing their validity.

By the time of Kepler's death, the whole astronomical scene had been radically changed, not only by his three laws, but also by the use of the telescope – introduced into astronomy at the beginning of the seventeenth century – to observe the sun, the moon and the planets. This is the story of Galileo. Kepler provides the essential link to this new world of science. His planetary geometry was near perfect, but he never discovered the physics that would explain it. It was sufficient to ascribe the perfection of his system to the design of an omniscient God (as exemplified by his theory of the five regular solids). In this sense Kepler remained true to the principles of ancient philosophy, which, in his day, had ruled for nearly 2000 years.

The science of light

The great sixteenth-century astronomers Copernicus, Tycho and Kepler, like all their predecessors going back to the furthest reaches of antiquity, took light for granted, something that no twentieth-century astronomer could possibly do. If, in the seventeenth century, there was one change in the world of science that counted above all others, it was that scholars for the first time took light itself as a subject for inquiry.

Even for the earliest forms of life on earth, light was the most essential of all natural phenomena. This truth must have dawned on humankind from the first days that its members walked the earth. Although for tens of thousands of years, the sun for most practical purposes was the only useful source of light, a great part of the mystery of fire was that it too produced light,[18] a question that will come up again in Chapter 8.

The decisive importance of light is paralleled by that of the eye in the living organisms with the power of vision. This, the essential power to react to the world around, depends on the physiology of the eye combined with that of the visual cortex of the brain. There is much truth in the precept, 'What the eye cannot see, the heart cannot grieve over.' Given the power of speech as a means of communication, it is perhaps no more than a half-truth; none the less, until the dawn of modern science almost any phenomenon reported in speech must originally have been observed, or at least have been capable of being observed, by the human eye. Since the introduction of the telescope, as reported in Sir Henry Wotton's letter to King James I – together with that of the microscope, which occurred at much the same time – modern science has changed all that to the point that the particle physics of today's post-modern science is concerned with phenomena which are difficult, if not impossible, to conceive of in terms of any visual image.

Light is first judged by its reaction to any medium through which it passes or to any surface on which it impinges. In everyday life any light reaches an observer through a transparent medium that is part of the earth's atmosphere. This is part of the air we breathe, but the question then arises as to the medium through which the light of the sun (or, for that matter, any other heavenly body) is transmitted outside the earth's atmosphere. Chapters 7 and 8 show how this question was only answered in the twentieth century, but at the beginning of the seventeenth century even the effect of the earth's atmosphere on the passage of sunlight was little understood. The effect of other transparent media, such as water, and more particularly glass, was much more important.

The key to this effect was refraction. This is the phenomenon by which the direction of a ray of light is altered as it passes from one transparent medium to another. Given that one medium can be air, and the other water, the phenomenon must first have been observed in prehistoric times. Even so, its usefulness and significance had to await the production, some time around the fourteenth century, of glass of sufficient purity to allow lenses to be made for the purposes of magnification. Their first use was in reading glasses, but in 1608 Dutch craftsmen produced instruments, based on two lenses, sealing the two ends of a long tube: these were the first telescopes, by which distant objects could be seen as if nearby. In late

September they were shown to the court in the Hague, where the Captain-General of the Republic was amazed by being able to see a clock in Delft and the windows of a church in Leiden. The instrument caught on immediately, and by the spring of 1609 Dutch peddlers were promoting it in northern Italy. The decision of an Italian professor, Galileo Galilei, to make his own version, when in May of that year he had heard of the new telescope, was to change the course of science, but this is running ahead.[19]

First we must look at basic theory, as discovered by the Dutchman Willebrord Snel (1580–1626), much less well known than Galileo. Mathematically, this can be simply stated by reference to Figure 2.2, which shows a ray of light passing from one transparent medium to another. The process of refraction is shown by the ray being bent at the point where this takes place.[20] The interface between the two media is represented by a plane, which is the simple case, such as occurs when light passes from air to water, or through a flat pane of glass. The rule established by Snel requires every homogeneous transparent medium to have a refractive index, n. Then, when the angle of incidence is i, that of refraction, r, and the refractive indices of the two media, n_1 and n_2, by Snel's law:[21]

$$n_1 \sin i = n_2 \sin r.$$

Now it may be that using trigonometrical functions I risk losing half my readers, but refraction is absolutely fundamental to optics

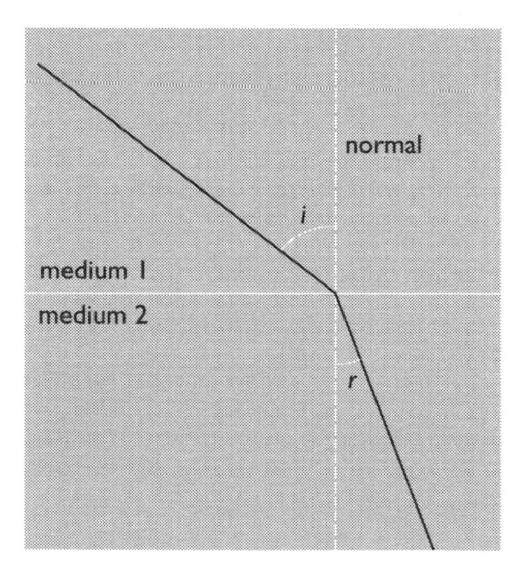

Figure 2.2 Snel's law.

(and to wave physics generally), and Snel's law is equally fundamental to refraction. Seventeenth-century physics would have got nowhere without sixteenth-century mathematics. The German mathematician, Rheticus (1514–76)[22] (who had helped ensure the publication of Copernicus' *De Revolutionibus*) led the field in trigonometry by publishing sine tables, some of which went to fifteen decimal places. Whatever way Snel came to his law, its correctness, combined with his own observations, would have been confirmed by Rheticus' tables.

Now Snel's law shows that it is only the ratio of the two respective refractive indices, n_1 and n_2, that counts. This allows us to take as 1, the refractive index of the ether, the medium though which light, in the sixteenth century, was seen to pass outside the earth's atmosphere. From this starting point, the refractive index of any other transparent medium can be calculated – a process with far-reaching results, as we shall see in due course. That of glass, critical for its use in scientific instruments, is about 1.5.

With Snel at the back of our minds, let us return to the lens, an optical instrument or component first developed some two or three centuries before his time. In its original form this was a curved, ground and polished piece of glass with two opposed surfaces, whose usefulness depended upon the way it refracted light. The basic form was biconvex, which meant that both surfaces represented a section of a sphere. (The extreme case, of a single spherical piece of glass, is a possible form, such as was found in the earliest microscopes.) The image of any object, as seen through a biconvex lens, is not only a distortion, but also a magnification of that object in the user's field of vision. The result, given the way the eye functions, is that the objective is resolved into details not observable with the naked eye. Anyone using reading glasses notices this effect whenever he opens a book. In the history of science, the breakthrough came when the telescope and the microscope made magnification by a factor measured in tens, and even hundreds, possible.

How, then, was this result achieved, and much more important, what were the consequences? At this point, the telescope and the microscope, the two fundamental instruments of magnification, part company. There may be counter-examples, but in the 400-year history of modern science, there has been remarkably little overlap between the disciplines that rely on the telescope and those that rely on the microscope. In both cases there proved to be limits to what could be observed, but in the twentieth century new instruments, such as the radio telescope and the electron microscope, were found to overcome them. The result (as will become clear in Chapters 8 and 9) is that the frontiers of post-modern science are defined either by the most distant celestial phenomena or the very smallest particles of matter. Now, however, we must return to the seventeenth century and look at the new cosmos revealed by the telescope and the new world of the microscope. We must then go on to look at the simplest of all optical instruments, the prism, and the elementary practical application of Snel's law. In the hands of Isaac Newton, this simple block of pure glass, with its flat sides, was to revolutionise our understanding of light – a story told at the end of this chapter. But first, to get the time sequence right, the telescope and the microscope must be considered.

Galileo and his telescope

Galileo Galilei (1564–1642) was a man destined to succeed. Indeed, at the end of his life, he was almost killed by his success. He was, however, a natural survivor, and few have ever equalled his achievements as a popular scientist. To begin with, the tide was always with him, and he knew it: he was a man born in the right place at the right time.

The right place was Pisa; the right time, the year 1564.[23] Pisa (known throughout the world for its famous leaning tower) had long been an important city in the Dukedom of Tuscany, whose capital, Florence (famed for the ruling house of Medici) was at the centre of the Italian Renaissance. Galileo's family came from Florence, and when he was nine years old he joined his father, Vincenzio, a music-master, who had returned to live and work there. The family, although not rich, was well connected. In 1580, Galileo, sixteen years old, with the financial support of a wealthy relative on his mother's side, returned to Pisa to study at the university, the best in Tuscany.

Galileo enrolled in the arts faculty, which meant, in practice, that he would study medicine, with mathematics as a possible subsidiary subject.[24] He left Pisa without a degree, exasperated – as he was to claim in later life – by the way that the works of Aristotle dominated every branch of learning. (Pisa, here, was no different from other European universities.) After some five years of study Galileo seems to have learnt nothing useful in either mathematics or medicine, at least not at Pisa. However, what he had missed in Pisa, he made good in Florence. There, Ostilio Ricci, who had played a major part in reviving interest in Archimedes, lectured on Euclidean geometry to the pages of the Tuscan court. Galileo took advantage of his own court connections to join the course, which taught him not only geometry, but also about measurement and perspective, both crucial in his future work. With this background he could have become a painter (for which he had definite talent), but Ricci set him on course to become an applied mathematician.

Galileo must have been an impressive student, for he was soon asked to lecture in both Florence and Siena, and then recommended by a friend to a chair in mathematics in Bologna. To strengthen his hand for Bologna, in 1587 he went to study with Clavius,[25] a professional astronomer and the leading mathematician at the Jesuit College in Rome. Clavius had worked hard to find a place for mathematics at the Roman College; his position was strong because of his work in preparing the new calendar introduced by Pope Gregory XIII in 1582. Galileo, it would seem, had chosen the right man, but in January 1588, Clavius rejected his proof of a theorem on centres of gravity. This disappointment was set off a month later by a recommendation, from a cardinal[26] no less, for the chair in Bologna.

Bologna (in a way familiar to any academic) was not thinking of making an immediate appointment (although another man was appointed in August). In June, however, Galileo was more than consoled by the acceptance by Guidobaldo del Monte (1545–1607) of the proof that Clavius had rejected. Guidobaldo, an expert on Archimedes, ranked high in Italian mathematics, and his support really counted, so much so that in 1589 Galileo was appointed professor of mathematics at Pisa – not bad for a man of twenty-five who had dropped out of college four years earlier. The young Galileo never allowed the grass to grow under his feet.

At Pisa, Galileo, still working with gravity, set about demolishing Aristotle: dropping weights from the top of the leaning tower may have been part of the exercise, although there is much doubt about this. In 1588, however, Galileo, once more with the support of Guidobaldo, moved to Padua, in the Republic of Venice, where the salary was better. There he came to enjoy the friendship and patronage of a nobleman, Giovanni Pinelli, who had both a vast library and a good collection of scientific instruments.

Among Galileo's courses at Padua was one on cosmography, covering the rudiments of both astronomy and geography. Just how his thinking developed during the early years in Padua is uncertain, but a letter, dated 30 May 1597, to his friend and colleague Jacopo Mazzoni shows that he had by then accepted Copernican astronomy. This was a most significant event: a copy of Kepler's *Mysterium Cosmographicum* had somehow come into the hands of Galileo. He needed only to read the preface to realise the importance of the book. He wrote immediately to congratulate Kepler, on 'his beautiful discoveries concerning the truth and promised to read the book'.[27] Galileo's letter, dated 4 August 1597, also confirmed his support of Copernicanism, but stated that he had not dared publish his reasons for fear of ridicule. Kepler was delighted, sent two more copies of his book, and encouraged Galileo about the prospects of Copernicanism.

That Galileo stayed in Padua for twenty-two years is largely due to his talent for having his salary increased in line with his family expenses. The greatest increase came after he had constructed his first telescope in 1609. This instrument, with 8× magnification, he promoted in August to those who ruled Venice, on the basis of its military, not its astronomical, usefulness. Before the end of the year Galileo achieved 20× magnification. His success was envied, and, given the instruments already marketed by the Dutch, it is not surprising that he was accused of plagiarism. No matter, his telescope was far superior, and it was not until 1630 that anyone produced telescopes with higher magnification.[28]

Following the first demonstration in August 1609, Galileo had within months used his telescope to look at the night sky. He discovered that the Milky Way consisted of thousands of tiny stars, that planets, but not stars, were proportionately enlarged by the telescope, that the moon had an irregular surface (contrary to the perfection required by Aristotelian physics) and then, with the new year, in January 1610, that Jupiter had four moons. In the following months his observation that the sun was 'spotty and impure' meant that he had discovered sunspots, and when it came to the planets he observed (without recognising them as such[29]) the rings around Saturn, and the moon-like phases of Venus. Everything that Galileo had discovered clashed either with Ptolemy or Aristotle, often with both. The phases of Venus were compatible with Tycho Brahe, but Galileo rightly saw that the Tychonic system had the worst of both Ptolemy and Copernicus. Much more important, the phases of Venus showed that the planets were opaque, and their light, like that of the moon, reflected sunlight.

Even before his first use of the telescope, Galileo, true to his nature, was playing off Venice against Tuscany. He made several attempts to persuade the Grand Duke in Florence to create a research professorship, submitting papers relating to such matters as the path of projectiles – a key question in ballistics. However, it was Galileo's discoveries with the telescope that won over the Duke, and having promised that he would continue to make new discoveries he was appointed Professor at the Court of Tuscany on 10 July 1610. Charm, ambition, good connections, perseverance, to say nothing of the brilliance of his achievements, had brought him where he wanted to be – the Court of the Medici.

This is the point to ask what Galileo, when he moved to back to Florence, had actually achieved. This question must be answered from three different perspectives: gravity, optics and astronomy. These we now know to be related in ways (described in Chapters 7 and 8) that Galileo could not have foreseen. As to gravity, Galileo's experiments consistently undermined Aristotle, but he made little progress towards a general theory. In optics, Galileo's main achievement was in making the best telescopes: he contributed nothing to the theory of lenses, and the rules governing refraction, upon which the theory depends, had already been establish by Snel. Galileo's breakthrough was the result of his being the first to use the telescope to look at the heavens, and what he then observed belongs to astronomy, not optics. (The true nature of starlight was not to be discovered until the development of spectroscopy in the nineteenth century – a story told in Chapter 6.)

So where does Galileo stand as an astronomer? Not very high, if we are to believe Thomas Kuhn's verdict that 'Galileo's astronomical work contributed primarily to a mopping up operation, conducted after victory was clearly in sight'.[30] How are the mighty fallen, or so one would think. But then Kuhn does recognise '...the greatest importance of Galileo's work: it popularized astronomy and the astronomy it popularized was Copernican'.[31] To see why this is historically so important – much more so than Kuhn seems ready to concede – we must look at Galileo's life after he returned to Florence in 1610.

The tide was with him: on 29 March 1611 he arrived at Rome, just after the dome designed for St. Peter's by Michelangelo had been completed. Galileo was the guest of the Tuscan ambassador, and a banquet was given in his honour by the Accademia dei Lincei. (This, the first of the European scientific academies, had been founded in 1603.) One of the guests at the banquet coined the word 'telescope' to describe the instrument with which Galileo had observed the heavens.

Galileo was also invited by the Jesuits to lecture at the Roman College, to a largely noble audience, which included three cardinals. The Church, it would seem, had endorsed Galileo's work, which continued at Rome in the midst of the social life of which he was at the centre.

After three months in Rome Galileo returned to Florence in triumph, and with hindsight he would have been wise to rest content with what he had achieved in Rome. None the less, he realised that his support of Copernicanism must mean a confrontation with the Aristotelians. Encouraged by his success in Rome, he moved the debate into the popular arena by publishing in Italian, and it soon became clear to the public that 'Galileo meant to discredit Aristotelian physics and cosmology wherever he could'.[32] The problem was that Aristotle, in the theology of Thomas Aquinas, which was accepted as canonical by the Church, was part of the canon. Dethrone Aristotle, and Catholic theology would be shaken in its foundations. By Galileo's day, the Counter-Reformation, the Church's reaction (formally stated at the Council of Trent (1562–63)) to Luther, Calvin and countless other Protestants defined its stand in the modern world. Like it or not, Galileo was playing politics close to the centre of power, in a world which was divided as it had never been before. It proved to be a dangerous game.

Although Galileo's friends, often in high places (such as Cardinal Barberini) advised him to stick to mathematics and physics, he was forced into theology, simply because too many still insisted that the theological argument against Copernicanism was decisive. Although this argument reflected the continuing dominance of Aristotle and Aquinas, a principle of 'inerrancy'[33] was also fundamental to it. Today we associate this principle, according to which the truth of the Bible may never be questioned, with the rejection of Darwinian evolution by the Christian right in America, but in Counter-Reformation Europe its focus was much more on biblical texts which supported Ptolemaic astronomy, and particularly the place of the earth at the centre of the cosmos. In this sense Aristotle (although he lived long before the Bible was first compiled) was a Christian theologian *avant la lettre*; this was the place assigned to him by Dante in the *Divine Comedy*.[34]

Galileo's problems started not in Florence, but in Pisa, where, on his recommendation, his friend Benedetto Castelli had been appointed professor of mathematics. In Pisa it transpired that while Galileo was little known for his discoveries, he was envied for his high salary. By chance, Castelli was invited by the Dowager Grand Duchess Cristina of Tuscany to attend her court at Pisa, where he was questioned about the compatibility of Copernicanism with Scripture. In the company, which included her son the Grand Duke and his wife, Castelli's theology, although strictly that of an amateur, carried the day, and so he reported in a letter, dated 14 December 1613, to Galileo. Galileo recorded his own thoughts in a reply, which quickly became known as the Letter to Castelli, and as such was widely circulated, often in inaccurate copies. In 1615, he produced an extended version, the Letter to the Grand Duchess, to be sent to the court at Pisa.

In Florence Galileo had made enemies of two Dominican friars of the Scuola di San Marco,[35] Tomasso Caccini and Niccolò Lorini. In December 1614, Caccini denounced Copernicanism from the pulpit of Santa Maria Novella, and worse still, in February 1615, Lorini, under the mistaken impression that the Letter to Castelli was a reply to Caccini, sent a copy to Cardinal Sfondrati, secretary of the Roman Inquisition. Then in March, Caccini himself, on his own initiative, went to Rome, to present his case to the Inquisition.

Galileo, suspecting his opponents' moves, tried to strengthen his own position by having a copy of the Letter to Castelli sent to Cardinal Bellarmino, founder of the Roman College, where Galileo had been so warmly welcomed in 1611. Bellarmino's reaction was to restate the position of the Council of Trent 'forbidding interpretations of scripture contrary to the Fathers', noting at the same time that this excluded a heliocentric cosmos. After all it was Solomon who had written that 'the sun rises and sets and returns to his place'.[36]

Although Bellarmino was ready to go a long way in accommodating Galileo, he still insisted that the earth stood still (the so-called geostatic principle) at the centre of the universe. If Galileo had accepted Copernicanism as no more than a convenient mathematical hypothesis for accommodating his telescoping observations, Bellarmino would have been content. Not only did Galileo refuse to do this, but he also entered the theological debate: it is accepted that if he had actually been able to prove the Copernican hypothesis, the theologians would have yielded, but no sufficient proof was ever produced.

Galileo, failing to make his case with Bellarmino, himself went to Rome to defend Copernicanism. There the egregious Caccini told him that he was still resolutely on the other side. Galileo reacted by having one of his supporters, Cardinal Orsini, approach Pope Paul V, whom he expected to be sympathetic. The Pope reacted by referring the question to the Inquisition, which was asked to state an opinion on two propositions:

1. The sun is the centre of the world and consequently immobile with local motion.
2. The earth is not the centre of the world nor immobile, but moves as a whole, also with diurnal motion.

These two propositions, with their awkward language, are a fair statement of the Copernican position. According to the judgement, which came on 23 February 1616, 'the first proposition was foolish and absurd in philosophy and formally heretical in that it expressly contradicted many sentences of scripture';[37] the second proposition was almost as objectionable.

Then, on 5 March, the Congregation of the Index, at the request of the Inquisition, published a decree 'to put an end to the spread of the false doctrine of the immobility of the sun and the mobility of the earth'.[38] The decree did not mention Galileo, but then, on 25 February, the Pope had already asked Bellarmino to advise Galileo to abandon Copernicanism. Persistent refusal could then lead to imprisonment.

Bellarmino confronted Galileo the next day: what actually happened when the two met is in dispute, but according to an unsigned account of the meeting, Galileo promised to obey 'a very solemn injunction to relinquish Copernicanism altogether, and not to hold, teach or defend it in any way, verbally or in writing'.[39] This was not the end of the road for Galileo, who continued to engage in dialogue in spite of the restrictions imposed upon him. He was taking sides in a battle between two institutions: one, the Accademia dei Lincei, was purely secular; the other, the Roman College, belonged to the Church. Galileo's most noted contribution, *The Assayer* (1623), was a devastating reply to a critique of his work by the Jesuit, Orazio Grassi, professor of mathematics at the Roman College. Grassi's reaction, three years later, was feeble in comparison.

The question is, what was the Accademia, in its support of Galileo, aiming at? It was also subject, at least by implication, to the Inquisition's injunction, but in the long run it wanted to demolish Aristotelian physics. And demolishing Aristotle would at the same time bring down the scholastic philosophy of the Jesuits – a result welcome to the Accademia, though not at the cost of offending the Jesuits. This was an extremely narrow path to tread, particularly at a time when the war then ravaging Europe threatened to undermine the authority of the Catholic Church.

Galileo, after his apparent success with *The Assayer*, decided to continue the argument with a much more comprehensive book, *Dialogue Concerning the Two Chief World Systems* – the Ptolemaic and the Copernican.[40] The *Dialogue*, published in 1632, takes the form of reporting a discussion, lasting four days, each devoted to a particular theme. Two of the purported participants, Giovanfrancesco Sagredo and Filippo Salviati, were old and influential friends of Galileo; the third, cast in the role of defending Aristotelian ideas, was an invented character, Simplicio.

The discussion is heavily biased, not so much against Aristotle, but against Scholastic Aristotelianism. Aristotle, himself, had built his philosophy on sense experience and reason: if then he could return to life, and see through Galileo's telescope, he would agree with Galileo. In the course of the discussion, Simplicio is consistently tied up in knots, while Galileo tries to convince 'the Aristotelians that their difficulties all arise from taking for granted what is in dispute, namely that the earth is fixed'.[41]

The *Dialogue* went a step too far: published in February 1632, it was referred to the Inquisition on 15 September, and Galileo was summoned to Rome to stand trial. The first interrogation was on 12 April 1633, and evidence presented four days later showed that the *Dialogue* taught and defended Copernicanism. The Letter to the Grand Duchess confirmed that Galileo was a Copernican. Galileo claimed that he had not accepted Copernicanism since the decree of 1616: the *Dialogue* was defended on the basis that it simply presented both sides of the question. Given the bias of its contents this, not surprisingly, was too much for the judges, who found him 'vehemently suspect of heresy'.

If Galileo were to abjure, the sentence would be formal imprisonment coupled with obligation to recite the penitential psalms weekly for three years. It was when he came abjure that Galileo is reported to added the famous words 'Eppur si muove' (And yet it does move) that to this day have been attributed to him. After the sentence, he was allowed to go to Siena, where he was the honoured guest of Archbishop Piccolomini: in practice imprisonment meant no more than exile, not in Siena or Florence, but to Arcetri, outside Siena, where he had a daughter in a convent.

Galileo had played, and lost, a dangerous game. Grassi attributed his fall to 'his inflated self-esteem and disregard for others'[42] – a not uncommon failing among leading academics – but then Grassi was hardly a friend. Galileo lived nearly ten years in Arcetri, but for some years before his death he was blind. Even so, he continued to write in the early years, and his last book, *Discourses Concerning Two New Sciences*, appeared in 1638. This had nothing to do with astronomy, so the question of Copernicanism did not arise again. On his death, on 8 January 1642, the Pope refused the Grand Duke of Tuscany permission to stage a public funeral or erect a commemorative mausoleum. Complete rehabilitation, by the present Pope, John Paul II, had to wait until 1992 – 350 years after Galileo's death.

Of the great scientists of the last 500 years, very few are better known than Galileo. But what did he really achieve? What can we add to Kuhn's judgement (quoted on page 51) that he 'popularized Copernican astronomy'? According to Kuhn, 'victory was clearly in sight', at a very early stage, probably even before Galileo first used his telescope in 1609 to look at the heavens. This is the benefit of hindsight: we know that Copernicus, who died a hundred years before Galileo, had found the right answer. It needed Galileo, however, to win the public round. The one thing that stands out from the historical record is that Galileo had a gift for maintaining a very high profile. Popes, Doges, Grand Dukes, and a wide range of scholars (including Thomas Hobbes and John Milton from England) all sought his company. Why then did the Church condemn him?

The easy answer, which is that contained in the judgement, is that the learning he propagated was heretical. This is too narrow. Galileo, as it must have seemed to the Princes of the Church, had the world at his feet, and it was a secular world. This, in the early seventeenth century, was the great threat to the Counter-Reformation Church: the world was acquiring a new dimension in which the magisterium of the Church would count for little. This was painfully clear from the success of the Protestant Reformation. Nowhere did this success count for more than in Holland, where the first telescopes came from. The condemnation of Galileo was essentially a rearguard action, doomed to failure. Galileo and all that he stood for were bound to win in the end. Even in Italy,

Giovanni Cassini (1625–1712), eight years old when Galileo was condemned by the Church, went on to important new discoveries in astronomy in the Galilean tradition. And Cassini was by no means alone in showing that the judgement of the Church would little impede the advancement of science. At the same time the Church had drastically overrated the theological implications of Copernican astronomy: as to fundamental doctrine, Galileo's rehabilitation in 1992 did not even show up on the Richter scale.

The microscopic world revealed

The invention of the telescope, and the observations made with it (particularly by Galileo), soon led to the idea that a similar instrument could be used to examine, in unprecedented detail, objects close to the observer. This explains the origins of the microscope, in the last four centuries the most versatile and widely used of all scientific instruments.

The microscope was also a Dutch invention, first appearing some time around 1620. Like the telescope, the earliest models used a system of two lenses, and such an instrument was actually used by Galileo to look at flies.[43] In contrast to the telescope, however, the magnification that could be achieved with a two-lens microscope led to few significant new discoveries. The greatest Dutch scientist of the age, Christiaan Huygens (1629–93) (inventor of the pendulum clock), long persisted with such compound microscopes, and by 1680 he had become convinced that their power depended on the smallness of the objective lens. In England, Robert Hooke (1635–1703) followed a similar path, using such a microscope for investigations in botany, chemistry, and many other branches of science. His *Micrographia*, in which he published his results, was a scientific masterpiece. However, in the context of the seventeenth century, both Huygens and Hooke were on the wrong track.

Since the 1660s, a compatriot of Huygens, Antoni van Leeuwenhoek (1632–1723), had begun to work with single-lens microscopes: these he had first used in his trade as a haberdasher to examine cloth fibres. His secret was simply to work with extremely small lenses, with very high curvature. Since, for every observation, he worked with a new instrument, the number van Leeuwenhoek used over the course of a long life was very large. Remarkably few survive, and these are valued museum objects. The reason simply is to be found in the minute scale of the lenses. The best of the surviving lenses is 1.2 millimetres thick, having a radius of curvature on both sides of roughly 0.7 millimetres and a magnification of approximately 270×. This may be the best lens van Leeuwenhoek ever made, but his correspondence hints at working with twice this magnification. Even so, 270× allows a resolution of 1.35 millimetres, allowing the observation of details measuring hardly a thousandth of a millimetre. No instrument attained better results until the nineteenth century.

Single-lens microscopes suffered from a number of defects, and it was some two centuries before they were cured. One was simply the quality of the glass available, but equally critical were two forms of aberration, spherical and chromatic. This meant, in effect, that the image view was distorted, except at the centre. The focal length of the small lens was inevitably very short, which meant not only that the object examined should be placed on a flat surface at exactly the right distance from the lens, but also that it should have next to no depth of its own. Anyone with experience of taking close-ups with a camera will be aware of the problem, but a remarkable amount of technology has been involved in ensuring that microscopic samples are thin enough to circumvent it. This, however, has nothing to do with aberration.

Spherical aberration comes from making lenses with surfaces consisting of a section of a sphere. The result is that light from the object does not converge on a single point, the divergence being greatest for that transmitted through the outer rim of the lens. The correct application of Snel's law (see page 46) requires a different, and mathematically more complex, curved surface. Without this, the only remedy is to mask the lens and restrict viewing to a small aperture – once more a process familiar to photographers. The price paid is in reduced illumination, where the margins, in any case, are very small.

Chromatic aberration is quite different, and depends on the fact that the parameters required to apply Snel's law vary with the wavelength of light, that is, according to the colours in the object viewed, as Isaac Newton would discover in 1666. The result of chromatic aberration is to disperse the image viewed, according to its colouring, impairing accurate observation. The only solution is to work, so far as possible, with one colour – another technique familiar to photographers.

Although, with microscopy, there was no obvious choice for the objects to be observed, from the very beginning the focus of all those using the new instruments was on living organisms. (Later the microscope would be important in geology, but this was not where interest lay in the seventeenth century.) The list of objects observed was as remarkable as the phenomena revealed with the magnification factor measured in hundreds.

Van Leeuwenhoek was undoubtedly the master: his observations transformed plant and animal physiology. He discovered protozoa in water (1674), bacteria in the tartar of teeth (1676), blood corpuscles (1674), capillaries (1683), striations in skeletal muscle (1682) and the structure of nerves (1717). In 1676, by examining faeces and decaying teeth, he became the first to observe micro-organisms[44] and in 1688, by examining the tail of a frog, he discovered the circulation of blood through capillaries.[45]

By examining human semen he was almost certainly[46] the first (1677) to observe spermatozoa, the most important, according to Christiaan Huygens, of the microscope's discoveries. He then showed that the males of all species, including fleas, lice and mites, produced spermatozoa, and went on to describe their copulation and life cycles. (Significantly, the German botanist, Joachim Camerer (1534–98), without the benefit of the microscope, had already identified pollen as the male element in plant reproduction, equating it to semen in animals.)

What sort of a man was van Leeuwenhoek? At a first glance his life is reminiscent of Galileo: he too combined an interest in mathematics with a talent for building his own instruments. Delft was hardly Florence, but it was a major cultural centre of the Dutch republic in the so-called Golden Century.[47] Like Galileo, van Leeuwenhoek capitalised on the enormous interest, both popular and aristocratic, in the discoveries made possible by new scientific instruments. Although he is now remembered for his microscopes, his interests were much broader, and in Delft he was considered an expert in navigation, mathematics, astronomy and philosophy. If van Leeuwenhoek concentrated on the microscope, it was probably because this was what most interested the best contemporary scientists and, on their recommendation, possible royal and aristocratic patrons.

Unlike Galileo, van Leeuwenhoek had something of an inferiority complex: he was worried about publishing his results, because his written Dutch was so poor, and he knew no other language. Haberdashers did not normally move in the best circles. In spite of being uncouth and scientifically illiterate, he was still socially ambitious: distinctions, such as election to the Royal

Society, in 1680, helped him, but the aristocratic Huygens brothers, Christiaan and Constantijn, always found him incurably bourgeois. None the less van Leeuwenhoek's visitors included Peter the Great of Russia, the Electors of Palatine and Bavaria, the future King James II of England and his daughter (and successor), the Princess Mary. At the same time the '... pursuit of his inquiries among the townsfolk of Delft and the sailors, fishermen and farmers of greater Holland enmeshed his researches in a dense and complex social network'.[48]

In the end it was not so much van Leeuwenhoek's lack of social graces that limited his scientific achievements, but his inability to read what his contemporaries – who wrote mostly in Latin or French – published. On the other hand, the remarkable results he achieved with his microscopes provided them with a resource of great value. Van Leeuwenhoek's drawings, such as that of the vessels observed in semen (Figure 2.3), were particularly fine. He was, however, not alone in the field of microscopy, which in his time was very much a Dutch preserve.[49] In human anatomy the best known rival was Regnier de Graaf, today known largely for his microscopic examination of reproductive organs. In 1672 he discovered the Graafian follicles, coining the word 'ovary' for the female gonad.[50] (The ovum itself was only discovered by K. E. von Baer in 1827). He had already shown in 1668 that the testis was a tangle of minute vessels.[51]

A greater rival to van Leeuwenhoek was Jan Swammerdam (1637–80) of Amsterdam. A workaholic loner besotted by religion, he was none the less 'a master anatomist, a brilliant experimenter in physiology, and an extraordinary naturalist'.[52] His researches into the anatomy of snails revealed 'many wonders and unheard of things that have perhaps never been imagined' – all the work of an 'inscrutable and incomprehensible God.[53]

Emotional stress led Swammerdam to direct his microscopic research to insects. This subject had not only interested Aristotle, but also led him to make statements which, characteristic of so much of his science, were completely erroneous. Contrary to Aristotle, Swammerdam, as a comparative anatomist, was able to show that insect physiology operated on the same principles as more advanced forms of life. In Protestant Holland, however, there was no papal Inquisition to be worried about.

Although Swammerdam's *Historia Insectorum Generalis* was already in the press in 1669, he was much impressed by a study

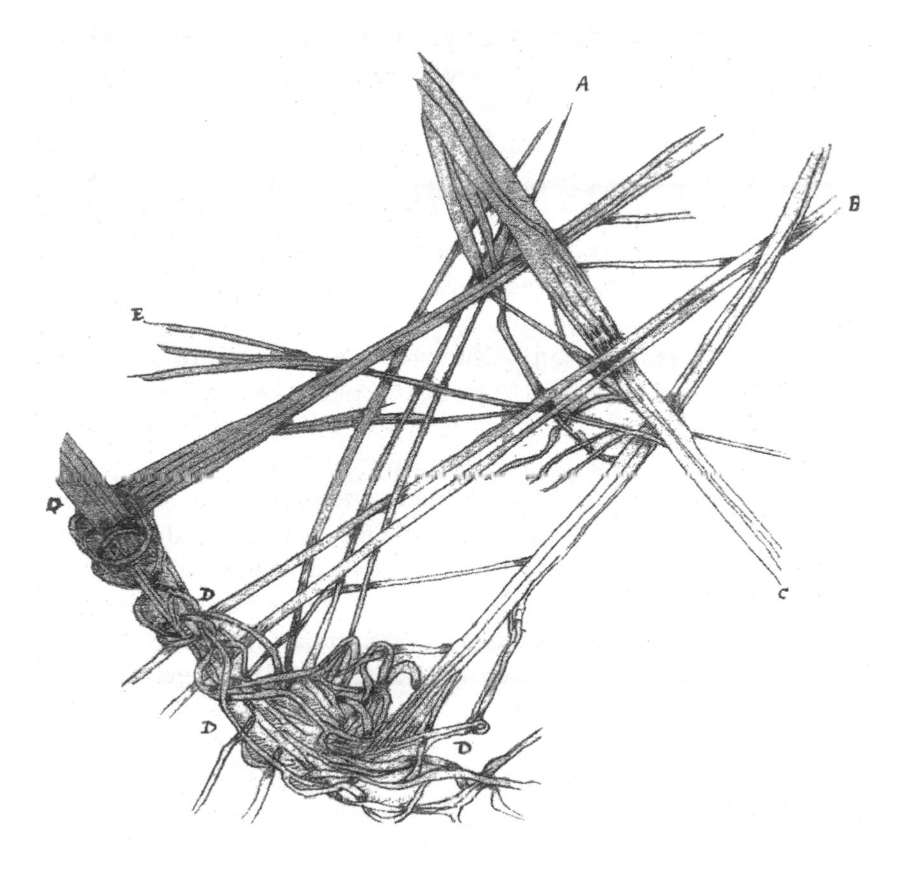

Figure 2.3 Vessels in semen reported by van Leeuwenhoek.

of the life history of the silkworm, published by the Italian Marcello Malpighi (1628–94) in the same year and illustrating both the moth and the caterpillar. Swammerdam improved on Malpighi by discovering the parts of the future butterfly in the mature caterpillar. He described the life cycles of many different types of insect, ranging from the mayfly to the honey-bee. He discovered the ova of the queen bee (until then thought to be a king), and later, in 1668, the testicles and penis of the drone. His method of classifying insects according to the metamorphoses occurring in their life cycles is still used.

As an individual, Swammerdam must be remembered for his obsession with '. . . the religious significance of scientific research'. In this, few have ever equalled him in his insistence that 'the only goal of scientific labors . . . was to demonstrate the attributes of God in all his works so that mankind would revere and glorify him all the more . . . In particular God was as great and almighty in his smallest creations as in the large, as astonishing in the louse as in Behemoth and Leviathan.'[54]

Although Swammerdam can never have been an easy person to get on with, this is also true of many other great scientists, including, as we shall see later, the young Isaac Newton. To the loss of science, Swammerdam surrendered entirely to his morbid religious beliefs. Science had to be abandoned, because it 'perverted by both the pleasure it provided and the ambition it aroused and served'.[55] Swammerdam, entering into his own 'dark night of the soul', represents a type familiar enough in the psychology of religion.

Although Swammerdam's position was extreme, the acceptance of discoveries made with the microscope still undermined the philosophical and theological establishment. At the same time, there proved to be crucial limits to the world revealed by the microscope. The fundamental problem was to come to terms with phenomena only observable with the new instruments. This compares with the problem that faced Galileo, but with this difference, that he already had Copernican astronomy at hand to explain what he observed with his telescopes. The microscopists had nothing comparable.

Philosophy was dominated by René Descartes (1596–1650), a Frenchman who, significantly, spent the greater part of his adult life in Holland. His perception of the real world was extremely economical: for Descartes, knowledge based on reason alone defined 'the realm of unseen particles and pores'.[56] Although Descartes found the microscope far more useful than the telescope, his philosophy still left it little scope, as was clear to his Dutch disciple, Theodoor Craanen, who insisted that 'the subtlety of nature surpasses our powers of thought'.[57] In Holland, Herman Boerhaave (1668–1738), inspired by Newton (who had already repudiated Descartes' astronomy), put an end to Cartesianism by his comprehensive anatomy, based on a 'schematic hierarchy of increasingly complex structural forms in the body . . . through an unknown number of repetitions'.[58] Even so, he and his followers had to accept the existence of complexities beyond the reach of microscopic investigation.

As for religion, the microscope was incompatible with the Calvinist doctrine that 'the revelation of divinity in God's creation had to be openly emblazoned before mankind'.[59] As late as 1705, one of van Leeuwenhoek's visitors 'having been shown some microscopic structure, still puzzled what its purpose might be if no eye could see it'.[60] Swammerdam, on the other hand, accepted all that the microscope revealed as the demonstration of 'the attributes of God in his works so that mankind would revere and glorify him all the more',[61] but then he was no orthodox Calvinist, and, as we have seen above, his fixation on religion led to his abandoning science.

In science, the microscope, sooner or later, would put an end to the lore of spontaneous generation, which, in the seventeenth century, was 'extensive, elaborate and richly specific'.[62] Its focus was on putrefaction and decay, in astonishing variety:

> Diverse kinds of mosquitoes and gnats derived from decaying matter in stagnant water . . . including the dew under certain kinds of leaves. Ancient tales about the origins of bees in the carcasses of oxen, bulls, cows, calves, and sometimes lions still echoed widely . . . Shellfish were ascribed to mud and slime, and snails to the putrefaction of fallen leaves, but the "common folk" had snails and mussels dropping from the sky as well, apparently engendered by ocean vapors, while frogs and tadpoles seem to have fallen with the rain everywhere . . . the broad range of animals that arose from spontaneous generation – from all kinds of worms and insects to species of fish, reptiles, occasional birds, and even rats – was emphasized repeatedly.[63]

For the educated public the basis for all this was a distinction, derived from Aristotle, that identified 'imperfect' animals, 'marked not only by their small size and lack of elaborate structure but by their stunted ability to propagate as well'. All this had been brought up to date, only a century before, by the German alchemist, Paracelsus (1493–1541), according to whom such fauna were 'monstrous, poisonous, usually short-lived and hated by those of their kind who had been properly born'.[64]

No part of this lore was acceptable to the leading microscopists: from 1687 van Leeuwenhoek started a ten-year campaign, during which he examined the life cycle and the sexual apparatus of grain weevils, grain moths, flees, lice, aphids and various flies, with the sole end of demonstrating the absurdity of spontaneous generation. In spite of van Leeuwenhoek, the question was not finally resolved until the nineteenth century, when experiments carried out by Louis Pasteur (1822–95) finally convinced the French Académie des Sciences that claims still being made for the observation of the phenomenon, under laboratory conditions, were false.[65]

Van Leeuwenhoek failed to take full measure of the fact that the microscope, while greatly extending the boundaries of the observable world, still encountered a critical threshold. At the end of the day, the newly revealed phenomena would still have secrets that even the best optical microscopes would never penetrate. Semen could be magnified hundreds of times to reveal spermatozoa, but there was no way that spermatozoa could be magnified, by a comparable factor, to reveal their true characteristics. Seventeenth-century science did not remotely conceive of today's world of molecular biology, with its chromosomes and genes.

Unfortunately, van Leeuwenhoek – in the way of so many of his predecessors whose work he had demolished – used his observations to found theories that they did not warrant. Here he was by no means alone. His obsession was that spermatozoa were the essential instrument of reproduction.[66] This view must be seen in the context of a general and well-established belief in the pre-existence of organisms. This was to be found at every level of society up to and including such distinguished families as the Huygens's. The principle was that every new organism, from its very first existence as an embryo, encapsulated the essential characteristics of all its predecessors. The first moment of creation determined the development of life into the indefinite future, a principle with powerful biblical support.

The problem was to identify the means of transmission. This explained the microscopists' constant interest in semen and plant seeds. The spermatozoon, once discovered, offered a ready-made solution to the problem. Implanted as one among millions in the favourable environment of the uterus it could become the embryo of a new individual: it was already programmed to develop into a new member of the species. On this principle the female role was simply to nourish the embryo, within the womb, in the period before birth. (Although according to today's understanding, every new being would then be a clone, this is still the common view in many traditional cultures.[67])

Van Leeuwenhoek, and the many who agreed with him, overlooked the significance of research into the development of the chick embryo, conducted by William Harvey (1578–1657) (who had earlier discovered the circulation of blood). Harvey's study of eggs,[68] at every stage of incubation, showed how the embryo developed from a microscopic pool of inchoate fluid into a viable organism – the newly hatched chick. He coined the word 'epigesis' to describe this process of 'unintelligible separation' (*divisio obscura*).[69] He too failed to realise how this process was started.

The role of the spermatozoon in fertilising the egg, and the equal contribution of each to the character of the embryo, fundamental to modern genetics and common to all biological species, would not be discovered until the nineteenth century. By the end of the seventeenth century the microscope, it seemed, had spent its force. According to Robert Hooke (1635–1703), who had been noted for his microscopic examinations, only van Leeuwenhoek continued to be interested.[70] This, at least, was how things were in the life sciences, for which the eighteenth century proved to be a dead period in microscopy.

On the other hand, the microscope proved to be almost equally important in chemistry (and related sciences such as geology), particularly in the study of crystals, as described in Chapter 6. Van Leeuwenhoek examined any number of crystalline substances, and noted the geometrical regularity in the shape of particles of alum, saltpetre and sal ammoniac, to name only a few of the substances he worked with.[71]

In 1747 Andreas Marggraf's (1709–82) demonstration that the crystals forming the juice of beet sugar were identical to those in cane sugar was the first instance of the use of the microscope for chemical identification. None the less it was more than fifty years before this seminal discovery was exploited commercially, and in 1776 the Comte de Buffon, known today for his monumental *Histoire Naturelle*, declared that the microscope had produced more error than truth.[72]

Microscopes really came into their own when J. J. Lister's (1786–1869) technical improvements in 1826 produced much superior resolution: this made possible the new science of histology – the microscopic study of tissue in living organisms – basic to modern pathology. This was very much a German interest, with Carl Zeiss's (1816–88) new achromatic compound lenses producing microscopes far superior to any that had gone before.[73] These were compound instruments, and the day of van Leeuwenhoek's single lens had passed.

In the nineteenth century also, with the precise measurement of the wavelengths (between 390 and 740 nanometres) of light in the visible spectrum, came the realisation that this placed a limit on the magnification possible with optical instruments, which no improvement in the quality of lenses, or the systems incorporating them, could overcome. Observations, at molecular level, now commonplace in biology, would have to await the development of new instruments. The first breakthrough came with Joseph Bernard's ultraviolet microscope in 1926, which allowed a virus to be seen for the first time, and after this the development of electron microscopes, starting in 1934, was to lead, before the end of the twentieth century, to levels of magnification measured in billions.[74] It is time, however, to return to the seventeenth century and the man who established the study of light at the centre of scientific research.

Isaac Newton's optics

In the history of science, Isaac Newton's study of optics provides a good introduction to one of the most creative intellects of all time. So let us look first at the man, and his background, before going on to his scientific achievements.

In appearance Newton was striking rather than handsome: there are any number of portraits and popular prints of the grown man, to say nothing of the odd statue. Invariably the eyes, above a beaky nose and a quizzical smile, are those of an inquirer. Nothing escapes the gaze of this man, but there is no suggestion of warmth. This was a man to admire rather than to know.

As a boy he was no different, perhaps because of a troubled childhood. He was born, on Christmas Day 1642, in the sheep country of Lincolnshire, in the manor of Woolsthorpe, which belonged to his mother's family, the Ayscoughs. His own father had died three months before, and, when he was three years old, his mother remarried, choosing for her second husband a prosperous but totally unsympathetic clergyman, the Reverend Barnabas Smith. The young Isaac was never to be welcome in the new household, which would soon include three further children. Instead, he was sent back to his Ayscough grandparents at Woolsthorpe, to grow up in a home that left him with few happy memories.

The same is true of Newton's days at grammar school in Grantham, the local market town, where he lodged with the apothecary. This is the period of the first recorded contacts, which note Newton's sharp intelligence and inventiveness, combined with absentmindedness in tasks – such as watching over sheep – that did not interest him. He had a passion for sundials, born out of his interest in noting the progress of shadows.

In the end, Newton's uncle, the Reverend William Ayscough, and

John Stokes, the schoolmaster at Grantham, saw that he had no future except at the university. He went, therefore, to Cambridge, to study at Trinity, 'the famousest College in the University'.[75] Neither the servants at Woolsthorpe nor the schoolboys at Grantham were sorry to see him go. It would be difficult to find a great man who had known so little love during his childhood: genius has its price.

When Newton entered Trinity in 1661, the college was a socially stratified intellectual backwater, teaching a curriculum based on academic Aristotelianism, and quite oblivious of the revolutionary scientific developments of the time. Galileo may have shaken the papal establishment, but not Cambridge. This was hardly a favourable climate for Newton, particularly since his status at Trinity, that of a sub-sizar, placed him at the bottom of the social ladder. He was also hard up, largely because of the meanness of his mother and of the family into which she had married. His only useful connection was with Humphrey Babington, one of the eight senior fellows who governed Trinity, and also Rector of Boothby Pagnell, a neighbouring parish to Woolsthorpe. This relationship became particularly close when Newton sought refuge with Babington during the Great Plague of 1665.

Newton's life in Cambridge was austere. He was not only poor, but suffered a religious crisis in which even the simplest indulgence – such as beer, cherries or custard – left him with an appalling sense of guilt, born of the conviction that he had let God down at every turn. A man, according to a contemporary, 'of the most fearful, cautious and suspicious temper that I ever knew',[76] Newton did not follow the normal course of studies, which would have led to a degree followed by Anglican orders – about the only profession open to a Cambridge graduate in the seventeenth century. Instead he set out on his own, and in his notebook he recorded under the title 'Quaestiones Quaedam Philosophicae' the books he should read. Later he added the slogan, 'Amicus Plato amicus Aristoteles magis amica veritas' (Truth is a greater friend than Plato or Aristotle).

Newton had not only read Galileo's *Dialogue*, but had discovered the French philosopher and mathematician René Descartes. Although Descartes is better known as a philosopher, he was also a remarkable mathematician and scientist. Like Galileo, he had a gift for attracting noble patronage, which was to lead him to spend some twenty years of his life in Holland, a country he left less than six months before his death in order to teach philosophy to the Queen of Sweden in Stockholm. Unlike Galileo, he was 'lofty, chilly and solitary'[77] – attributes which remind one more of Newton. As a mathematician, Descartes was condescending, if not downright hostile, but then his *Geometry* carried the subject beyond the range of Euclid's theorems that had defined it for nearly 2000 years. His principle innovation was to define geometrical figures according to coordinates x and y on a graph, related to each other by an algebraic equation, so that, for example, $x^2 + y^2 = r^2$ is a circle with radius r; he also introduced the use of a superscript, such as 2 in the above example, to indicate higher powers.[78]

Newton mastered Descartes' book with almost no grounding in Euclid, and certainly no help from anyone at Cambridge. As always, he was self-taught, and the path he followed would soon lead him to establish the foundations of modern mathematics – a story told in Chapter 4.

Whereas, however, Newton found new inspiration in Descartes' *Geometry*, he had problems with Descartes' theory of light, which was to be found in another book, *The Dioptric*. This work followed another, *The Treatise on Light*, which was inspired by the observation, in 1629, of parhelia, or sun haloes, at Rome. Then, in 1633, Descartes, hearing of the Inquisition's condemnation of Galileo, suppressed publication, even though a similar fate was unlikely for a resident of Protestant Holland. But then Descartes, still a devout Catholic, would do anything for a quiet life, and *The Dioptric* contained little that would upset the Church.

Descartes, like Aristotle, got immense mileage out of pure thought[79] (although, historically, he himself played a major part in demolishing Aristotle). Descartes' approach led him to the principle that there could be no empty universe, while, at the same time, there was only one kind of space: essentially, therefore, matter and space were identical. The universe, as we perceive it, was the result of the circulation of matter around countless vortices, of which the sun was one. Centrifugal forces then led particles of agitated matter to reach the human eye, either directly, or, more generally, by reflection: this could be either from points representing other bodies in the solar system (such as the planets) or from that part of the world directly surrounding the observer.[80]

A significant part of Newton's 'Quaestiones' is devoted to objections to this theory, which are as valid today as they were in 1664. With Descartes knocked out of the ring, there was no alternative to an experimental approach that owed little to other scholars. Newton was particularly interested in the phenomenon of colour, and to begin with he experimented with his own visual perception, often risking permanent damage to his sight. He looked at the sun, 'until all pale bodies seen . . . appeared red and dark ones blue', and later he 'slipped a bodkin "betwixt my eye & the bone as near to the backside of my eye as I could" in order to alter the curvature of the retina and to observe the colored circules that appeared as he pressed'.[81]

While keeping up the 'Quaestiones' and conducting his experiments Newton knew that to continue at Cambridge he must be elected a scholar of Trinity. Even so, when he came to be examined by Isaac Barrow, professor of mathematics, he proved to know little of Euclid's geometry, while no questions were asked about Descartes. Isaac Barrow is recorded as being unimpressed, but Newton still became a scholar, and Barrow went on to promote his cause, to the point that when Barrow resigned from the Lucasian professorship in 1669, Newton succeeded to the chair.

There is a paradox about the young Newton: on the face of it, he was his own worst enemy, leading a life calculated to antagonise those whose support was essential if he was to continue as a scientist. He was unsociable, his hours were irregular, and he hardly took any notice of the syllabus upon which he would be examined. In the judgement of J. M. Keynes,[82] 'Newton was profoundly neurotic of a not unfamiliar type, but . . . a most extreme example. His deepest instincts were occult, esoteric, semantic – with a profound shrinking from the world, a paralyzing fear of expressing his thoughts, his beliefs, his discoveries in all nakedness to the inspection and criticism of the world.' Not surprisingly, he never married: as a Tom Stoppard character noted (in the play *Arcadia*), sex was 'the attraction which Newton left out'.

For all his shortcomings, Newton became first a scholar, then a fellow of Trinity, and at the age of 26, Lucasian professor. The only possible explanation is that even the backwoodsmen at Cambridge could not fail to recognise his genius. Even as a professor, Newton continued to try his luck, by making it clear that he would not take up Holy Orders as the statutes required. The prospect did, however, lead him to devote time to theology (in which he achieved no distinction) which was then lost to science. Mercifully, in 1675, Barrow helped him get the necessary royal dispensation.

Newton remained at Cambridge until 1696, and although he was to live until 1727, all his creative work had been done at Cambridge. Here we must return to the work which, years after it had been carried out, took its final form in *Opticks*, published in 1704. Colour was the constant leitmotiv, but Newton no longer tortured his eyes with experiments designed to elucidate the physiology of perception, although these had led to significant results about the nature of primary colours. Instead he investigated the heterogeneity of light with optical instruments, mainly the prism. At an early stage Newton had worked with lenses (whose use, occasionally, would remain necessary): he had learnt from Descartes that, following Snel's law, elliptical or hyperbolic surfaces would cure spherical aberration, but no device he designed ever produced the precision needed for his experiments. The prism, essentially a glass block with plane surfaces, was much easier to produce, and with this Newton dissected light into its elementary components.

Newton was not the first to use a prism for optical research: not only Descartes had preceded him, but also two noted and older British scientists, Robert Hooke (1635–1703) and Robert Boyle (1627–91). All worked with pencil beams of sunlight, directed by prisms on to a flat white surface, but only Newton had the wit to work with a surface 22 feet away from the prism – a distance at least five times that of his predecessors.

The result was exactly what Newton's theory required. The spectrum, which contained the colours of the rainbow (actually produced by the same optical effect and first recorded in antiquity) in the familiar order, red, orange, yellow, green, blue, indigo, violet, was five times as long as it was broad. This phenomenon, which no theory current in Newton's day could explain, was the result of the unprecedentedly large distance separating the prism from the spectrum. Newton contended, correctly, that the refractive index of glass is not constant for all light, but increases from one end of the spectrum to the other. (This follows from the fact, not discovered until the nineteenth century, that the wavelength of light, from red to violet, decreases from 7.8×10^{-7} to 3.1×10^{-7} metres; this change also corresponds to a change in the velocity of light, in a given medium, across the spectrum.)

Newton's discovery opened the way to any number of experimental refinements, all of which confirmed the fundamental theory, that light is heterogeneous, being compounded according to its actual colour, from components from across the spectrum. In what Newton called his '*experimentum crucis*', a second prism, introduced so as to interrupt the rays from the first, did not produce any new colours. In a number of ways he was able to cast spectra from three prisms so that they partially overlapped, with a point where all colours coincided: the fact this point was white, showed how refraction within the prisms worked both ways. A similar result followed from using a lens to focus light of different colours, obtained from a prism, on to a single point.

All this was a far cry from exploring one's own retina with a bodkin. For the first time, colour was reduced to an objective phenomenon, to be produced from a single source – in Newton's day, invariably sunlight – by replicable experimental procedures. Without knowing it, Newton had found a research procedure with untold possibilities, and, with the development of new instruments, capable of being extended into realms whose existence he could never have conceived of.

For Newton, however, there was still the problem of the colour of solid bodies. The solution he arrived at, although simple enough, is a remarkable achievement in the history of science. Newton lived in a world in which the sun was the main source of illumination. (What would he have made of a place like Las Vegas, which only comes alive at night, and where the accepted meaning of 'dark' is 'closed'?) True, even in Newton's time, there were oil lamps and candles, but the everyday world was still essentially sunlit: at night the sky was lit by stars, but their power was too weak, by a factor of millions, to provide useful illumination. The sunlit world, both indoors and outdoors, was full of colour, and the great painters of the day had become masters in reproducing it. But what was it that a red pigment, used, say, by Newton's near contemporary Anthony van Dyck, had in common with red in the subject portrayed? Newton's answer was that a red object reflected red light, and absorbed light of any other colour. Subject to the physiology of the observer's optical system, the balance between absorption and reflection for all the colours in the spectrum determined the actual colour perceived in any part of the field observed. Stating this simple principle was one thing; proving it, another. Newton achieved this result by painting red and blue patches on a piece of paper, and then focusing beams of red and blue light, obtained from the sun's spectrum, upon them. This showed that light of the same colour accentuated that of the patch, but that of a different colour weakened it. The trick was to produce a beam of pure red or blue light. Now with a beam of light passing from a medium, say glass, to another, say air, with a lower refractive index, the angle of incidence i is always less than that of refraction r. As a result (see Figure 2.2), as i increases, it reaches a critical point at which r reaches 90°. Then, by Snel's law,

$\sin i = 0.66$, or $i = 41°$. Beyond this critical point, the light is reflected within the medium with the higher refractive index. Since, however, the index for any medium increases very slightly across the spectrum, from red to violet, there is a narrow range of values for i in which light beyond a certain point is reflected within the medium, whereas below that point it will pass outside it. This process can be used to produce red light in air, from a white beam passing through glass. With the help of a prism, the blue light can also be used as it passes to the outside medium, say air, through another face. With unprecedented accuracy Newton worked constantly with prisms, often in conjunction or combined with lenses (see Figure 2.4),[83] to produce results which abundantly confirmed his theoretical analysis.

Newton also continued his work with telescopes: his problems with lenses had turned his mind to mirrors, and in 1668 he built a reflecting telescope, having invented his own alloy to cast and grind the mirror. He also built the tube and mount, and although the instrument was only six inches long, its magnification was nearly 40×, better than anything possible with a 6 foot refractor. The earliest surviving of Newton's letters mentioning the new telescope dates from February 1669, but it only came to the notice of the Royal Society towards the end of 1671: the fellows immediately asked to see it, and Isaac Barrow brought it to them from Cambridge before the year was out. It caused an immediate sensation.

Henry Oldenburg, the secretary of the Royal Society, lost little time in writing to Newton to thank him for the telescope, telling him, in the same letter, that he had been proposed for election – which followed on 11 January. Oldenburg, anxious to prevent any 'Usurpation of forreiners'[84] – a very real risk in the climate of the day – communicated all the details of Newton's instrument to

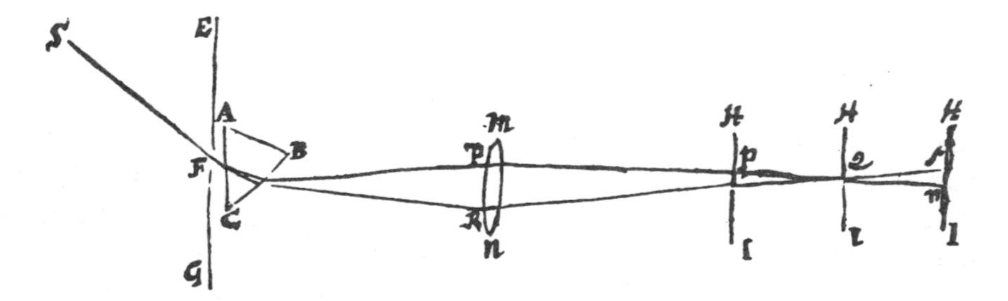

Figure 2.4 Newton's diagram of a prism and lens system for analysing the sun's spectrum.

Christiaan Huygens in Paris. This was good strategy: if Huygens, recognised as Europe's leading scientist, particularly expert in the field of optics, accepted Newton's priority, his judgement would be decisive.

Huygens, who had been elected to the Royal Society in 1663, shared its enthusiasm for the 'marvellous telescope of Mr. Newton', while the inventor himself expressed his surprise 'to see so much care taken about securing an invention to mee, of wch I have hitherto had so little value'. At the same time he promised 'to testify my gratitude by communicating what my poore & solitary endeavours can effect towards the promoting your Philosophicall designes'.[85] He kept the promise on 6 February 1672, by posting for publication in the *Philosophical Transactions*, a letter that the head-note summarises in the following words:

> A Letter of Mr. Isaac Newton, Mathematick Professor in the University of Cambridge; containing his New Theory about Light and Colors: where Light is declared to be not Similar or Homogeneal, but consisting of difform rays, some of which are more refrangible than others: And Colors are affirm'd to be not Qualifications of Light, deriv'd from Refractions of Natural Bodies, (as 'tis generally believed;) but Original and Connate properties, which in divers rays are divers: Where several Observations and Experiments are alleged to prove the said Theory.

The letter represents what in Newton's day was a new way of practising science. By submitting his findings to the Royal Society, Newton was among the first to initiate discussion and debate within a community of scholars, who communicated with each other in the pages of a scientific journal, in this case the *Philosophical Transactions*. As T. S. Kuhn[86] has pointed out, 'through the discussion, in which all the participants modified their position, a consensus of scientific opinion was obtained. Within this novel pattern of public announcement, discussion, and ultimate achievement of professional consensus, science has advanced ever since.'

Like much of good science, Newton's explanations are simpler than the alternatives with which they had to compete: their subject matter is also as fundamental as any in the whole of science. He described his discovery of the true nature of light as 'the oddest, if not the most considerable detection, which hath hitherto been made in the operations of nature'. This may sound arrogant, but to quote Kuhn[87] once more, 'An innovator in the sciences has never stood on firmer ground.'

Newton's achievements in the physics of optics were not, however, all gain. When it came to the actual nature of light, he worked out a corpuscular theory, according to which light must consist of a stream of infinitesimal particles, whose velocity would depend on the medium through which they travelled. Now although both Galileo and Descartes had been able to determine crude lower limits for the velocity of light, it was only in 1676 that the Danish astronomer, Ole Roemer (1644–1710), by observing the eclipses of Jupiter's inner moon over a period of six months, noted a 22 minute time discrepancy, that could only be explained by equating it to the time that light would take to travel a distance equal to the length of the diameter of the earth's orbit round the sun. Because, in the seventeenth century, there was no accurate measure of the distance between the earth and the sun, velocity calculated on this basis was subject to a substantial error: none the less the principle was sound and significant, also, for proving false any theory requiring the instantaneous transmission of light.

Newton's corpuscular theory of light, combined with Snel's law, required the ratio $\sin i : \sin r$ to be the same as $v_r : v_i$, where v_r is the velocity of the refracted light, and v_i that of the incident light (see Figure 2.2). This would require light to travel faster in a dense medium, such as water. Huygens, on the other hand, developed an elaborate wave theory of light (from which Snel's law could be deduced mathematically): according to this theory the ratio $\sin i : \sin r$ would be the same as $v_i : v_r$, the inverse of that required by Newton. This, although correct, does not prove the correctness of Huygens' wave theory, which is in fact seriously flawed. The accurate measurement of the velocity of light in the laboratory had to wait until the nineteenth century, and it was only in 1850 that Jean Bernard Foucault (1819–68) in Paris produced the first accurate comparison between the velocities of light in air and in water. This was then seen as a decisive confirmation of the correctness of a wave theory of light (which, in the twentieth century, would have to give way to the quantum theory pioneered by Albert Einstein, as related in Chapter 7).

Huygens had also observed how two crystalline blocks of transparent Iceland spar, on being revolved relative to each on their common axis, successively blocked and allowed the transmission of light. This was the result of polarisation, but the correct explanation of the phenomenon, which is that a polarising medium such as Iceland spar only allows light waves to be propagated in one plane, came only in the nineteenth century. This, the principle behind today's Polaroid sun-glasses, if known in Huygens' day, would have confirmed his theory of light.

Light interference

There was, however, one phenomenon, light interference, known to both Newton and Huygens, but understood by neither of them, which at the beginning of the nineteenth century seemed to vindicate Huygens and discredit Newton. This was the result of work done by Thomas Young (1773–1829) in England and Augustin Fresnel (1788–1827) in France. The essential apparatus is illustrated by Figure 2.5. This shows a cross-section of three screens, of which the first has one narrow slit and the second two, while the third reflects the image projected by a beam of light. The image will only appear if the slits are sufficiently narrow, but will then vary according to the distances between the screens: in the right setting it will take the form in Figure 2.5. Light, coming from the left, then projects as the pattern shown in the figure.

This phenomenon can only be explained by a wave theory of light. However, waves take two forms, one transverse, such as can be observed with a vibrating string, and the other, longitudinal, which is the form of sound waves propagating by a sequence of air compressions and dilations emanating from their source. Huygens'

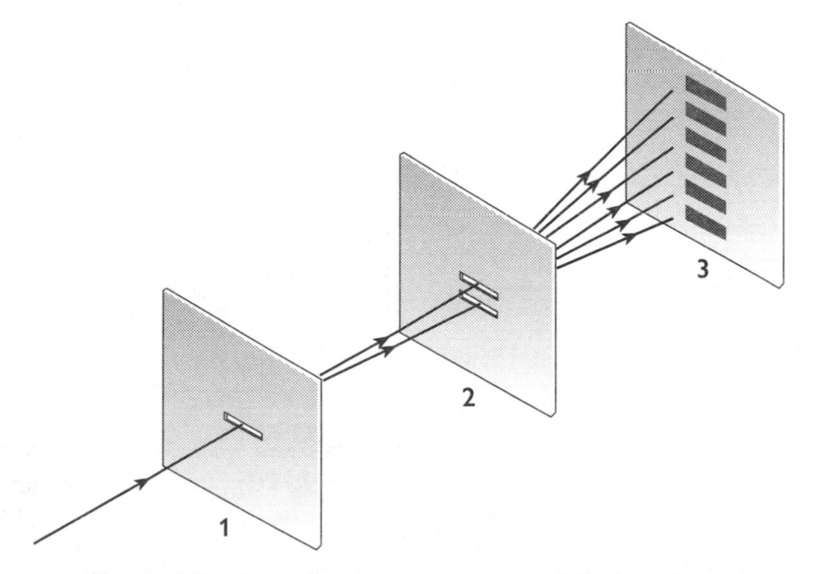

Figure 2.5 Apparatus for demonstrating light interference.

theory was based on longitudinal waves, but Fresnel showed that only transverse waves could explain double refraction, a phenomenon occurring when light passes through certain crystals. Until the end of the nineteenth century some still held that light waves could be both transverse and longitudinal, but quantum theory now allows only transverse waves.

Neither Young nor Fresnel had an easy ride in their chosen field of optics. Young's theory was suspect in England because it was too anti-Newtonian, and Fresnel in France had to contend with all the political upheavals occurring during his lifetime. He and his brother Leonor were, however, popular in Britain, not for the wave theory of light but for the lenses they developed for lighthouses. These, after much local controversy, were adopted by successive generations of Stevensons for all the Scottish lights.[88]

Young's and Fresnel's wave theory came at just the right time, since the new science of spectroscopy provided the means of measuring the wavelength λ for different colours of light. With the increasingly accurate measurement of the velocity v of light, this enabled the frequency ν to be calculated according to the elementary formula $v = \lambda\nu$, which is still fundamental.

3

Science, technology and communication

Travel and navigation

THE SCIENTIFIC revolution of the seventeenth century unfolded in a continent where communications had little improved since the time of the Romans: the Romans' skill and enterprise in civil engineering, still to be seen in roads and aqueducts, was almost completely lost to medieval Europe. Such improvements as there were, notably in the second millennium, came from the Islamic world.

The two centuries before 1600 had, however, seen considerable gains in mathematical notation, and above all the spread of knowledge by means of books. By the end of the sixteenth century, the Dutchman, Willebrord Snel, making use of new resources in trigonometry, had established triangulation as a means for land surveying that could extend over any distance.

The principle was simple: any area could be surveyed from a fixed baseline, whose length and orientation were accurately known. This would then be the base of a triangle, whose apex could be any point visible from the two ends of the baseline. The triangulation point thus became the basis of all land surveys.

The method required only an accurate means of measuring the angles between the baseline and the two other sides of the triangle. The theodolite, with plain sights, dates from the sixteenth century, but once the telescope, invented in the early seventeenth century, was introduced by Jonathan Sisson, it soon became a standard component.[1] It is mounted so as to allow both horizontal and vertical rotation, with the number of degrees in both cases shown on graduated circles. A perfectly horizontal base is essential for accuracy: this is achieved by a system of adjustable screws working in conjunction with a spirit level.

If triangulation was the means for making maps of unprecedented accuracy, these did little to improve transport and communication. In seventeenth-century England commerce depended largely on rivers, so that unlikely places such as Cambridge were important ports. There were no all-weather roads, so that pack-horses were the only reliable year-round means of transport over land. Wheeled transport, outside towns, depended on dry weather. The position was no better in continental Europe, let alone the New World.

The seventeenth century did, however, witness the beginnings of a process that would transform navigation by sea, which, with the discovery of the New World, had become every day more critical. Although, on the other side of the world, Pacific islanders crossed thousands of miles of ocean, using a system of navigation in which the rising and setting of known stars played a part, the canon of accurate astronomical records first established by Tycho Brahe in the sixteenth century went far beyond the position of stars as lodged in the memory of sailors.[2]

At the turn of the eighteenth century, mathematical tables, accurate telescopic observation and Newtonian celestial dynamics provided, in principle, all the means necessary for precise navigation on the open sea, so long as the weather allowed the sighting of heavenly bodies. Two methods were possible: one required measuring the angle between a star and the moon, the other, while only requiring the altitude of a star above the horizon to be measured, also depended on the accurate measurement of time at sea.

Both methods looked like practical propositions, particularly after the publication, in 1725, of John Flamsteed's catalogue of 3000 stars, and the invention in 1731 of a double-reflection quadrant (prototype of the sextant). The first method faced two problems: one, practical, was measuring the necessary angles at sea, and the other, mathematical, was the lack of lunar tables. The second method failed simply for the want of an accurate chronometer: its use was otherwise much simpler.

In mid-century both methods found a solution to their problems, but that for the second of them ensured that it would be accepted as standard into the indefinite future. This was the invention of John Harrison's (1693–1776) H4 chronometer. Its story is told in Dava

Sobel's *Longitude*,[3] subtitled 'The True Story of a Lone Genius Who Solved the Greatest Scientific Problem of His Time'. Harrison may have been a lone genius, but the problem he solved belonged not to science, but to technology. He made a clock that would keep time at sea with unprecedented accuracy.

The book's title shows why this was so important. Because the earth rotates on its axis once every 24 hours, the observed altitude of any heavenly body, at any given time, uniquely determines the longitude of the observer – that is, the number of degrees east or west of a standard meridian. There are two coordinates here: the altitude, which can measured from the horizon at sea by a quadrant, and time, which in principle can be told from a clock, or in the technical jargon of navigators, a chronometer.

The Longitude Act of 1714 offered a prize of up to £20,000 for a method to determine longitude to a prescribed degree of accuracy. The means were left open to all competitors, but from an early stage it was clear that a sufficiently accurate sea-chronometer would qualify for an award. John Harrison, born in Yorkshire, grew up in Lincolnshire. This was also Isaac Newton's home county, but the great man himself was sceptical about the chances of a 'watch', noting that 'by reason of the motion of the Ship, the Variation of Heat and Cold, Wet and Dry, and the Difference of Gravity in different Latitudes, such a watch hath not yet been made'.[4] This was the ultimate challenge to John Harrison.

Before he was twenty, Harrison, the self-taught son of a carpenter, started making pendulum clocks out of wood. His local reputation allowed him to pursue this craft, and with his brother, James, he built up a sound business: they improved the accuracy of their clocks first by inventing a pendulum which maintained a constant time across changes in temperature, and second by devising a new 'grasshopper' escapement. Their clocks, tested against the motion of the stars, were accurate to within a second a month – a level unequalled elsewhere.

It is not surprising that Harrison, somewhere around 1727, began to think of the £20,000 prize. The problem (which he saw immediately) was that no pendulum-clock could keep time in a rough sea. In 1730, Harrison, after three years working on plans for a timepiece without a pendulum, went to the Greenwich Observatory to show them to Edmond Halley, the Astronomer Royal. Halley, somewhat sceptical, referred him to George Graham, FRS and England's leading scientific instrument maker. Graham, impressed, sent Harrison home with a generous loan to help him develop his first model.

Five years later this led to the H1, Harrison's first chronometer. He submitted it (accompanied by a strong recommendation from Graham) to the prize Commissioners, and was granted a sea-trial – a voyage to Lisbon and back. When land was first sighted on the return journey, the chronometer's time correctly showed that it must be the Lizard, and not the Start, as the ship's master had reckoned – both points being on England's south coast. The master was so impressed that he wrote out a certificate praising the chronometer's accuracy.

Armed with this Harrison returned to the Commissioners. Instead of pressing his claim for a West Indies trial – essential for a prize award – he mentioned the defects of H1 and asked for time and modest financing to correct them. This was the beginning of a laborious process, leading first to H2, then H3 and finally, in 1759, H4. This was a completely new model, weighing only three pounds and entirely contained in a brass case five inches in diameter.

Harrison was delighted with what he had achieved, claiming 'that there is neither any Mechanical or Mathematical thing that is more beautiful than this my watch or Timekeeper for the Longitude'. He was right, and ready once more to face the Commissioners.

Among them he encountered a formidable opponent, Nevil Maskelyne – a clergyman as well as an astronomer (which may explain his having been described as 'rather a swot' and 'a bit of a prig'). In the scientific establishment, Maskelyne, who would become Astronomer Royal in 1762, was an insider who, following Newton's lead early in the century, had long backed the lunar method. He had good reason: as early as 1754 a German map-maker, Tobias Mayer, had compiled the necessary tables and sent them to London. Sea-tests showed that they had the required accuracy, and, although Mayer died in 1762, the Commissioners awarded his widow £3000.

This was not all. Also in 1762, Maskelyne, committed to Mayer, published an English translation of his lunar tables, the *British Mariner's Guide*. This then led to the first edition of the *Nautical Almanac*, published by Maskelyne, as Astronomer Royal, in 1766. If Harrison's H4 was to win the battle, years of hard work would go for nothing.

It is no wonder that Harrison encountered opposition, exacerbated by his reluctance to tell the secrets of H4's mechanism.

Finally, in 1764, a sea-trial with Harrison and his H4 sailing to Barbados and back proved his claims, and he was awarded £10,000: the rest of the award had to wait until 1773, when he was eighty. The chronometer was an immediate success, and the necessary tables, confusingly named the *Abridged Nautical Almanac*, were published. Captain James Cook took an H4 with him on his second long voyage of exploration (1772–75), and related how 'our never failing guide, the Watch' performed triumphantly. The lunar method fell out of use.

If the H4 revolutionised travel by sea, inland travel also made great progress. In the first half of the eighteenth century, rivers were dredged and locks were built, and, in the second half, canals completed the system of inland waterways. And at last, something was done about the roads, so that a network of stagecoaches, allowing relatively comfortable long-distance travel, developed.

Steam and hot air

All this went hand in hand with the industrial revolution, and the new wealth it created. Starting as early as 1712, the use of steam engines for pumping water out of the new deep mines was to lead to a long line of research relating to different kinds of energy. (In Italy, Torricelli had already shown that because of atmospheric pressure, water will not rise more than 33 feet in a suction pump). Towards the end of the century, the much improved steam engines invented by James Watt (1736–1819) were to transform British industry, and on Tyneside wooden rails were use to run coal-wagons down to the river.

At the same time the French moved in quite a different direction. On the 4 June 1782, a hot-air balloon, designed by two brothers, Joseph and Etienne Montgolfier, made its first ascent from the city of Annonay. In the town square, where the event took place in front of the Deputies of Vivarais (and many other people), the final words on the commemorative plaque are 'ICI EST NEE LA NAVIGATION AERIENNE'.

This is something of an exaggeration: the balloon was unmanned. In the event, two other men, Pilatre de Rozier and the Marquis d'Arlandes, were the passengers in the first manned flight, which took place in November 1783: this covered a distance of 7½ miles in something under half an hour.

Travel by balloon has made remarkably little progress since 1782: the balloon used for the successful circumnavigation of the world in

1999 was still the same basic Montgolfier model. And if practical air travel had to wait until the twentieth century, the scientific potential of the balloon was realised immediately. Lavoisier saw that hydrogen, a gas much lighter than air, could provide the essential lift, apparently disregarding the risk of fire.[5] Balloons had the most to offer to meteorology, where they are still used for measuring atmospheric pressure, wind speeds and other variables at high altitudes. In the United States, James van Allen, in 1946, found a way to use rockets to launch balloons, which he then used for taking high-altitude measurements of cosmic rays.

The historical importance of the Montgolfier balloon is that it opened an entirely new area to scientific research: at the heights reached by balloons, any number of non-terrestrial phenomena can now be observed, particularly in the field of radiation from space,[6] such as X-rays, where the earth's atmosphere prevents observation at ground level. Physics, however, was nowhere near this stage at the end of the eighteenth century.

Uniform standard measures

If, in spite of the Montgolfiers, scientific research was to remain essentially earthbound throughout the nineteenth century, the infrastructure upon which it depended was rapidly transformed. This was the result of four inventions, in the 50-year period 1790–1840. In chronological order these were a standard unit of length, steam locomotion, electric telegraphy and the postage stamp.

In 1788, a year in which there were 2000 units of measure current in France (most of them used only in one locality), a commission of six scientists was set up to consider how to establish a uniform system. Its members, who included Coulomb, Laplace and Lavoisier, could hardly have been more distinguished. It would start work in 1789, exactly a thousand years after Charlemagne had established uniform measures throughout his empire (some of which still survived in Britain).

The Commission's first decision was to make a completely new start, with some constant of physics as its base. There were two possibilities (neither of which would have been open to Charlemagne). One was to make use of Christiaan Huygens' discovery that the period of oscillation of a pendulum depended only on its length (so that, for instance, the standard could be the length of a pendulum with a period of one second). The other possibility was to base the standard on the length of a meridian (that is, a great circle passing through the two poles). The National Assembly could not make up its mind: on 8 May 1790 it decided for the pendulum; on 30 March 1791, for a quarter of a meridian (that is, the distance between a pole and the equator). At the same time Lavoisier had devised a means for accurately determining the weight of a prescribed unit volume of water: this would then provide a new measure of weight, linked to that for length.

At the end of the day the pendulum was rejected, partly because it lacked charisma – but also for the good scientific reason (already known to Newton) that gravity varies slightly over the world's surface.[7] The problem, then, was to measure the meridian: the only practical way to do this was to find a meridian, running precisely from north to south and joining two coastal locations. The difference in the two latitudes (determined astronomically) then provides the means for measuring the length of the quarter-meridian.[8]

Conveniently France proved to be the only country in the world where a meridian could be found satisfying the requisite conditions; even more conveniently it could be chosen to pass through the Paris Observatory. In fact, as shown on the map in Figure 3.1, the meridian so chosen intersects the coast of the Mediterranean just inside Spain, but with a little diplomacy French surveyors could be allowed to start their work there.

This was exactly how the operation was planned: two surveyors would map the line of the meridian by means of the triangulation process established by Snel two centuries earlier. One would start at the north end, and the other at the south, to meet, by pre-arrangement, somewhere in the middle. And in 1791, Lavoisier, who had become Treasurer of the Academy, arranged for the necessary finance.

Two astronomers, Pierre Méchain (1744–1804) and Jean-Baptiste Delambre (1749–1822), were appointed to the task and equipped with a new instrument, superior to the English theodolite, invented by the Chevalier de Borda in 1780.[9] The two could hardly have been more different, as would be reflected in the way they carried out their work and surmounted the many obstacles encountered: Méchain, who would work north from the coast near Barcelona, was pessimistic and withdrawn, while Delambre, who would work south from Dunkirk, was optimistic and outgoing.

The distances to be covered by each were measured in *toises*, then the unit most commonly used (but due to be superseded as a result of

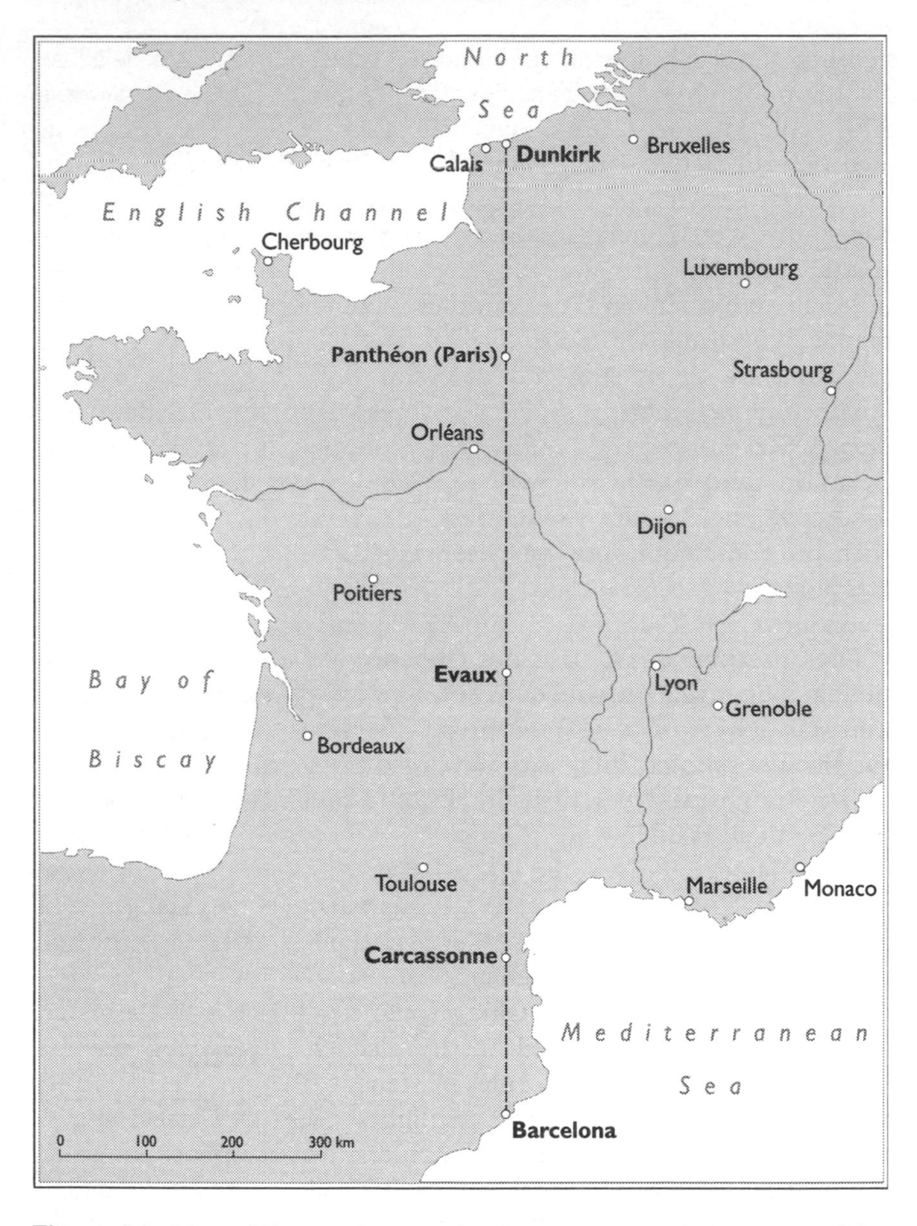

Figure 3.1 Map of France showing the five astronomical stations used for establishing the metric system.

the task in hand). Because the Spanish sector was almost unknown, Méchain was assigned much the shorter distance, 170,000 toises, where Delambre got 380,000. The two would then meet in the small town of Rodez, somewhere south of the Dordogne.

The *modus operandi* was to carry out successive triangulations by sighting standard signals, in the form of large coloured boards, placed on local high points, sometimes natural (e.g. the summit of a hill), sometimes man-made (e.g. the top of a bell-tower). These then defined stations for locating succeeding triangulation points. In addition, there would be five astronomical stations, located by star-sights as with sea navigation. Two of these were the terminal points, Dunkirk and Barcelona, a third was the Panthéon in Paris, and the two remaining, Carcassonne in south, and Evaux in central France. The result was that there would be four separate stages in measuring the distance by triangulation.

The time was hardly propitious for such an undertaking: the French Revolution did not make life easier for Méchain and Delambre, and suspicious local people, without any idea of what was going on, obstructed the work when their help was needed. With the rudimentary infrastructure of the time, many triangulation points were almost inaccessible – and things were worse when the weather was bad.

The operation was carried out with two baselines, each 12 kilometres long. This distance had to be measured with extreme accuracy; otherwise the whole project would be worthless. There was also the problem of finding two areas, along the meridian, each with a straight road across perfectly flat terrain. In the north this was the main road between Melun and Lieusaint, just south of Paris. Delambre built two stone pyramids, 25 metres high, at each end: even so, 500 trees had to be cut down to clear the line of sight between them. Equally thorough preparations were needed for the southern baseline near Perpignan.

The actual measurement, taking some seven weeks in the early summer of 1798, was carried out by placing, successively end to end, four identical platinum rules of standard length. Endless care was taken to protect them from sunlight, to ensure perfect alignment and fit between two successive rulers. Using a system devised by Lavoisier (who by this time had lost his head to the guillotine), a copper ruler, with a different coefficient of expansion, was used for corrections taking into account changes caused by heat in the length of the platinum standard. Some idea of the care taken is shown by an average rate of progress of 20 metres per hour. At the end of the day, when the two baselines were compared as a result of the triangulations carried out across the distance separating them, the error was of the order of 3 centimetres over a distance of 12 kilometres – an astonishing degree of accuracy.

Méchain and Delambre were busy for more than six years, but while they were still at work, the Commission in Paris was also involved. First, it had to decide on new names for the measures, and then how they were to be related. To ensure that the new system could be used internationally, new terms were coined from Latin and Greek roots (following the practice, recently adopted, for the newly discovered chemical elements). The key units, named mètre, litre and gramme, could be subdivided into smaller units, defined by Latin suffixes, milli-, centi- and déci-, and consolidated into large units, with Greek suffixes, déca, hecto and kilo. At the same time the liquid measure, the litre, was defined as 1 cubic décimètre, so that the weight of a litre of water would then define a kilogramme.

In 1798, the year of completion, Napoleon, who would become first consul of France a year later, had led French armies in conquests that radically changed the political alignment of Europe. Talleyrand, the French Foreign Minister, acting on the principle of *carpe diem*, convened what was effectively the first ever international scientific congress. Its agenda had one main item: the adoption of the new metric system.

The powers invited to the congress were either neutral or allied, the latter consisting largely of recently constituted French puppet states, such as the Dutch Batavian Republic. England, which on 1 August 1798 had destroyed the French fleet at the battle of Aboukir, was not invited, nor were Prussia and the United States. The English-speaking world, with its archaic system of weights and measures, is still paying the price. The rest of the world has had the benefit of the metric system for more than 200 years.

Steam locomotion: the first powered transport

George Stephenson (1781–1848), known to the Victorians as 'The Father of Railways', was born in poverty and never attended school. Significantly, the Stephenson family lived in a coal-mining village, close to the River Tyne in the north of England. The young Stephenson, therefore, grew up in what for its day was a world of advanced technology. Coal was the only source of energy for industry, and much of it was consumed in the process of mining and transporting it.

Every day, wagons carrying coal, drawn by horses but running on wooden rails, passed the home where Stephenson grew up. There his earliest ambition was to become an engineman, overseeing the operation of a steam engine, and after three years as a fireman, he was appointed a 'plugman' responsible for the operation and repair of a pump engine at a local mine. This was a good job for an illiterate eighteen year old, but Stephenson saw that without learning to read and write he had little future. He therefore took lessons (also in arithmetic) three nights a week after work.

During his twenties Stephenson, in a number of different jobs, succeeded on at least two occasions in making repairs to machinery where other craftsmen had been defeated. On the second occasion, his repair to the pump at the Killingworth Colliery allowed the mine to reopen after it had been flooded for a year. There, in 1810, Stephenson, twenty-eight years old, was appointed engine-wright, and almost from the first day his innovations substantially reduced the working costs of the colliery.

In the following years any number of steam locomotives were invented, but either their boilers blew up or they chewed up the rails: one such had even been introduced in the neighbouring Wylam colliery. Even so, an efficient, economic and reliable locomotive engine had yet to be invented. This was the task that Stephenson set himself in 1814, a year in which there was already considerable interest in the possibility of steam-powered railway travel.

Stephenson's first successful steam locomotive, the *Blucher* (1814), was used for hauling coal-wagons over short distances, such as that between Killingworth and the Tyne. Its success was due not only to better design, but also to better machining: Stephenson, who had noted how other locomotives had failed for poor workmanship, set unprecendently high standards for precision.

Local success at Killingworth led a prominent Quaker, Edward Pease, to propose a railway over the twenty miles between the mining centre of Darlington and Stockton, the seaport for shipping coal to London. Stephenson accepted this unprecedented challenge and the railway, when it opened in 1825, carried passengers as well as coal.

Although revolutionary, this was not quite enough. Opponents of railways questioned the capacity of iron wheels to run on iron rails, and doubted whether trains could cope with gradients in hilly country. Stephenson had the vision to persevere, and in 1831 the Liverpool and Manchester Railway, built over difficult terrain, opened for traffic – and boasted the *Rocket*, the most advanced and best known of all Stephenson's locomotives. The world was convinced and Stephenson's fame was made: the future of transport lay with railways, whose speed and capacity for traffic far exceeded any previously known means of transport. The process of development started almost immediately, with 2000 miles of new lines laid in the 1830s, and another 3000 in the 1840s.

The rest of the world soon followed: France opened its first railway in 1833, Germany, in 1835. The first Dutch railway, from Amsterdam to Haarlem, which opened in 1839, was soon to have the distinction of proving the train's usefulness as a scientific instrument.

In 1842, an Austrian physicist, Christian Doppler (1803–53), professor in Prague, stated a principle according to which the observed frequency of a wave is higher or lower according to whether its source approaches or recedes from the observer. The rule was tested in 1845 by having a locomotive draw an open carriage with several trumpeters through a station on the Dutch railway. The pitch of the trumpets, as heard by observers on the platform, lowered immediately as the train passed by. The drop in frequency, related to the speed of the train, accorded precisely with Doppler's principle.

The 'Doppler effect', as recorded in 1845, is of extreme generality, so that it also applies to electromagnetic waves. The principle, applied to the so-called 'red shift' observed in the spectra of distant stars, today provides strong evidence for an expanding universe.

The electric telegraph

Before the end of the 1830s, it became clear that if railways were to be safe and efficient, some means must be found for communicating, along the line, both the impending approach of trains and their successful arrival (which would leave the line free for new traffic). The solution was Charles Wheatstone's electric telegraph, the first practical use of a current provided by a battery. This was first installed commercially in 1838, when the Great Western Railway laid a line between Paddington and Slough, opening a new era in railway signalling.

The receiving terminal consists of a compass needle placed between two poles of an electric magnet. The sending terminal consists of a handle with three positions. In the central position no current flows, so the electromagnet at the receiver is not energised and the compass needle also remains in the central position. Turning the handle either way, clockwise or anticlockwise, then switches on a current, the direction of which is such that the compass needle turns the same way. With three possible positions at both terminals, the receiver mimics the sender. With each position coded for a specific signal, 'line blocked', 'line clear', 'train on line', this became the basic railway block-instrument, still occasionally used to regulate the passage of trains over successive sections of track.

A more elementary use of electromagnetism allows a bell-code to be sent by telegraph. A key operated by the sender sends a current to an electric magnet connected to the clapper of a bell at the receiving end. Well into the twentieth century the bell, combined with the telegraph, sent trains safely and at high speed from one end of the country to another.

In the years 1832–35, Samuel Morse, in the United States, invented an alternative electric telegraph, for which he devised the famous Morse code, introduced in 1838. The receiver consisted of an electric magnet, fixed so as to attract a metal armature. Immediately the circuit was completed, it was broken by the armature being attracted to the magnet. The magnet lost its power, the armature was released, and the circuit was restored. In this way the cyclical process could continue indefinitely: the result was a buzz, which could be regulated by the adjusting screw.

Morse saw that by using a key to interrupt the circuit he could send a code consisting of short and long pulses – the once familiar dots and dashes – so that the two combined could represent all the letters of the alphabet. Soon, skilled operators could send and receive thirty words a minute, over any distance, so long as an electric cable could be laid across it. The first link ever, laid down in 1845, was between Baltimore and Washington. Helped by undersea cables, the system would cover the whole world before the end of the century.

The postage stamp

The postage stamp, first proposed by Rowland Hill (1795–1879) in 1837, and introduced in Britain in 1840, provided the means for an administrative reform of an ancient institution. Written communications had been sent over long distances ever since writing on clay tablets first developed some thousands of years ago. By 1840 the post office was already well established as a national institution, but every letter had to be accounted for separately and paid for on receipt by the addressee. This was slow, inefficient and extremely expensive.

Rowland Hill, who came from a family of social reformers, had seen how the industrial revolution had brought many young men to the new cities, separating them from families that remained in the countryside. At the same time also, popular education was making literacy much more widespread, while the new railways could carry post at a speed and on a scale never possible before. Putting two and two together, Hill saw that if a letter could be prepaid, by affixing a stamp bought from a post office, without the two transactions otherwise being related, the saving in costs would be enormous, the more so if the system attracted a large number of new users.

Hill's belief in the new system was justified beyond all expectations. The reformed postal service operated on a scale hundreds of times greater than anything that had gone before. Its success was such that within ten years it was being adopted world-wide, with postage stamps in countless different designs being printed in their millions. Hill's invention was perfect almost from the start: the only thing he failed to conceive of were the perforations in a sheet of stamps – his stamps had to be cut out with scissors.

For communication across the international community of scientists the new postal service soon became essential. Papers and periodicals could be circulated at low cost, as they still are, even in the present era of fax and e-mail. Since 1840 technology has been relatively unsuccessful in improving the system. Stamps are still sold over the post-office counter, and post is still delivered by hand. Mechanical sorting is standard, but must still deal with objects of many different forms and sizes. Bar-codes, instead of written addresses, and invisible codes, which can be read electronically, incorporated into stamps, add to efficiency, but communication by letter is by twenty-first-century standards extremely labour intensive.

Today's big question is how far e-mail and the internet will take over communication over long distances. One critical point here is the high level of investment required of all who use the system: computers are expensive, and prohibitively so in developing countries (where even telephones are scarce). The internet café may help solve the problem at a popular level and is certainly useful for travellers who can access their e-mail over the world-wide web.

Bell's telephone

This story has run too far ahead, and the thread must be taken up once more in the last quarter of the nineteenth century. In 1876 Alexander Graham Bell (1847–1922) was granted his first patent for a telephone: the instrument applies principles of both acoustics and electricity.

The impact of sound on a diaphragm (which is best made of fairly elastic metal) can transform a sound wave into movement, with the same mathematical profile: following the induction principle established by Faraday (see page 115), connecting the diaphragm to a magnetic core, free to move up and down in the inside a coil of wire, will generate an electric current, with the same profile (but 90° out of phase).

This can then be transmitted over a distance to a similar instrument, where the current received causes the diaphragm to vibrate in phase, causing sound waves identical to those impacting on the diaphragm of the sender. In principle no outside source of power is needed. In practice telephones work on a low-voltage direct current supplied by batteries: the magnetic system of the sender is then replaced by a pack of carbon granules, whose electrical resistance varies in phase with the movement of the diaphragm. The variations in the current are then taken up by the diaphragm at the receiving end, with the same result as before.

Edison's electric light

The carbon microphone, which has decisive technical advantages, was the invention of Thomas Edison (1847–1931), one of the most prolific inventors ever. His most useful invention was the electric light bulb: the underlying principle is basic in the transformation of energy from one medium to another. An electric current generates its own energy, according to an elementary formula: volts × amps = watts, all units familiar in any household connected to mains electricity.[10] This energy can be transformed into mechanical power, heat or light, for countless different uses.

The problem with light is that although a wire carrying an electric current can become incandescent, in normal circumstances the radiated energy will take the form of heat – as with an electric fire (where the glow of the elements is a very poor source of illumination). To be useful as a source of light, the temperature in the wire must be much higher: the problem then is that the wire will immediately be subject to combustion – simply as the result of the chemical processes investigated by Lavoisier (see page 152) a century before Edison. (The process is instantaneous as can be seen when the glass of a light-bulb is broken.)

Edison saw that placing a wire with very high resistance inside a glass bulb containing a vacuum could solve this problem. Without oxygen there would be no combustion, so that there was no upper limit to the temperature of the wire: this could be raised to a point of incandescence where the electromagnetic waves emitted were mainly of light frequency. The ideal effect comes from a long thin wire, which is the essence of the coiled coil light filament. Edison's invention (1879) preceded Hertz's discovery of electromagnetic waves, so his understanding of the underlying theory of his invention was inevitably one-sided. Even so, it was still sufficient, although following Ramsay's discovery of the inert gas argon in 1894, this for technical reasons came to replace the vacuum in the light bulb.

Hertz's waves

Hertz developed his apparatus (see page 214) for generating electromagnetic waves in the 1880s, and in the 1890s Rutherford in England and Guglielmo Marconi (1874–1937) in Italy showed how they could be transmitted over a distance. While Rutherford went on to pure science, Marconi applied his knowledge and, with his invention of radio, became a rich man. He found the practical means of generating Hertzian waves, at fixed frequencies, in such a way that they would transmit over long distances. With a simple waveform, little more was possible beyond making and breaking the electrical circuit that generated it. Even so, this was sufficient for the Morse code, adapting the basic principle established by Morse himself in 1830s.

Marconi sent the first radio signal across the Atlantic in 1901. The potential of the new medium for communicating with ships at sea (and later with aircraft) was recognised almost immediately, and in 1910 it was used to intercept Dr. Crippen, a notorious murderer, when he was escaping from England to Canada on the SS *Montrose*. This was probably the first use of radio to arrest a criminal.

Sound radio, as we know it, is based upon waves at low audiofrequencies (that is, those of audible sound) being imposed upon a continuous Hertzian carrier wave at a high radiofrequency. Some feature of the carrier wave then varies in step with the audiofrequencies in the sound wave in such a way that the receiving instrument can separate the two. The audiofrequencies are then converted into an electric current activating a diaphragm, as in a telephone. This is the loudspeaker. The technology is very complex, as can be seen by dismantling any radio, or now any television set. Until about 1950 it was based on circuitry connecting thermionic valves (which looked like complicated electric light-bulbs), but these have now been replaced by semiconductors, commonly known as transistors (which are much smaller, and key components of any computer). The related science belongs to solid-state physics, now a very active research field.

Hydrocarbon fuels

Transport in the twentieth century was revolutionised by the invention of engines running on hydrocarbon fuels – gasoline, diesel oil and paraffin. The molecules in a pure hydrocarbon consist exclusively of hydrogen and carbon atoms: of these benzene (see page 189) is the most fundamental. Fuels come from naturally occurring petrochemicals, consisting mainly of hydrocarbons, but with other chemical components, such as nitrogen, oxygen and sulphur. Refining produces first gasoline (with five to eight carbon atoms), then paraffin (with eleven or twelve) and finally diesel oil (with thirteen to fifteen).

Each of these three fuels has its own characteristic engine, the distinctive features being determined by the form of combustion. Gasoline, vaporised by means of a carburettor mixing it with air, is ignited by an electric spark and explodes to drive a piston in a cylinder: this explains the term 'internal combustion' (IC). Paraffin burns under pressure in a process of continuous combustion, so producing a jet of hot gases that can be used directly as a means of propulsion. Diesel fuel, in the form of vapour, explodes as a result of being compressed in a cylinder, hence the common designation 'compression ignition' (CI): as a result of the gas laws (see page 141) the temperature then increased to the flashpoint at which combustion occurs.

The IC engine was largely the invention of Karl Benz (1844–1929) whose first car appeared in 1885: the engine is still standard for automobiles, but the CI engine of Rudolf Diesel (1858–1913) is more suited for heavy vehicles, locomotives and boats, where it is more than twice as efficient as comparable steam engines. The much lighter IC engine, on the other hand, was much more suitable for aircraft, and was standard from the Wright brothers' first airplane until the invention of the jet engine driven by paraffin. Today's light aircraft still use the IC engine with gasoline as fuel. The advantage in every case was the extreme light weight both of the engine and the fuel it consumed. The need for a tender of coal and water was completely bypassed. A steam-driven aircraft would never have left the ground.

Powered flight

Apart from the essential IC engine, Orville (1871–1948) and Wilbur Wright (1867–1912) designed their first airplane according to aerodynamic principles of wing theory developed in the second half of the nineteenth century, mainly in Germany. The main problem, the shape of the cross-section of the wing, can be solved with the use of the mathematical theory of complex variables. The basic principle, established by the Swiss mathematician Daniel Bernoulli (1700–82), is that air must flow more rapidly over the top, than under the bottom, of the wing, so that the pressure of the air above is less than that below the aircraft, thereby creating 'lift'. For this to happen the aircraft must be moving forward: this, originally, was the function of the propeller, first developed in 1842 by Isambard Kingdom Brunel (1806–59) for his ship, the *Great Britain* – the fluid dynamic principles being the same. (The greatest engineering problem facing the Wright brothers was finding the right metal for a propeller shaft that would withstand the stress caused by torque.)

A jet aircraft substitutes the thrust of hot gases for that of the propeller. The distinction is critical for performance. The propeller is most efficient in the relatively dense atmosphere of lower altitudes, but this at the same time offers the most resistance to the forward movement of the airplane. The jet engine is most efficient at high altitudes where this resistance is at a minimum, the only limit to height being the need for some atmosphere to provide lift – even so the U2 reconnaissance airplane can fly at heights of up to nearly 30,000 metres.[11]

The power of a jet engine is essentially that of a rocket, so that the technology goes back a long way in history. In China, projectiles powered by rockets were used in the Sung-Chin wars, almost a thousand years ago.[12] The rocket was superior to any comparable war engine, in that it continued to deliver its power after launching. The basic science consisted of finding a fuel (gunpowder in Sung China) that would burn at an optimal rate. This is the essential chemical problem in all ballistics. Cordite, first developed in the nineteenth century as a propellant for shells, became standard for this purpose, simply because of its unprecedentedly high performance.

Rockets and space

In the Second World War rockets came into their own as a means of sending explosive warheads over long distances. This is the origin of the ballistic missile, although the term only became current in the 1950s. The most successful rocket, the V2, first used in August 1944, was German. In the years following the war, its inventor, Wernher von Braun (1912–77), would become the dominant figure in rocket technology, not only for missiles, but also for space travel.

The political history of the rocket in the twentieth century is at least as remarkable as its scientific development. Following the First World War, scientists in Germany, Russia and the United States already saw the rocket as a means for going beyond the earth's atmosphere into space. In Germany the leading figure was Hermann Oberth (1894–1989), and he, together with others sharing his enthusiasm for rockets, organised the *Verein für Raumschiffahrt*, the 'society for space travel'.

In 1928 Wernher von Braun became a member while still at school. He was a prodigy born into a wealthy, cultured, intellectual, aristocratic family: his father, a minister in the government that Hitler drove out of office in 1933, was intolerant of any form of corruption. Out of office, he survived the twelve years that Hitler was in power by successfully maintaining a low profile.

This was hardly true of his son. The younger von Braun, having enrolled in the Technische Hochschule in Berlin (where he saw his first rocket motor) soon realised that only the German military had the means to develop rockets, which also had the advantage of not being subject to the armament limitations imposed by the Treaty of Versailles in 1919. Von Braun's first employment, at the *Raketenflugplatz*, or 'rocket airfield', Reinickendorf, was soon noticed by the military, and he was offered a senior position in development and research. The year 1934 saw von Braun, twenty-two years old, gain his Ph.D. for a thesis entitled 'Combustion Phenomena in Liquid Propellant Rocket Engines'. This came from the Friedrich-Wilhelm University in Berlin, where von Braun had lectures from three Nobel prizewinners, Erwin Schrödinger, Max von Laue and

Walter Nernst[13] – giving him a useful connection with the scientific establishment. In the same year the rocket on which he had been working had its first successful altitude test.

Beginning in 1935, a new site, located near a remote fishing village, Peenemünde, on the Baltic Sea, was acquired for the military programme, and von Braun became its scientific director. This was to be the most successful of all the German weapons programmes in the Second World War. Von Braun's skill as an engineer, and his understanding of the problems of rocket performance, were focused on a single weapon, the A4 rocket. This had its first successful launch on 3 October 1942, when it reached a height of 85 kilometres over a range of 190 kilometres.

Although Hitler, following a revelation in one of his 'infallible' dreams, had predicted failure, full production was ordered. Then, on 17 August 1943, the RAF bombed Peenemünde, causing such damage that Hitler immediately ordered production to be transferred to an underground site, at the same time entrusting it to Heinrich Himmler's secret police. A disused mine in the Harz Mountains (far from Peenemünde) was converted into an underground factory, to be known as Mittelwerk, running on slave labour.

Von Braun, remaining in Peenemünde, soon made clear that his operation was not part of Himmler's empire. In March 1944 this led to his arrest on the charge that he never intended the A4 as a war weapon, but only for space travel – an incident that would soon become a useful part of his curriculum vitae (even though he was released within two weeks).

Mittelwerk in all produced nearly 6000 rockets,[14] of which 3200 were launched operationally. The first cities to be targeted, on 8 September 1944, were Paris and London.

Then, in January 1945, with Peenemünde facing the advancing Soviet armies, von Braun was ordered to evacuate his entire operation to Bleicherode in central Germany. The first train left a month later, and then in April came a final move to Oberammergau. With the end of the war fast approaching, von Braun found a safe house in the nearby village of Oberjoch. When the Americans arrived, he sent his younger brother, Magnus, who spoke near perfect English, to make contact and explain what he had offer to the West.

Von Braun, summoned for interrogation, was able to retrieve technical documents from Peenemünde, which he had hidden in a mineshaft. An American expert, after examining them, reported that 'one of the greatest scientific and technical treasures in history is now securely in American hands'.[15] Von Braun also told the Americans how to locate some 1000 people who had worked at Peenemünde but had escaped to the West. He also pointed out that the Mittelwerk production line would be part of the Soviet Zone of Germany. The Americans, in nine days, took away 341 freight cars containing parts and machinery, and shipped the whole lot to the United States. Von Braun was also beginning to sell the idea that he and his team should go as well.

By this time, with the Pacific war coming to its end after the first operational use of atomic bombs, the Americans had no doubt that they wanted von Braun. In the end he moved to the United States, with 126 of his colleagues from Peenemünde. The 127 Germans were first set to work in Fort Bliss, Texas, where they remained, officially, prisoners of war. Then in 1950 they became recognised immigrants, and in 1955, after a further and final move to Huntsville, Alabama, they all became American citizens in one grand ceremony.

The operation at Huntsville[16] was organised into thirteen divisions, and as late as 1962 all of them were still headed by one of the 127 original Germans (the last of whom would only retire in the 1990s); it was only in 1973, with a new director, that the Germans ceased to dominate at Huntsville. On 10 October 1957, the Soviet Union launched Sputnik, the first successful satellite, and, on 12 April 1961, Yuri Gagarin became the first man in space. These Soviet achievements were a considerable spur to work at Huntsville, and within months the Americans had drawn level. Then, in the course of the 1960s, with the Apollo Moon Project, led by von Braun, the United States drew ahead, to achieve the first successful moon-landing on 16 July 1969.

By this time the usefulness of rockets to science was beyond doubt. Even so, what was their essential contribution? The answer lies mainly in the character of the earth's atmosphere. For one thing, as Fred Whipple, professor of astronomy at Harvard, noted, the optical system of an earthbound telescope, is 'like a dirty basement window'.[17] Something can be done by building observatories at the top of mountains, in a relatively dry and dust-free location, such as Mauna Kea in Hawaii, where a whole cluster occupies the summit. But going into space, as with the Hubble telescope, provides a spotlessly clean window.[18]

Besides a telescope, many other kinds of instrumentation can be part of the payload of a rocket-powered vehicle in space, so that measurements can be taken of cosmic and X-rays, the frequency of light from the sun, to say nothing of the temperature, pressure and composition of the atmosphere at high altitudes. To give one specific example, the measurements taken by Geiger counters on the *Explorer I* satellite,[19] launched on 31 January 1958, led to the discovery of the Van Allen radiation belts[20] – now essential to atmospheric studies. The was only a beginning: the further spaceships such as the Voyager (illustrated in Figure 3.2) can reach, the greater becomes the wealth of information, particularly about other planets.

These discoveries were not von Braun's: he merely provided the technology. For as Fred Whipple said, 'Wernher was an engineer,

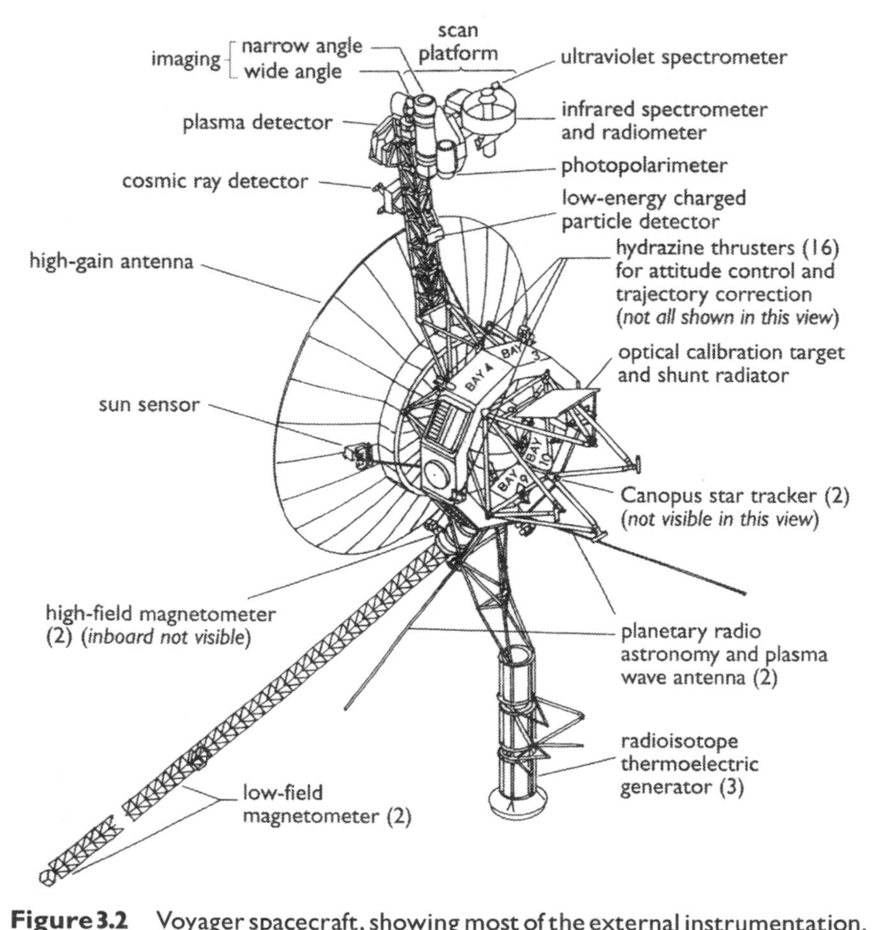

Figure 3.2 Voyager spacecraft, showing most of the external instrumentation.

not a scientist, in the eyes of scientists.' The technology had many dimensions, and von Braun was master of all of them. His particular strength lay in the use of liquid fuels[21] (as shown in his 1934 Ph.D. thesis). In addition, he was the first to develop booster and multi-stage rockets – so that, for example, the Saturn launched the first moon rocket, while the Apollo carried it on to the moon. The choice of materials, both for the rocket shells, and everything inside, was always critical; electronic systems had to be devised, with increasing dependence on computers; communications with the NASA[22] Johnson Space Center at Houston, Texas, had to be assured.

Few men in the last fifty years have attracted so much praise as von Braun. A good-looking man, blond and heavy-set, he looked people straight in the face, to capture them with his charm and charisma. He enjoyed sailing, flying, scuba-diving, but also played the piano (having had, as a boy, lessons from the composer Paul Hindemith) and the cello. He was variously described as a scientist, engineer, philosopher, humanitarian, politician, diplomat, realist and visionary – all in all, too good to be true.

His professional knowledge was essentially second-hand and derivative, but he coordinated and managed brilliantly, and people always saw that he was a winner. Throughout his life he made his own luck, and others paid the price. While he was at Peenemünde, the rockets he designed were being made by slave labour, by people working under appalling conditions, at Mittelwerk. Once these rockets were operational, they killed several thousand people: during the Second World War they were the most successful new weapon at the disposal of Hitler. It is not for nothing that von Braun's A4 rocket, once operational, was renamed V2 – with the V for *Vergeltungswaffen*, a weapon of retribution, which was precisely its appeal to the Nazi leaders.

The same easy conscience that allowed von Braun to prosper under Hitler stayed with him in the United States, where his rockets could carry an unprecedentedly destructive payload. They were never used to deliver the payload (just as the V2 did not save Germany in the war), but still von Braun's legacy is not all gain, in spite of the long familiar list of rocket-powered vehicles in space – Apollo, Mariner, Mercury, Galileo, Magellan, shuttle, Skylab, Viking, Neptune, Voyager – all evoke the memory of one man.

Even so, as Carl Sagan once said of von Braun: 'He was willing to use nearly any argument and accept any sponsorship as long as it could get us into space. I think he went too far... The modern rocket, which he pioneered, will prove to be either the means of mass annihilation through a global thermonuclear war or the means that will carry us to the planets and the stars. This dread ambiguity, which faces us today, is central to the life of Wernher von Braun.'[23]

And then, going back to the 1960s, Tom Lehrer reminds us how

When the rockets go up
Who cares where they come down
That's not my department,
Says Wernher von Braun.

4

Discovering electricity

Static electricity

THE FIRST serious study of magnetism and electricity was *On the Magnet*, published by William Gilbert (1544–1603) in 1600. The book was the result of long talks with navigators and experiments with lodestone. This, in Gilbert's day the only known material with the power of magnetic attraction, was the basis of a primitive compass, consisting of a raft, upon which the stone was placed, floating in a bowl of water.

The magnetic properties of the lodestone recorded in the book were discovered by systematic experiment, a procedure almost unknown to science before Gilbert's pioneering work. The book was also revolutionary in comparing magnetic with electrical attraction. This was a long-recognised phenomenon, in which hard substances, notably amber, after being rubbed by certain cloths, attracted light objects, such as paper, feathers and chaff. In the course of his researches, Gilbert, who called this property 'electric' (after *elektron*, the Greek word for 'amber'), found it was shared by glass, sulphur, wax crystals and several different gems.

The versorium (see Figure 4.1), an instrument designed and named by Gilbert, is also pictured in his book. This is the first known pictorial representation of any electrical instrument. The needle swings on its pivot towards any electrically charged body, which makes it the first example of an electroscope, a standard piece of laboratory apparatus. This follows from the fact that before Gilbert's day St. Elmo's fire (sometimes observed at sea in stormy

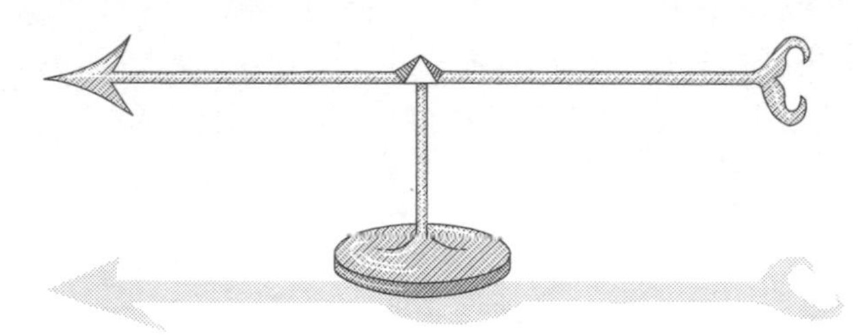

Figure 4.1 Gilbert's electroscope.

weather) and the torpedo fish (described by Aristotle) were the only recognised electrical phenomena.

In 1660 Otto von Guericke constructed the first electrical machine: this consisted of a large rotating sulphur globe, which by brushing against cloth attracted light material. Although this was little more than a mechanism for demonstrating an already-known phenomenon, von Guericke was still the first to see man-made electrical sparks and recognise them for what they were. In the early 1700s, Francis Hauksbee in England, with apparatus based on the same principle but with a rotating hollow glass globe, produced a luminous glow with the friction generated when he placed his hand on it. This was another electrostatic phenomenon.

All these various instruments operated by accumulating a limited electric charge. In about 1745, in the Dutch city of Leyden, a means was found of storing these charges. The so-called Leyden jar was simply a substantial glass container, with separate layers of metal foil on the inside and outside surfaces. The inside was charged by a metal chain connecting it to a charged body, which then lost its charge to the jar. This apparatus, the first example of what is now known as a capacitor, allowed a very considerable electrostatic potential to be built up between the inner and outer surfaces of the jar. The power produced by releasing it proved sufficient to kill small birds, ignite alcohol, light candles, explode gunpowder or inflict on ordinary mortals a considerable electric shock. This effect delighted King Louis XV of France, when he had 700 monks hold hands in Paris, with the two at the ends of the chain using their free hands to discharge a Leyden jar. But, for all that, it remained a key piece of apparatus (now to be seen mainly in museums) until the end of the eighteenth century.

This century knew, however, one scientist interested in electrical phenomena who stood head and shoulders above all his contemporaries. This was Benjamin Franklin (1706–90), unrivalled for his international reputation in many different fields – science and invention, publishing and politics. Until the twentieth century he was probably the greatest name in American science.

Franklin, a second-generation American, had to leave school when he was ten to help his father, a tallow chandler and soap boiler, in his business in Boston. Because he loved books, his father decided that he should become a printer's apprentice. Before he was eighteen he broke the indenture and moved via New York to Philadelphia, which he would make his home for the rest of his life.

Success as a printer came so fast to Franklin that Governor Keith offered to sponsor a trip to London to buy equipment and make business contacts. On the ship Franklin discovered that the governor had arranged for neither funds nor any useful introductions. None the less, after two years usefully spent in London, Franklin returned home, still only twenty years old.

Back in Philadelphia, Franklin set up his own printing shop, started a newspaper, the *Pennsylvania Gazette*, and published *Poor Richard: An Almanack* – a book which made him famous – and all this before he was thirty. He soon had a finger in every pie, and with his amiable disposition and considerable talents, became a major figure in the colony. In 1751 he became a member of the Pennsylvania Assembly, which in 1757 sent him to London as their agent. By this time, his scientific research had won him in 1753 the Copley Medal of the Royal Society and in 1756 election as a Fellow, with a number of articles already published in its *Philosophical Transactions*. In America, Harvard, Yale and William and Mary awarded him honorary degrees. All this happened in the first forty years of a man who would live to be eighty-four. Just to read about Franklin leaves one breathless. What then did he achieve as a scientist?

The answer to this question is to be found in the title of his book *Experiments and Observations on Electricity, Made at Philadelphia in America*. This book, first published in 1751, when Franklin was forty-five, soon became a standard work, not only in English, but in French, German and Italian translations. It was the product of intensive work over a six-year period, 1745–51, during which Franklin purchased any electrical apparatus he could get his hands on. His experiments with both glassy and waxy substances led him to conclude that all bodies contained charges in a neutral state of electrical equilibrium. Friction could result either in a gain of electricity, in which case the charge was positive, or a loss, in which case it was negative. Equilibrium could be restored by discharging the charged body (whatever the sign of its charge) – a process familiar from working with Leyden jars. This was the 'one-fluid' theory that secured Franklin's election to the Royal Society.

At an early stage Franklin's experiments focused on how to discharge charged bodies: he discovered that a pointed conductor would produce a steady discharge at a distance of 6–8 inches, while, for a blunt conductor, it could do so for no more than an inch – at which distance there would be a sudden discharge with a spark. Such experiments led Faraday and his collaborators to the concept of 'electrical fire... not created by the friction, but collected only' with each body containing its own 'natural quantity'.[1]

This was essentially a theory of the zero-sum conservation of charge. Franklin succeeded in giving Leyden jars both positive and negative charges, and showed that the force itself was stored in the glass of the jar with the charge being proportional to its surface area. Basically all that was needed was a plate of glass separating two metal plates: applying this discovery, Franklin produced a glass and lead battery consisting of eleven such condensers (as they would now be known) connected in series.

The fact that both positive and negative charges were possible led Franklin to the discovery that bodies with a like charge repelled each other, whereas those with unlike charges attracted each other. The phenomenon led Franklin to develop an elaborate theory of 'electrical matter' and 'electrical atmospheres', but although it could explain the repulsion between positively charged bodies, or the attraction between bodies with unlike charges, it could not accommodate the repulsion between negatively charged bodies.[2] On the other hand, according to the theory, a charge in one body could be induced in another, and Franklin achieved this result experimentally.

At the end of the day, Franklin, the scientist, is remembered for his experiments with lightning. He was the first to conceive of the idea that this was an electrostatic phenomenon, in principle the same as the discharges produced by experiment. This was towards the end of the 1740s, and in November 1749 Franklin made a list of twelve observable similarities between the two phenomena. In July 1750, his *Opinions and Conjectures* stated his thoughts on this subject. Following his work with discharges through pointed conductors, he was already warning people not to be near 'prominences and points', such as church spires, during a thunderstorm.[3]

To confirm his theory Franklin devised two well-known experiments, one with a sentry-box and the other with a kite. The former required a sentry-box, constructed on an insulated stand, to be placed on a high building. A rod, 20–30 feet long, pointed at its far end, would then rise vertically through the roof of the sentry-box. When thunderclouds gathered, the man inside (who was advised to be well insulated), brought a wax handle close to the rod. The experiment succeeded when it attracted sparks from the rod, which could only have received its electric charge from the clouds.

The actual experiment was carried out for the first time in France in May 1752. The king, who had already shocked the monks, and his court were present. The occasion, which gave 'complete satisfaction',[4] was widely publicised and the experiment was repeated in many different places. Franklin had always made clear that the wax handle should be earthed with a metal wire to protect the man in the sentry-box. In St. Petersburg, a local scientist called Richmann omitted this precaution and was electrocuted. By this time Franklin had long enjoyed widespread international renown.

The kite experiment is even better known, particularly to American schoolchildren. It was also carried out in 1752, but this time by Franklin himself: he flew an ordinary kite in a thunderstorm, having attached a key to the wet string by a dry silk ribbon. Electric sparks once again showed that the key had an electric charge, which could only come from the kite. Always a practical man, Franklin, with his new understanding of lightning, invented the lightning conductor, a pointed metal rod which could be fitted to tall buildings or the masts of ships to conduct electricity safely to earth. This prevented any number of fires from being caused by lightning. (Franklin's versatility is shown by the fact he also invented the rocking-chair and bifocal lenses.)

Given the part Franklin played in the American Revolution – he was, with Thomas Jefferson and John Adams, one of the three authors of the Declaration of Independence – it is not surprising that he is remembered mainly as a statesman. Even so, his international reputation was acquired first as a scientist, and, as such, among the many he came to know was an Italian, Alessandro Volta (1745–1827), who at the turn of the nineteenth century was to publish an invention more important for the future of electricity than anything achieved by Franklin. It is to Volta's story that we now come.

The electrodynamic revolution

When Alessandro Volta (1745–1827) was born to a noble family in the beautiful lakeside city of Como in Lombardy, it was then part of the Austrian Habsburg empire: in 1859 it became part of Garibaldi's Italy. Volta was well-born, but his family fortunes were at a low ebb. This was no doubt the fault of his father, who, after being a Jesuit for eleven years, left the order at the age of forty-one to marry a lady, also of good birth but twenty-two years younger. Their family was brought up in the shadow of the Church: of five sons three became priests; of four daughters, two became nuns.

Volta, although backward when very young, suddenly showed great intellectual promise when he was seven, and when he left school at sixteen he shone particularly in chemistry, Latin, French and Italian. He combined these skills in a poem with 500 verses about Joseph Priestley, the English discoverer of oxygen and the author (in 1767) of a history of electricity. The family wished to steer the boy into the Church or the law, but trying both he liked neither. Instead he followed a career in physics and chemistry, working hard and neglecting both food and dress in a way reminiscent of the young Isaac Newton.

Volta's diligence was rewarded by his being appointed, in 1775, Rector of the Royal School in Como. In that same year, working in the school's laboratory, he developed his electrophorus, or carrier of electricity, which like Franklin's battery, could store a continuously increasing electric charge. The device became very popular, but it was not entirely original, and is now no more than a museum piece.

In 1778, Volta accepted a chair at the ancient University of Pavia, where Columbus had once studied. There he first worked on gases (following in the footsteps of Priestley and Henry Cavendish), but in 1782 he devised the 'multiplying condenser', an unprecedentedly sensitive instrument, capable of detecting negative electricity in steam, smoke and the gas produced by immersing iron in dilute sulphuric acid.

The early 1780s Volta devoted to travel, and in Paris he met not only Lavoisier and other notable French scientists, but also Franklin – the greatest figure in his own field of electricity. His election to the Académie des Sciences indicates his standing. In England he finally met Priestley, together with other Fellows of the Royal Society (which would elect him a Fellow in 1791).

In these final years of the eighteenth century, Luigi Galvani (1737–98), another Italian, but a physiologist, was investigating the anatomy of frogs. Observing how dead frogs convulsed when fixed to an iron fence with brass skewers, Galvani concluded that he had discovered 'animal' electricity produced by the frogs' muscular and nervous system. This result came to the notice of Volta, who showed that the effect observed had nothing to do with animals, but depended solely on the difference between the two metals and the nature of the substance that separated them. This result came from Volta's probing his own anatomy, particularly tongue and eyes, with strips of two different metals, such as zinc and silver, and noting the neurophysiological effects in a way that once more recalls Newton. The more Volta continued these experiments, the more he became convinced that the electricity came from the difference between the two metals, so exit Galvani (leaving us with the word 'galvanise').

Volta's insight led him to the invention of the pile, the first instrument ever to produce a continuous electric current. This was no more than the prototype of the power cell, still in everyday use, in countless different forms (including batteries), some 200 years later. Because of its fundamental importance, Volta's underlying principle must be described in some detail.

Volta discovered that a pile (his own word) consisting of two different metals with outside terminals, and separated by a layer of some non-metallic substance, could produce a continuous electric current. Take any single-cell battery apart, and this is what you find, but the same result can be achieved by sticking a copper pin and a zinc pin into a potato – to produce a current strong enough, say, for a hearing-aid.

To understand Volta's invention some basic terminology is essential. It also helps to look ahead in time, and adopt today's accepted explanation of the operation of Volta's pile. In modern terms, this consists of two metal electrodes, separated by an electrolyte of a suitable chemical composition. Now, as explained in Chapter 6, reaction is the key to chemistry: all metals are reactive, but at widely different levels. This means that they can be placed in order, to constitute the so-called reactive series. In simplified form, the lowest point is that of platinum, followed, in terms of increasing reactivity, by gold, silver and copper. The practical result, for these four metals, is that they occur in natural deposits, which means that all have been known since prehistoric times. Scientifically, it does not matter that they are embedded in complex geological strata. (In practical terms, needless to say, this can mean that extraction is extremely expensive when the source is deep underground, and contains only relatively small quantities of the metal. Gold and silver are quite easily refined: the problem is their scarcity.[5])

Continuing up the scale, there is a middle range of the reactive series defined by lead, tin, iron and zinc. These base metals are relatively common – iron, indeed, measures nearly 5% in terms of abundance.[6] The problem is that they occur in compounds, with each metal having its characteristic ore. Because the ores are either oxides, or capable of being reduced to this form by heating, the metal can always be extracted by burning carbon – originally as charcoal, more recently as coke. This explains why they have been so long known to humankind. Copper, although occurring in pure form, is best included in this group, since pyrites, its most common source, needs refining in much the same way as zinc or lead.[7]

In ascending order, aluminium, magnesium, calcium, sodium and potassium all belong to the top range of the reactive series. Although extremely common, they are so highly reactive as to occur only in compounds. None had been isolated when Volta invented his pile, but Davy (as described in Chapter 6) used the pile to develop the process of electrolysis, which then made isolation possible. So let us look at the pile again, first as invented by Volta, and then as developed by Davy.

The electric potential of the pile – today measured appropriately in volts – is determined by the distance separating the elements used in the electrodes in the reactivity series. From the base metals available to Volta and Davy, the best choice for the electrodes is copper and zinc, simply because they are furthest apart in the reactivity series. Since the electrolyte may be a compound, there are any number of possibilities. Volta's choice of a copper sulphate solution – which worked well enough – was largely practical. With the cell in use as a source of power, the zinc electrode, with higher reactivity, releases electrons in a process known as ionisation, and positively charged zinc sulphate begins to form around it in the electrolyte. This process, known as oxidation, is ultimately self-defeating, in that the cell becomes exhausted. At the same time the copper electrode gains electrons, in a process confusingly known as reduction.[8] The flow of electrons is then the electric current produced by the cell. The electrode losing electrons is known as the anode, and that gaining them as the cathode (although these terms were only coined by William Whewell half a century later). These correspond to the positive and negative terminals of any battery, and this was what mainly interested Volta and, derivatively, Davy.

Considering the widespread practical use of the electric cell, it is surprising that, apart from Davy's electric arc,[9] the first actual instance dates from the 1830s, a generation after Volta, when Charles Wheatstone (1802–75) developed the electric telegraph. This was first installed commercially in 1838, when the Great Western Railway laid a line between Paddington and Slough, opening a new era in railway signalling. The first commercial telephone line, also dependent on batteries, came only in 1877. The revolution in telecommunications could not, however, have come earlier, since it depended on a series of experiments in electricity and magnetism (described later in this chapter) which did not begin until the 1820s.

Despite the lack of practical uses, Volta's fame following his invention of the pile spread throughout Europe, but although he lived another twenty-seven years his achievements in the nineteenth century in no way equalled those of the eighteenth. His pile immediately became indispensable in the laboratory and, within a generation, outside it. With such an achievement, it is understandable that he was content to sit back and enjoy his fame.

Electromagnetism

Michael Faraday (1791–1867) was popular and attractive, diligent and hardworking, talented and creative, to a degree unequalled in the history of science. His own detailed record of his work, covering much of the nineteenth century, shows how he laid down the empirical foundations of the theory of electricity and magnetism, which, since his day, has defined much of science and technology. A brilliant communicator, Faraday, for some forty-odd years from his base at the Royal Institution in the heart of London, explained to the public the scientific developments that would change their lives. Electricity and magnetism, field theory, optics, chemistry, and metallurgy were all part of his programme.[10]

Faraday was born in London into a humble family of religious dissenters. The gospel proclaimed by the Sandemanian sect was fundamentalist, austere and demanding, and Faraday only in part kept his religion separate from his science. William Barratt once described Faraday's religion as 'God's revelation to man of the Divine purpose' and his science as 'man's revelation of the divine handiwork'.[11] Faraday was always loyal to the Sandemanians; when he married it was to another member of the sect, and he later became an elder. Today the sect, whose last member died in 1992, is largely remembered for having had Faraday as a member.[12]

At school, attended until he was thirteen, Faraday acquired little more than basic literary and mathematical skills. He then had the good fortune to be apprenticed, aged fourteen, to a bookbinder: George Riebau must have been one of the kindest masters in a relationship more commonly associated with oppression and exploitation. He not only allowed his apprentice the freedom of his library, but also permitted experiments in the back of his shop.

In 1810, following advice given in Isaac Watts' *Improvement of the Mind*, Faraday, aged nineteen, started attending weekly meetings of the City Philosophical Society. In his first year he presented a paper defending the 'two-fluid' theory of electricity (which he would later renounce). He also started to correspond with other scientists. Within three years his reputation led Sir Humphry Davy to appoint him as his assistant at the Royal Institution; this was almost immediately followed by the continental tour described in Chapter 6.

Returning home, Faraday was reappointed to the Royal Institution, with a long official title. He was promoted in 1821, and in 1825 he became Director of the Laboratory, a position he held until his death, with only one increase in the modest salary (although the fees for his numerous lecture courses, and a civil list pension from 1835, added substantially to his income).

The Royal Institution proved to be the ideal base for Faraday's work as a scientist – with a large laboratory in the basement (now open as a museum), the lecture hall (one of the finest in London) on the ground floor, and the upper floors constituting his private home – and all this in Mayfair. From 1825 to 1862, more than a hundred Friday evening discourses allowed Faraday to present his, and sometimes others', new discoveries to the public. According to John Tyndall (who would eventually succeed him), the 'Friday evening discourses were sometimes difficult to follow. But he exercised a magic on his hearers which often sent them away persuaded that they knew all about a subject of which they knew but little.'[13] And these were only part of a lecturing programme, in which courses were offered to many different audiences, including children.[14]

A final break with Davy was the price to be paid for Faraday's success at the Royal Institution. This was the result of Faraday being proposed for the Royal Society in 1823. By convention, Davy as president could not sign the certificate, but, having not been consulted either, he opposed the nomination. The fact that Faraday became secretary of the Athenaeum in the same year, and director of the Royal Institution in 1825, suggests that Davy still respected his talents.[15]

The break, none the less, was final. For the future of science this was unimportant: Davy, who died in 1830, achieved little in the last years of his life. It is still a tragedy that Faraday's unparalleled charisma never succeeded in healing the rift between them.

For all his gifts as a communicator, Faraday, experimenting in the basement laboratory, worked alone. As he once explained, 'I do not think I could work in company, or think aloud, or explain my thoughts at any time.' This is typical of the man: his brilliant public demonstrations of his experiments were the result of endless preparation and rehearsal. The only clue to the thought processes that led to them is in the seven volumes of his laboratory diary. Paradoxically for so public a man, his essential contributions to science are not all that transparent.

The flow of discovery, through so many different channels, produced far too many results for any short summary. None the less, by focusing on electricity and magnetism, Faraday's key contribution to science becomes apparent.

In the years before 1823, when Faraday was still collaborating with Davy, remarkable discoveries relating electricity to magnetism were made by two near contemporaries, André Ampère (1775–1836) in France and Hans Oersted (1777–1851) in Denmark.[16] Ampère, a theorist, had an obsession to prove the identity of magnetism with electrodynamics. He conceived of the idea that magnets consisted of particles around which electric currents circulated and tried to justify this experimentally. Oersted, inspired by the philosopher Kant, pursued the idea that nature's forces had a common origin. Here he was moving along the same lines as Ampère, but his discovery in 1820 that a compass needle was affected by an electric current represented an experimental breakthrough never achieved by Ampère.

On 1 October 1820 the Royal Institution heard of Oersted's discovery, and in May 1821 Faraday achieved the reverse effect, by using a permanent magnet to deflect an electric arc. At about the same time the editor of the *Annals of Philosophy*, a friend of Faraday's, asked him to write an account of everything achieved so far in the new science of electromagnetism – this was the beginning of a lifelong obsession.

For this new assignment Faraday started with 'a sketch' and, in order to test the empirical basis for hypotheses put forward by those already in the field, repeated their experiments. Ampère's experiments failed this test, which led Faraday to question his theory (which was supported by that of Oersted) that there were two electrical fluids, one negative and one positive, but 'equally positive in their existence and possessed of equal powers'.[17] On similar empirical grounds, Faraday also questioned Ampère's concept of electric currents circulating around particles in magnets.

Although Faraday was convinced that Ampère was wrong, and that he himself was right, he was still loath to cross swords with a man, sixteen years his senior, with an international reputation. This may explain why Faraday never published the sketch under his own name, and Ampère probably never read it, for there is no sign in his work that he profited from its criticisms.[18] The fortunes of the two men could hardly have been more different. Ampère's father had been executed in the French revolution, his first wife died young, and his second marriage failed disastrously. His fervent religious life oscillated between intense mysticism and sceptical despair. His writings, although containing brilliant scientific insights, were often incoherent. Faraday, in contrast, was born under a lucky star, and the world around him knew it. The only shadow cast on his life was when he fell out with Davy.

In disagreeing with Ampère, Faraday relied on experiments carried out with apparatus developed in his own laboratory. What he called a 'helix' (and what is now known as a 'solenoid') was no more than a long length of wire wound round a core, like a thread round a bobbin. He then discovered that the helix, when activated by an electric current, behaved like a magnet. According to Ampère, the magnetic properties of the helix should have been the same as those of a magnet in the same form. To test this, Faraday, having magnetised a hollow steel cylinder so that opposite poles were at each end, used a compass needle to test Ampère's claim. It failed the test, demonstrating that Ampère's theoretical concept – that magnets consisted of particles around which electric currents circulated – was mistaken.

The difference between Faraday and Ampère, although largely a matter of temperament, had a significant historical background in the competition between English and French scientific principles. This went back to the seventeenth-century conflict between Newton's corpuscular theory of light and Huygens' wave theory. In the early nineteenth century, Augustin Fresnel's (1788–1827) Paris experiments with the interference, diffraction and polarisation of light, coupled with new ideas in mathematics, led to general acceptance of wave theory, although, as related in Chapter 7, twentieth-century quantum theory would see Newton vindicated, at least in part. Fresnel's English contemporary, Thomas Young (1773–1829), also used wave theory to explain interference, but his results were unacceptable for being too anti-Newtonian. When Faraday first surfaced around 1820, Fresnel was on the crest of a wave. Ampère's standing in France was helped by his friendship with Fresnel, who shared his house for some years. Even so, when it came to the crunch in electromagnetism, Ampère had to admit defeat by Faraday.

Faraday's criticised Ampère first for relying too much on *ad hoc* hypotheses, and second for his lack of predictive success. This was the result of Ampère's insistence that 'simple facts [*faits primitifs*] cannot be immediately observed, but can only be deduced from observations with the aid of mathematical calculation'.[19] A crucial experiment with copper rings, in which the first and second results contradicted each other, was swept under the carpet by Ampère. He had effectively surrendered his position as an experimenter (and incidentally confirmed the justness of Faraday's criticisms). After 1822 Ampère's work became austere and mathematical, to the point that James Clerk Maxwell would later call him the 'Newton of electricity', but it was Faraday who pushed him on to this new path (in which his work would prove to be much more durable).

Faraday was still collaborating with Davy when he first confronted Ampère, and it was Davy who improved upon a key experimental result of Ampère's by showing how iron filings, scattered on a circular disc, arranged themselves in a pattern of concentric circles, when a wire carrying an electric current passed through its centre. Although the experiment was Davy's, it was Faraday who realised its significance for creating lines of force.

As in the experiment with the helix and the magnetised steel cylinder, Faraday, in an experiment on 3 September 1821, reversed the parameters by substituting for Davy's wire a crank free to rotate on a vertical axis. He then passed a current through the crank, and approached it with a bar magnet. Taking different lines of approach, he discovered that the crank would move in a direction perpendicular to that of the approaching magnet, its direction, clockwise or anticlockwise, being determined by which pole of the magnet, north or south, was brought close to it. The immediate result was for the crank to swing round and strike the magnet. By quickly removing the magnet, to allow the crank to continue rotating, the process could be repeated, by bringing the magnet forward again every time the crank completed a rotation. In effect Faraday had created the first electric motor. In theory the sleight of hand would have been unnecessary if Faraday had available a magnet with only one pole, in which case the crank would rotate for so long as the current passed through it. Such a magnet, however, is a physical impossibility, as Faraday realised.

In 1822 Faraday solved the problem with the apparatus shown in Figure 4.2. On the right-hand side the wire replaces the crank, with its bottom end, in a bath of mercury, free to revolve round the bar magnet. The mercury, being a conductor, allows the electric circuit to be completed: when this happens the wire revolves, its direction depending on the polarity of the magnet (of which the lower pole is insulated from the electric circuit). The left-hand side shows how the apparatus still functions when the parameters are reversed, so that the magnet is free to revolve around a fixed wire.

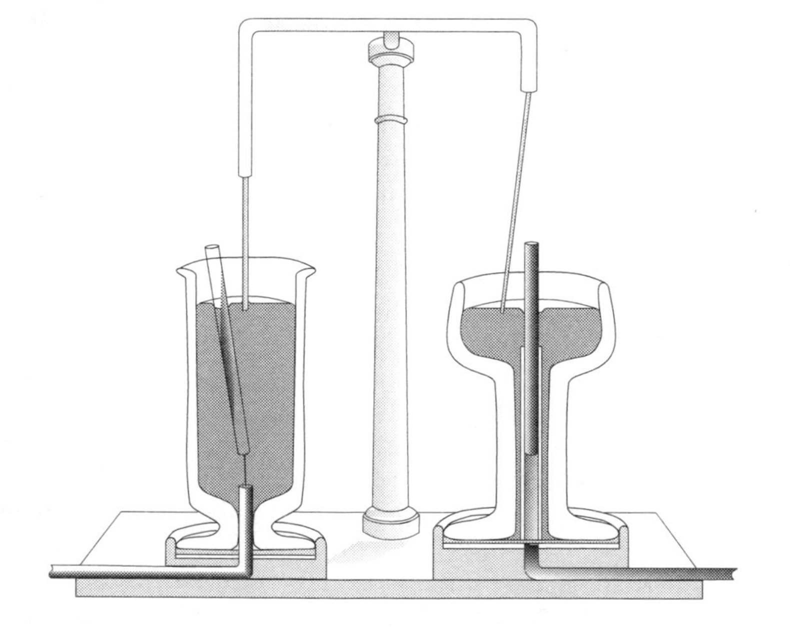

Figure 4.2 Faraday's rotation apparatus.

In the following years Faraday, knowing the magnetic effects of an electric current, went on to consider the possibility of the reverse effect, that is, an electric current produced by magnetism. Although during the 1820s Faraday produced no useful results, in Paris Dominique Arago (1786–1853) experimenting in 1825 with a moving magnet and a copper plate, kept hope alive. Then, on 29 August 1831, Faraday's 'induction ring' experiment produced results that would change the course of history.

His apparatus could hardly have been simpler: it consisted of two wire coils, *A* and *B*, wound round opposite sides of a soft iron ring. It was as if two helices were bent to form semicircles, and then welded together. Faraday's notebook describes how this was used:

> Charged a battery of 10 pr. plates 4 inches square. Made the coil on *B* side one coil and connected its extremities by a copper wire passing to a distance and just over a magnetic needle (3 feet from iron ring). Then connected the end of one of the pieces on *A* side with battery. Immediately a sensible effect on needle. It oscillated & settled at last in original position. On breaking connection of *A* side with Battery again a disturbance of the needle.[20]

Faraday found that the effect varied according to the number of turns in the coils (which the design of the apparatus allowed to be varied). The experiment also worked just as well when a cardboard tube replaced the iron core.

Less than a month later, on 24 September, new apparatus, substituting magnets for the *A* coil, produced the same effect (so that electromagnetism produced by the current in a helix was equivalent to that of a permanent magnet – a result of great significance to Faraday). This result was confirmed on 28 October, when Faraday 'made a copper disc turn round between the poles of the great horseshoe magnet of the Royal Society'[21] (which on 24 November received a report of the results of all three experiments). In the following year, 1832, further experiments showed that the electricity Faraday had produced by induction had the same chemical, magnetic and other effects as that produced by voltaic cells or any other method. The key to understanding the inductive process was that it was the process of change in the electromagnetic field that produced the current. This is what Faraday had failed to realise in 1825.

In principle, this insight was sufficient for the invention of a machine for generating electricity, and the first prototype, Hippolyte Pixii's (1808–35) magnetogenerator, appeared in 1832.[22] Faraday himself never produced a serviceable generator, but from 1840 onwards, Charles Wheatstone (1802–75), with whom he collaborated in the 1830s, produced improved magnetogenerators capable of supplying a steady current. (In the 1850s, Wheatstone, together with Werner Siemens (1816–92) and others, developed the self-exciting generator, with a field produced by electric magnets whose current was itself tapped from the generator; to this day this is the standard model to be found in any power station. Faraday doubtless saw a number of prototypes, but actual operational use had hardly begun when he died in 1867. Even so, the fundamental principle remained that of electric induction, discovered by him in 1831.)

Faraday's collaboration with Wheatstone followed from the latter's being invited to present his research into acoustics in Faraday's programme of Friday evening discourses at the Royal Institution. Wheatstone's working life started with his father, who made musical instruments. He also taught the flute to Princess Charlotte, daughter of the Prince Regent (later King George IV). In 1821, working in his father's premises, Wheatstone invented a giant magic lyre, which reverberated to music for harp, piano and dulcimer coming from unseen players in the room above.[23] (He also invented the mouth-organ and the accordion.)

Concerts on Wheatstone's gigantic musical gizmo were a sell-out at five shillings a ticket – in that day a colossal sum. Faraday (who was very musical) may first have met Wheatstone simply by attending one of the concerts, probably in 1823. No matter, the two plainly hit it off, not only in music but in physics – in which both electricity and acoustics are involved in the propagation of waves. On 15 February 1828, Faraday presented a discourse on resonance, based on material supplied by Wheatstone (who was much too shy to be a good lecturer).

The joint series continued on and off for twenty years, and in the 1830s Wheatstone, inspired by Faraday, turned his researches to electricity. In 1834 he conducted a series of experiments with electric sparks, using a half-mile-long circuit containing three small gaps to be bridged by sparks when the current was turned on. A device based on a revolving mirror should have indicated the order in which the sparks occurred, but it led Wheatstone to the conclusion that there were two electric fluids flowing in opposite directions in the circuit. This was more or less the point reached by Ampère in 1820, and demolished by Faraday. On the other hand, Wheatstone's apparatus allowed him to estimate the velocity of electricity in the circuit at 288,000 miles per second. With hindsight, this proved to be a remarkable result, when it became clear, towards the end of the century, that the velocity was the same as that of light in a vacuum – 186,000 miles per second.

Before leaving Faraday, there is one final question to answer. What, at the end of the day, was the underlying nature of the unseen forces whose observable effects were revealed by Faraday to a degree unparalleled in the history of science? In modern terms we are asking: what was the nature of Faraday's field theory, and how did it develop in the course of his life? According to a recent study, there is no clear answer:[24] all that we have to go on are occasional comments in the voluminous records of Faraday's lectures and experiments.

A good starting point is a very early comment, made in 1816, relating to three apparently distinct kinds of attraction:

> The attraction of gravitation, electrical attraction and magnetic attraction . . . appear . . . to be sufficient to account for all the phenomena of spontaneous approach and adherence with which we are acquainted . . . The Science of Chemistry is founded upon the cohesion of matter and the affinities of bodies and every case either of cohesion or of affinity is also a case of attraction . . . The attraction of aggregation and chemical attraction are actually the same as the attraction of gravitation and electrical attraction I will not positively affirm but I believe they are.[25]

To begin with, Faraday saw force as a property of matter, which fits in well with the lines of force around a magnet that various experiments can reveal. Later, when he came to work with electromagnetic induction in 1831, this position could no longer be held. Even later, in 1845, he discovered an interaction between magnetism and light, when he showed that a powerful magnet could change the plane of polarisation of light. Although he looked for it, he never found a link between electricity and gravity: this would have to wait for Einstein.

The breakthrough into field theory

The heart of the problem is that a field is only palpable when there is something present to react to it. A compass provides unmistakable evidence of the earth's magnetic field, and also a means for investigating it in detail. But the field is there even without the compass. Faraday was a genius in designing laboratory apparatus for revealing forces that otherwise had no observable effect, often by producing the forces themselves in novel ways. Although he worked in many different branches of science, his most important legacy is in electricity and magnetism and, above all, the relation between them. More than anyone else he established the empirical foundation for their development into realms whose existence he hardly conceived of. The breakthrough required a mathematical theory of the electromagnetic field that went far beyond Faraday's own capacity as a mathematician. This was pre-eminently the achievement of James Clerk Maxwell, who entered the scene a generation later. The mathematical equations established by Maxwell (and presented in Chapter 7) were essential to any further progress in electromagnetism, for they not only covered every phenomenon discovered by Faraday, but also opened the way to developments which Faraday could never have foreseen. At the same time, Faraday's work in electricity and magnetism opened the way for the power generation of electricity and workable electric motors, both of which were well under way before he died in 1867. Within a generation of his death electric power had become commonplace in the Western world.[26] The debt owed to Faraday by both science and the world at large is immeasurable.

5

Energy: science refounded

Heat and work

THIS CHAPTER describes the way in which mechanical systems, and the laws that govern them, developed so as to reach a stage in which they were recognised as being one aspect of the much more general concept of energy. Until well into the nineteenth century, the physics of mechanical systems was that of Isaac Newton's *Principia*. For Newton, with his interest focused on celestial dynamics (as described later in this chapter), gravity was the only force that counted. This was true also of his contemporaries, such as Christiaan Huygens, even if they disagreed with him on theory. Optics, the other major interest of their day, had its own independent domain.

In this world of seventeenth-century science there was hardly any place for heat, although it was, of course, part of everyday technology in such long-standing applications as the potter's kiln or the blacksmith's forge. In the eighteenth century, however, technology, mainly as applied to the steam engine, gradually forced the scientists to take heat more seriously, a process greatly helped by the invention of reliable thermometers, with appropriate units for the measurement of temperature, in the first half of the century.

By the end of the eighteenth century, the enormous success of James Watt's steam engines – the driving force of the Britain's industrial revolution – was beginning to interest scientists, particularly in France. First (as related in Chapter 6) the chemists, led by Antoine Lavoisier (1743–94), were beginning to understand the processes of combustion, by which heat was generated as a result of chemical reactions. Then, in the 1790s, Count Rumford (1753–1814), as a result of noting the immense amount of heat generated by boring cannon, produced a formula relating work done to heat generated – a very important step in the history of science. (His *An Experimental Inquiry concerning the Source of Heat Excited by Friction*, presented to the Royal Society in 1798, was so well received that he decided to leave Bavaria for England, where he founded the Royal Institution.[1])

In the nineteenth century, the action moved back to France, where Sadi Carnot (1796–1832) and others, relying more on engineering principles than on physics or chemistry, developed a scientific theory of heat engines, to the point that Jean-Baptiste Biot (1774–1862) included one of James Watt's designs in his *Traité de physique*. At the end of the day, however, the conversion of energy, with electricity and magnetism included in the equation, was dominated by British scientists, notably William Thomson (1824–1907) – later Lord Kelvin – James Joule (1818–89) and Michael Faraday. This is the story of the present chapter, which also includes a discussion of the place of mathematics in its relation to what, following the usage introduced by Kelvin, is now called physics. In this sense, the chapter describes the transition from natural philosophy (familiar from the title of Newton's *Principia*) to classic nineteenth-century physics. (Chapter 7 then relates the transition, at the end of the century, to modern physics.)

Newton's dynamics

Newton, in a letter to Robert Hooke, dated 5 February 1676, wrote: 'If I have seen further it is by standing on the shoulders of giants.'[2] His indebtedness to others is well recorded, yet his universal law of gravitation represented a statement of principle of the widest possible application, unprecedented in the history of science. After Newton, the principle, fundamental to Aristotelian physics, that the motion of bodies outside the sublunary sphere was governed by laws different from those applying within it, had lost all credibility. It was accepted that the same three laws of motion governed both celestial and terrestrial dynamics. What then did these add up to? And why did they lead Newton inexorably to the statement of the universal law?

These are Newton's three laws of motion.

1. Every body continues in its state of rest, or of uniform motion in a right line, unless it is compelled to change that state by forces impressed upon it.
2. The alteration of motion is ever proportional to the motive force impressed; and it is made in the direction of the right line in which that force is impressed.
3. To every action there is always opposed an equal reaction: or, the mutual actions of two bodies on each other are always equal, and directed to contrary parts.

The first law establishes the fundamental concept of inertia: quite simply, in a domain containing a number of different bodies, each and every one will continue in its own way, in a straight line, unless some combination of forces causes it to deviate from it. In two dimensions one can picture a billiard table, of unlimited size, with balls rolling across it without ever colliding. The measure of the inertia of any one body is equivalent to its mass. Galileo had also conceived a model with balls rolling over a plane, but in his book the plane had to be finite, so that he had to consider what would happen when a ball rolled over its edge At this point 'a downward propensity due to its own weight'[3] takes over.

At this point it is clear that Galileo recognised the principle of Newton's first law: it is less certain that he realised that the downward propensity of a body in free fall represented an instance of Newton's second law. Now according to Aristotelian physics, the ball, on rolling off the edge of the table, would simply have changed direction, and gone into free fall at a constant speed, proportional to its weight.

This fallacy had also been exposed by Galileo (even without experiments from the Leaning Tower of Pisa), but for Newton the fact that a body in free fall accelerates (as Galileo had demonstrated) means that it must be subject to an impressed force: this is what the second law requires. The question is, what is the measure of this impressed force?

The combined operation of Newton's first two laws of motion is inherent in what Galileo discovered about the trajectory of projectiles, and Newton was ready to concede that Galileo, without making the point explicitly, knew both laws. The first, and key part, of the second law, relating to the alteration of motion, cannot, however, be traced back to any result of Galileo's.[4] With Newton it led to the simple equation: $F = ma$, where F is the force, m the mass of the body, and a its acceleration. The problem still remained as to how this equation relates to falling bodies. The solution came as a result of looking beyond terrestrial limits, to consider what forces might act on a planet to maintain its elliptical orbit round the sun.

In the course of the year 1683, this was discussed in London at informal meetings between three remarkable scientists, Robert Hooke, Edmond Halley and Christopher Wren. By the end of the year, Halley had concluded that this force varied inversely with the square of the planet's distance from the sun. He had arrived at this result by substituting Kepler's third law (see page 44) into a formula for centrifugal force recently published by Christiaan Huygens. (In fact Wren had already suggested this as a possible solution in a conversation with Newton in 1677.[5])

At a meeting of the Royal Society in January 1684, Hooke not only agreed with Halley, but claimed 'that upon that principle all the laws of celestial motion were to be demonstrated and that he himself had done it'.[6] Pressed by Halley and Wren he failed to produce a solution: with no progress in London, Halley finally went to Cambridge in August 1674 to consult Newton. Halley asked what curve would fit the orbit of a planet, if the inverse square law determined the force attracting it to the sun? Newton answered that it would be an ellipse (which would accord with Kepler's first law, which physics had yet to prove), but searching through his own papers Newton failed to find his proof.

However, in November 1684, Newton sent Halley a nine-page treatise, 'De Motu Corporum in Gyrum' (On the motion of bodies in orbit). This not only contained the proof that Newton had failed to find in August, but also that of the more general case, showing that the inverse square law always produces a conic orbit, in practice an ellipse or a hyperbola, according to the velocity of the body. (The elliptical orbit is characteristic of satellite bodies, such as planets; the hyperbolic, of bodies which appear from and disappear back into outer space.) This was not all: Newton also derived Kepler's second and third laws (noted on page 43) from the inverse square law.

Newton was not content to let matters rest with his 'De Motu'. From corresponding with John Flamsteed (1646–1719), the Astronomer Royal at Greenwich, where the observatory had been founded in 1675, Newton was able to confirm that not only comets but also the satellites of both Jupiter and Saturn satisfied Kepler's third law. Whatever force it was that attracted planets towards the sun also attracted to them their own satellites. Newton went even further: knowing that Jupiter and Saturn were approaching conjunction, records supplied by Flamsteed enabled him to confirm that their orbits round the sun were disturbed by their mutual attraction – in itself an instance of what would become his own third law of motion.

The last stage in Newton's work with celestial dynamics was to bridge the gap separating it from gravity, as a terrestrial phenomenon. The moon provided the link. Observation confirmed that it too was subject to the inverse square law. Why not then the apple falling from the tree? The problem was essentially one of scale. Celestial dynamics is concerned with the attraction between bodies whose dimensions are extremely small in relation to the distances separating them. On earth the opposite is true: the apple falls a few feet on to a body whose dimensions are measured in thousands of miles. From what part of the earth, then, must distances be measured so that the inverse square law can apply? In 1685, Newton proved, mathematically, that a 'homogeneous spherical shell, composed of particles that attract inversely as the square of the distance, attracts a particle external to it, no matter at what distance, inversely as its distance from the center of the sphere'.[7] The physical equivalent to this proposition is that the attraction between any two bodies is the same as it would be if the entire mass of each of them was concentrated at its centre of gravity. The moon and the apple can then be treated as being attracted by a mass equal to that of the earth concentrated at a point, in the one case, some 238,900 miles, and in the other, some 3964 miles, away.[8]

In 1685 the dimensions of the earth had been known since antiquity, astronomers had measured the distance of the moon and its period of revolution with unprecedented accuracy, and Christiaan Huygens' invention of the pendulum clock made possible the accurate measure of g, the acceleration of free fall at the earth's surface. If Newton was right as seeing gravity as a single universal property of all matter, then there must be a precise mathematical correlation between the moon and the apple. The result he obtained was correct to the last inch (or one part in 400) of Huygens' determination of g.

Newton had now established not only the logical foundation for universal gravitation but also its essential empirical basis. The gravitational attraction F between two masses m_1 and m_2, separated by a distance d, could now be expressed in a single formula:

$$F = Gm_1m_2/d^2.$$

It made no difference what the masses were: m_1 could be an apple, and m_2 the earth, but m_1 could just as well be the moon. It was fundamental to the law that G should be constant, but how then should it be measured? This was a practical problem that Newton never solved, quite simply because he could find no observation that could provide the necessary measurements. (See page 295 for the use by Henry Cavendish[9] (1731–1810) of the recently invented torsion balance to estimate G, and then the use of this result to estimate the mass of the earth: this was the starting point for estimating, with the help of Kepler's third law, the masses of the sun, the moon, and the planets and their moons.)

Newton left unsolved one problem which mathematicians long after his day proved to be insoluble. The mathematical principles of gravitation were derived from the mutual attraction between two bodies, in accordance with the third law of motion. Newton fully realised that in any system comprising more than two bodies the motion of any one of them was determined by the attractive force of all the others. In principle, therefore, his proof that the inverse square law required an elliptical orbit could not hold in the actual circumstances of the solar system, comprising any number of planets and satellites, to say nothing of comets or asteroids.

The vast mass of the sun, more than a thousand times that of the largest planet, Jupiter, and more than six million times that of the smallest, Mercury, provided the basis for a practical solution to the problem. To take the case of any one planet, the attractive force of any other will be always be small in relation to that of the sun, and mathematical methods going back to Newton's day have long been sufficient for making the necessary corrections to the planet's respective orbits. This explains Newton's request to Flamsteed for accurate measurements of the positions of Jupiter and Saturn when they were in conjunction. Even at this point, at which Saturn's distance from the sun was more than twice that from Jupiter, the attraction of the sun was still some 186 times greater – and that at the time of an event occurring only once in 150 years. In 1846 the planet Neptune was discovered following discrepancies observed in the orbit of Uranus. These were far beyond the observational capacity of any telescope available in Newton's day, but the discovery of Neptune may none the less be seen as 'the ultimate triumph of Newtonian dynamics'.

Newton's work on gravitation, and its application to both terrestrial and celestial dynamics, found its final form in his

Philosophiae Naturalis Principia Mathematica, first published in 1687. Although this was to a considerable degree a synthesis of Newton's own work (such as had appeared in the 'De Motu') and that of his contemporaries, it was an unparalleled achievement in the history of science. It did not come easily to Newton, and quite likely would never have been published were it not for the dedication of one man, Edmond Halley, who during the time it was being written was Clerk to the Royal Society.

Although Newton, in the middle years of the 1680s, sensed that he had the power to bring about a revolution in science, he was often daunted by the difficulties he faced, which included claims by others, notably Robert Hooke, that they were the original authors of some his own basic principles. Halley, who never failed in his support, persuaded the Royal Society, to whom the *Principia* was dedicated, to be its publisher, which meant in practice that he took all the responsibility upon himself. The manuscript, when first presented to the Society, was described as giving 'a mathematical demonstration of the Copernican hypothesis as proposed by Kepler, . . . making out all the phænomena of the celestial motions by the only supposition of a gravitation toward the center of the sun decreasing as the squares of the distances therefrom respectively',[10] but, as I have shown, this tells only half the story. What is more, the *Principia* annihilated any number of competing principles, postulated by such well-known figures as Kepler and Descartes.

The book took England by storm, although it was, and still is, seen as extremely difficult. Dr. Babington, Newton's first patron at Trinity, was one of those who said 'that they might study seven years, before they understood anything of it', and a Cambridge undergraduate, passing Newton in the street, observed 'there goes the man that writt a book that neither he nor any body else understands'.[11] Halley, in his review for the *Philosophical Transactions*, was in no doubt about its significance: 'This incomparable Author . . . has shown what are the Principles of Natural Philosophy, and so far derived from them their consequences, that he seems to have exhausted his Argument, and left little to be done by those that shall succeed him.'

On the continent things were less simple: Newton had sent both Huygens and Leibniz copies of the *Principia*, but both rejected its central concept as absurd. Even so, Huygens noted 'the beautiful discoveries I find in the work he sent me' and in his correspondence with Leibniz it was constantly mentioned. Huygens questioned only the physics, not the mathematics, and although in doing so he may have harmed his own reputation, it can be said now that he had found Newton's Achilles' heel. We now know from Albert Einstein that the *Principia* did not pronounce the last word on gravity, which today is one of the four basic forces recognised by particle physics. It is also, paradoxically, much the weakest, but that is another story, to be told in Chapter 7.

Mathematics and the scientific method

Mathematics has been part of science since the time of the earliest known records, as shown by the eclipse lore of ancient Mesopotamia.[12] The correct relationship between mathematics and science has long been the subject of debate. For some, science must begin with a 'direct appeal to facts' followed by 'strict logical deduction from them afterwards'. The words quoted are those of Sir John Herschel[13] (1792–1871), a noted astronomer who did not hesitate to apply mathematics to his results.

Strictly speaking pure mathematics provides the only means of strict logical deduction. This paradoxical conclusion follows from Bertrand Russell:

> Mathematics and logic, historically speaking, have been entirely distinct studies. Mathematics had been connected with science; logic with Greek. But both have developed in modern times: logic had become more mathematical and mathematics has become more logical. The consequence is that it has now become wholly impossible to draw a line between the two: in fact, the two are one.[14]

Russell characteristically overstates his case, but once logical processes are admitted that do not satisfy the strict criteria of pure mathematics, the path is open to all kinds of deductions leading to results that prove demonstrably false. Stick to the straight and narrow path of mathematics, and false results can only follow from false premises. This is the GIGO principle familiar to any computer freak: garbage in, garbage out. Any number of important scientific discoveries are the result of applying this principle, just as any number of false conclusions have followed from failing to do so.

Russell, in referring to Greek, is pointing his finger mainly at Aristotle, often known as the 'father of logic'. Russell sees Aristotle's logical system 'as definitely antiquated as Ptolemaic astronomy', conceding, however, that 'Aristotle's logical writings show great ability, and would have been useful to mankind if they had appeared at a time when intellectual originality was still active'.[15]

One crucial shortcoming was Aristotle's overreliance on deduction, the dialectical process by which a single original premise can lead to a whole series of propositions. The alternative is induction, the process by which a proposition rests upon a convincing number of observed instances of the phenomenon to which it applies, without there being any significance counter-instances. Aristotle did in fact ask how one knew 'the first premises from which deduction must start',[16] but still he based almost the whole of his science on a combination of false premises and illogical deductions from them.

Francis Bacon (1561–1626) is well known among logicians for taking the opposite position, and one would think, therefore, that he would be commended by Russell. His fault, however, was that he failed to realise that observed instances of a phenomenon, however numerous, do not necessarily suggest a correct hypothesis to explain it.[17] Swammerdam, with his microscope, observed the spermatozoa of different species on countless occasions, and, although his hypothesis that one single means of sexual reproduction was common to a remarkably wide variety of different species was not only correct but essentially new, he never formulated any hypothesis about the fertilisation of the ovum – an essential part of the process.

Essentially, to be useful to science, a hypothesis must transcend the observed facts, and then allow a series of deductions leading to some phenomenon – which, intuitively, may be quite unrelated – that can be confirmed by further observation and experiment. Mathematics' claim to be central in this process is that its deductive methods are faultless. This was the merit of both Newton's and Einstein's theories of gravitation. Newton's mathematics required certain disturbances in the orbits of Jupiter and Saturn (which were observed by Flamsteed), and Einstein's a certain deviation of starlight due to the sun's gravity (which was observed at the time of the solar eclipse of 29 May 1919[18]).

The historical problem is simply that science does not stand still: Einstein could not exist without Newton, yet their theories of gravitation are incompatible. Einstein's mathematics – absolutely essential to his theory – was a development, at several removes, of Newton's calculus, without which Newton himself would never have established his universal theory. How then can Newton be reconciled with Einstein?

Now there was no fault in Newton's mathematics (apart from the fact that it was presented in a way that it made it difficult to understand). On the other hand, it was quite insufficient for Einstein, for whom mathematical methods only invented after Newton's death were indispensable. As to the observations relied upon by Newton, their main fault was insufficient accuracy, unavoidable with the telescopes available in his day. Between Newton and Einstein, there was not only considerable improvement in the power of resolution, which in 1838 had allowed stellar parallax to be observed for the first time, but the usefulness of telescopes was decisively enhanced by the ability to record observations photographically – with the first successful result in 1850.[19]

Mathematics does not admit of a single mistake: theorems proved in antiquity have lost none of their validity.[20] Ideally, the process of mathematicising a science reflects a 'one-to-one correspondence between the body of induced empirical knowledge and the formal system in terms of which it is cast'.[21] Today's mathematical physics goes a long way in constructing analytical systems containing formulae, connected by lines of logical – that is, mathematical – reasoning, that do not necessarily correspond to anything real. The process does not affect the truth of the phenomena that underlie it. Newton's law of gravitation had, as its inevitable mathematical consequence, the truth of Kepler's three laws. It does not tell us, however, what gravity is; nor, for that matter, does Einstein.

Both Newton and Einstein produced formal systems, based on the state of the art in the mathematics of their day (which Newton, at least, himself largely created), that corresponded more exactly than any competing systems with the underlying phenomena observed by astronomers. In Newton's case the correspondence was such that for some 200 years the mathematical system not only kept pace with new discoveries but told astronomers what they stood for. As already noted on page 124, the process culminated with the discovery of Neptune in 1846.

Thereafter, physics began to produce results, particularly in the accurate measurement of the speed of light, which went beyond the range of any Newtonian system (although Newton's work on colours can be said to have foreshadowed them). At the same time, Einstein's theory of general relativity, which required that light was subject to gravity, was the result of mathematical techniques unknown in Newton's day. This still does not mean that the subject of gravity has been closed by Einstein: the gravity scene is extremely busy, but the actors mainly produce more mathematics, occasionally throwing out a result that can be tested by observation or experiment. The key fact is that the mathematicians have driven the practical scientists to a point where they must produce results of unprecedented accuracy, relating to phenomena observable only with the help of instruments belonging to today's big science, which can be anything from a particle accelerator contained in a 26 kilometre tunnel to a telescope mounted on an earth satellite.

It would be impossible to trace the development of mathematics in relation to science in any but the broadest outlines. One point to be stressed is that improved notation has often been as important as new methods. It is easy to think of Arabic numerals, the decimal point, the equals sign, trigonometrical functions, logarithms, and indices to indicate powers such as 5-squared or 4-cubed as belonging to mathematics since the dawn of history, but they were only introduced to the Western world in the last 800 years. The familiar graph, with x and y axes, dates only from the eighteenth century, and it was only in the seventeenth that Descartes (as already noted on page 68) found a way of describing geometrical figures in terms of algebraic coordinates.

Pythagoras' theorem well illustrates the transformations that can be achieved with such innovations. On the face of it, the theorem, which states that the square on the side of the hypotenuse of a right-angled triangle is equal in area to the sum of the squares on the two other sides, belongs to Euclidean geometry, and the textbook proof requires the three squares actually to be drawn, to produce the familiar result: in algebraic terms, $a^2 = b^2 + c^2$. This was useful in antiquity when integral solutions, such as $5^2 = 4^2 + 3^2$, allowed surveyors, working with knotted ropes, to produce right angles.

The move from geometrical diagrams to algebraic symbols opens the way to much further development. For one thing, the statement of the theorem can be generalised to cover all triangles by stating it in the form $a^2 = b^2 + c^2 - 2bc \cos A$ (noting that Pythagoras proved only the special case when $A = 90°$, so that $\cos A = 0$). This is only a beginning: the original theorem can also be stated in the form $a^2 - b^2 = c^2$, so that $(a - b)(a + b) = c^2$. The substitutions $a + b = 2kp^2$ and $a - b = 2kq^2$ then provide two simultaneous equations that can be solved to provide a general formula for all right-angled triangles:

$$a = k(p^2 + q^2), \qquad b = k(p^2 - q^2), \qquad c = 2kpq,$$

which mathematically is the first step into a quite different world – and one of little relevance to hard science.

The invention of the differential calculus in the seventeenth century transformed the power of mathematics as an instrument of science. The question of who was the inventor is one of the most contentious in the whole history of science. Russell notes how 'the ... dispute as to priority was unfortunate and discreditable to all parties',[22] but it still makes very good history.[23]

There are two contenders: Newton and Leibniz. Newton claimed to have invented his direct method of fluxions in November 1665, and the inverse method in May 1666.[24] These are equivalent to Leibniz's differential and integral calculus, which he invented in Paris in the autumn of 1675, although publication only came in 1684. Newton's claim is only supported by his own unpublished letter, but his *Treatise on the Methods of Series and Fluxions*, written in 1671 but not published in English until 1736, is well attested and establishes his priority over Leibniz. Even so, Newton's notation in its final form did not come until some twenty years later. The difference between the two is shown in Figure 5.1.

Leibniz's notation won the day at a very early stage, and is still in continuous use: this is largely because it proved to be adaptable to the requirements of calculus, which developed very rapidly in the eighteenth century. Newton's notation, even its final form, was still tied to its origins in physics, which related change to time, which in the hard sciences is often the key variable. And it is primarily in physics where Newton's dotted notation still survives today.

The difference between Newton and Leibniz, fundamental in defining the interface between science and mathematics, is well summed up in Newton's 'Account of the Book Entituled

$$\begin{aligned} \dot{\overline{xy}} &= (x + o\dot{x})(y + o\dot{y}) - xy \\ &= ox\dot{y} + oy\dot{x} + o^2\dot{x}\dot{y} \\ &= o(x\dot{y} + y\dot{x}) \end{aligned} \qquad \begin{aligned} \mathrm{d}(xy) &= (x + \mathrm{d}x)(y + \mathrm{d}y) - xy \\ &= x\,\mathrm{d}y + y\,\mathrm{d}x + \mathrm{d}x\,\mathrm{d}y \\ &= x\,\mathrm{d}y + y\,\mathrm{d}x \end{aligned}$$

Figure 5.1 Newton's (left) and Leibniz's (right) notations for differentials.

Commercium Epistolicum',[25] published by the Royal Society (of which he was then President) in 1715:

> It must be allowed that these two Gentlemen differ very much in Philosophy. The one proceeds upon the Evidence arising from Experiments and Phænomena, and stops where such Evidence is wanting; the other is taken up with Hypotheses, and propounds them, not to be examined by Experiments, but to be believed without Examination.

The properties of matter, both at rest and in motion, have been central to science at least since Aristotle's *Physics* propounded fundamental theories relating to both states. He was, needless to say, mistaken on every possible count, but it was only in the seventeenth century that the process of correcting his mistakes began, and it was not until the nineteenth century that it was completed. This only confirms the judgement of Bertrand Russell that Aristotle's authority, in science, was 'a serious obstacle to progress. Ever since the beginning of the seventeenth century, almost every serious advance had to begin with an attack on some Aristotelian doctrine.'[26]

Time, motion and matter

When it comes to matter, rest and motion present quite different problems, and the fields in which the state of art is applied differ widely. The absence or presence of time, as an essential factor, defines the difference between the two. Physically speaking, the science of matter at rest – at least until the early eighteenth century – was concerned only with the dimensions of length and mass, and measuring instruments were designed subject to this limitation. The measurement of time only became critical with the study of matter in motion.[27]

Matter at rest then has two different aspects. The first, which takes the properties of any sort of matter, wood, metal, glass, or whatever, for granted, is concerned with how they can be used in fields such as architecture and engineering. The second aspect is concerned with how matter is constituted: this, today, defines the realm of chemistry (considered in Chapter 6), but until well into the seventeenth century it belonged to alchemy, and the principles governing it were those laid down by Aristotle.

For building massive structures, whether cathedrals, castles or aqueducts, there was essentially a choice between two materials, wood and stone, with brick, a manufactured product, replacing stone, where economic factors made this desirable. This recalls the world of medieval craftsmen – carpenters, stonemasons, bricklayers – who over the course of centuries perfected their art, to achieve such marvels as the chapel of King's College, Cambridge (only completed in 1515), where state-of-the-art working with stone enabled the construction of a building whose walls seem to consist mainly of glass. The medieval church builders did make some advances, such as developing flying buttresses to support ever higher walls, or using wood-vaulting to allow ever greater open spaces in the interior – to produce in the end such remarkable structures as the dome of Brunelleschi's duomo in Florence (which was then clad in brick) – but still there was little advance on such remarkable, and much older buildings, as the Parthenon in Athens (433 BC), the Basilica of Hagia Sophia in Constantinople (537), or the temple of Tôdaiji in the Japanese Nara (dating from 752 and still the largest wooden building in the world). And the craftsmen were not using any science, not even Aristotle's, in any modern sense.

In seventeenth-century Europe metals only appeared in the tools – saws, drills, hammers, and so on – used in large-scale construction, or as small components, such as nails, screws and bolts. Cranes and hoists were still made of wood, with cables of rope. This was in complete contrast to the wealth of relatively compact metal instruments, from timepieces to telescopes. These, at the cutting edge of scientific discovery, came into their own in the course of the century, largely as a result of the radical improvements made by leading scientists such as Galileo, Huygens and Newton.

In contrast, the first cast-iron bridge was only completed in 1789, when it had the advantage of being close to the earliest coke-fired iron smelter, established at Coalbrookdale in 1709. (Here China was well ahead with iron-chain suspension bridges dating from the eighth century, while Thomas Telford's (1757–1834) suspension bridge across the Menai Strait (1825) was not built until nearly a thousand years later.) By the eighteenth century, however, the industrial revolution was well under way, harnessing the power of steam after Thomas Newcomen's (1663–1729) invention of a steam engine in 1698. All this required a much deeper understanding of the material properties of metals subject to stress and strain, particularly in cases of high temperatures and pressures.

Chemistry: the beginnings of a science of matter

The foundations were laid in the seventeenth century by Robert Hooke (1635–1703) and Robert Boyle (1627–91). The latter, often known as the father of chemistry, published in 1661 *The Sceptical Chemist*, which repudiated both Aristotle's four elements and the fundamental principles of contemporary alchemy (which Newton, who certainly knew Boyle, never abandoned). At the very least, then, Boyle cleared away some cobwebs. Even so, in chemistry, he never got the fundamentals right. That would have to wait another hundred-odd years, as related in Chapter 6.

Even though his success in chemistry never equalled Newton's in the fields of gravity (described later in this chapter) or optics, Boyle's place in the history of science is secure. Like his contemporary, Christiaan Huygens in Holland, he was always at home in ruling circles. The child of an earl, he followed school at Eton with a three-year grand tour that took him to Paris, Geneva and Florence, to return to England to find his father dead, and himself heir to a country estate. He set up a laboratory in Oxford, where, in addition to his research into materials, his invention of a successful air pump fitted in well with the 'experimental philosophy' of the Royal Society, which he had helped found in 1660. Like Newton, Boyle was obsessed with religion: as a Governor of the Society for the Propagation of the Gospel in New England, he circulated translations of the scriptures at his own expense, also endowing the Boyle Lectures 'for proving the Christian Religion against notorious Infidels'. He lived with his sister in London from 1668, and religion occupied him until his death twenty-three years later. Like Newton, he never married.

New dimensions: temperature and pressure

Boyle comes across as a humourless workaholic, writing up his work in a prolix style that invited caricature, particularly by Jonathan Swift, better remembered as the author of *Gulliver's Travels*. Today Boyle is best remembered for Boyle's law, familiar to all physicists, according to which the pressure and volume of gas in a closed container are inversely proportional. His work with gases fits in well with the general study of atmospheric pressure, which had been pioneered by Evangelista Torricelli (1608–47), close friend to Galileo in the last year of his life,[28] who in 1644 invented the mercury barometer – an instrument still in general use. The invention followed research by a remarkable Frenchman, Blaise Pascal (1623–62), who himself invented the tram, the wheelbarrow, the syringe, the hydraulic press and a calculator for his father to use for his business accounts. To test the underlying principle of the barometer, father and son carried two glass tubes containing mercury to the top of the Puy de Dôme, noting the way that the height of the columns decreased with altitude. Like Boyle, Pascal fell prey to religion, and spent the last seven years with his sister in a retreat. Although this produced his most famous work, the posthumous *Pensées* (1669), the gain to religion was certainly a loss to science.

As for Boyle, his law led to the invention of the manometer, the standard instrument for measuring gas pressure, and the aneroid barometer, both still used today. The principle, similar to that of the mercury barometer, was adapted by Daniel Fahrenheit (1686–1736) to produce an accurate alcohol thermometer in 1709 and a commercially successful mercury thermometer in 1714. Unlike the case of the barometer, where the height of the column of mercury supported by the atmosphere can be used to provide the units of measurement, a thermometer, to be of any use, had to be calibrated by reference to two fixed points, determined by states of matter known (or at least presumed) to occur at a constant temperature. This is easier said than done, but Fahrenheit believed that melting ice and the human body provided two such points, which then became 32° and 96° on his scale. In 1742, Anders Celsius devised a scale that substituted the boiling point of water for body temperature, with 0° for the freezing point and 100° for the boiling point of water.

Fahrenheit was the first to apply a physical phenomenon, that is, the expansion of matter when heated, to the measurement of temperature: mercury, a metallic element occurring in pure form as a liquid at normal everyday temperatures, could be contained in a glass tube capable of being calibrated so as to measure its expansion with heat. The principle was similar to that already adopted by Torricelli for his barometer. For the first time then in the history of science, two new units of measurement were introduced within a period of less than a hundred years.

This was a considerable breakthrough, although the instruments designed by Torricelli and Fahrenheit would prove to have substantial limitations: mercury, after all, has its own freezing and boiling points, which critically limit the usefulness of any instrument based upon it. In practice, other phenomena known to physics can be applied for measurements outside these limits, but once again this is running ahead to Chapter 8.

Gravity and Newton's concept of mass

Finally, science owes to Newton the concept of mass, which is a fundamental property both of any material aggregate, or of any separate part of it, in any part of the universe. (Although at the atomic level the concept must be revised to reckon with quantum phenomena, this goes far beyond anything known in Newton's day). For Newton it was sufficient that 'the quantity of matter is that which arises conjointly from its density and magnitude ... This quantity I designate by the name of body or of mass'.[29] Physics now provides endless examples of the relevance of mass: to give but one, it determines the quantity of heat required to raise the temperature of a body by a prescribed amount. For Newton, however, mass was important, not as a property of matter at rest, but as an essential component in his second law of motion – for however a body moves, its mass remains constant, or, in the language of the seventeenth century, 'mass is indifferent to motion'. (Once again we have a principle not acceptable to modern physics: by Einstein's general theory of relativity, mass increases with velocity, tending to infinity as velocity approaches the speed of light. This is just one point where Newton has had to yield to Einstein, as explained in Chapter 8.)

The link between mass and the motion of any body leads us to look at the laws governing motion. From the late sixteenth century, for a period not far short of a hundred years, the study of moving objects was at the cutting edge of science: if the opening moves were made by Galileo, and the end-game played by Newton, many other noted scientists played their part. Some events, such as Galileo observing different objects falling from the top of the Leaning Tower of Pisa, or Newton contemplating an apple falling from a tree, have long been part of scientific folklore, although the evidence for them is questionable. This does not matter, for any number of experiments and observations satisfying the demands of scientific rigour supported Newton's general theory of gravity and his three laws of motion, considered earlier in this chapter.

Where then did Galileo start? Although Galileo's early studies on motion are undated, the first of them, 'De Motu', probably dates from about 1590, when he was professor of mathematics at Pisa.[30]

This work is based on 'observations, solid proofs and arguments that very many of Aristotle's conclusions about motion, hitherto held to be perfectly clear and indubitable, were wrong'.[31] Galileo was not the first to take this line: starting with a small group at Oxford in the fourteenth century, scholars had questioned Aristotelian theory on its two weakest points, the movement of projectiles and the acceleration of falling bodies.[32]

As to projectiles, Aristotelians required a continuous force acting throughout the trajectory, but where did this come from after they had left the projector? The Aristotelian answer was 'the commotion which the initial movement had produced in the air',[33] but then what would happen to a projectile launched in a vacuum? Galileo borrowed from his predecessors the idea of an impressed force, or impetus, imparted to the projectile by the process of launching. It continues in the body 'as heat stays in a red-hot poker after it has been taken from the fire'.[34] A ball thrown up in the air has impressed on it sufficient force to overcome its weight, but at a certain point the inherent heaviness of the ball gains the upper hand (in the same way as the poker begins to cool) and according to basic Aristotelian principles, brings it down to earth.[35]

This leads to the second problem, that of the acceleration of falling bodies. According to Aristotle there should no acceleration, but simply, for any one body, a constant velocity proportional to its weight. It is this principle which the experiment carried out from the Leaning Tower, if it ever took place, would have disproved. However this may be, Galileo did experiment with rolling balls of different sizes down a calibrated plane slope, showing first that they all took the same time to reach the bottom, regardless of their weight, and second that their velocity steadily increased during their descent. In a letter written in 1604 he stated that 'the increase of the speed of a falling body is proportional to the distance from its starting point': this was a mistake, corrected in 1609, to accord with the rule that the distance covered by a falling body is proportional to the square of the elapsed time. The correct theory is laid out in full in Galileo's *Discourses and Mathematical Demonstrations Concerning two New Sciences*, published in 1638.

The *Discourses* also describe Galileo's theory of the trajectory of a projectile. This subject, of great importance to the science of war, was what first aroused Galileo's interest in motion before the end of the sixteenth century.[36] His demonstration, probably in 1605, that the path of the trajectory was a parabola not only refuted a key principle of Aristotelian physics but showed how, in effect, the trajectory of a projectile could be explained by combining two rules: the first was that governing a falling body, and the second that governing the unimpeded motion of a body in the absence of external forces. In this second rule Galileo anticipated Newton, who was to state it – as noted on page 121 – as the first of his three laws of motion.

Although the rules governing different forms of motion as worked out by Galileo were correct, they were essentially mathematical formulae rather than any general statement of scientific principle. This would have to wait for Isaac Newton, whose *Philosophiae Naturalis Principia Mathematica* was not published until 1687, nearly forty-five years after Galileo's death. This monumental study, generally known simply as the *Principia*, is of unequalled importance in the history of science. As Edmond Halley, who was largely responsible for ensuring its publication, was to write:

> This incomparable Author having at length been prevailed upon to appear in publick, has in this Treatise given a most notable instance of the extent of the powers of the Mind; and has at once shown what are the Principles of Natural Philosophy, and so far derived from them their consequences, that he seems to have exhausted his Argument, and left little to be done by those that shall succeed him.

So complete was the work of Newton, in every aspect of the study of motion, that it is easy to overlook the course of its development from the time of Galileo. Any such survey must focus on two very able men, René Descartes and Christiaan Huygens. Of these, Descartes is now much the better known, but this is because of his work as a philosopher and mathematician. We have already seen how the optics that he derived from his fundamental concept of space-matter was mistaken, and the same is true of his understanding of motion. None the less, Descartes improved upon Galileo's conception of uniform motion by expressing it in a law of rectilinear inertia according to which 'a moving body acted upon by no force will move in a straight line'[37] – essentially Newton's first law of motion.

The dynamics of Huygens' pendulum

Christiaan Huygens was, next to Newton, the greatest scientist of the day. As a man, Huygens was almost everything that Newton was not. In his year of birth, 1629, the Dutch republic, although still at war with Spain, was well into the 'Golden Century' of unprecedented wealth and fortune. Although not a noble family, the Huygens's had been at the centre of affairs ever since Christiaan's grandfather had seen a secretary to Prince William the Silent (1533–84) – generally regarded as the first of the new nation's founding fathers. Christiaan's father, Constantijn, held a similar office, and his older brother, also Constantijn, was well known as a poet. From 1637 onwards, the family lived in a grand house in the very centre of the Hague. A more civilised ambience was not to be found anywhere in the world.

The young Christiaan, good-looking in an effeminate sort of way, was gifted and precocious, with a love of music, languages and, above all, mathematics – so much so that his loving father called him, 'My Archimedes'. He was composed and amiable, and had good manners and beautiful handwriting. Good with his hands (like Newton), he built himself a lathe when he was thirteen and went on to grind lenses, an occupation he kept up for his whole life.

As a young man he knew, through his father, Rembrandt and Descartes, and later in life he would know Newton. He first published in his early twenties, and in 1655 he made his first visit to Paris, then regarded as the centre of the scientific world. In 1656 he invented the pendulum and discovered Saturn's rings. His discoveries made him famous: in the early 1660s new visits followed to Paris and London, where he became a member of the newly founded Royal Society. In 1666, King Louis XIV of France invited him back to Paris, where he was given spacious apartments in the Royal Library, to organise a similar institution, the Académie des Sciences. In 1673 he dedicated his major work on the pendulum clock, the *Horologium Oscillatorium*, to the king, even though France was then at war with the Dutch republic. In all innocence, Christiaan Huygens seemed not to notice. The pendulum clock would prove to be important, not only for keeping time, but for providing a means of measuring the acceleration g of a body in free fall. Galileo, experimenting with balls rolling down a slope, calibrated by lines separated by equal intervals, had demonstrated, in a rough and ready way, that the times taken to reach successive dividing lines increased according to a simple square law. If the times could be accurately measured, and proper allowance made for friction, then Galileo's apparatus, suitably refined, did provide a means for measuring g.

Antoni van Leeuwenhoek, microscope pioneer, 1632–1723.

Christiaan Huygens presents his pendulum clock to Louis XIV, King of France.

Isaac Newton as a young man, 1642–1727.

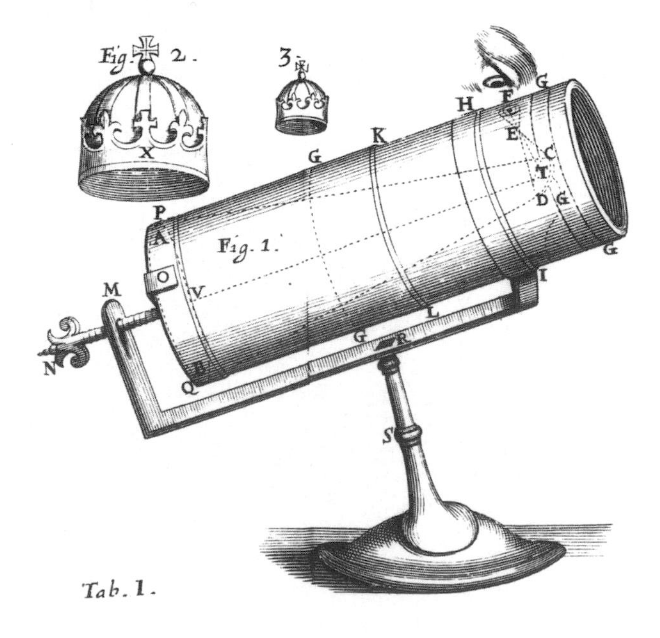

Isaac Newton's reflecting telescope.

Benjamin Franklin, 1706–1790.

Michael Faraday, 1791–1867.

Alessandro Volta demonstrating his electric pile to Napoleon, Emperor of France.

Antoine Lavoisier,
1743–1794.

Dimitry Mendeleyev,
1834–1907.

Left: James Clerk Maxwell, 1831–1879.

Below left: Marie Curie in her laboratory, 1867–1934.

Above: Niels Bohr, 1885–1962.

Right: Ernest Rutherford, 1871–1937.

Nuclear Physics Congress, Rome 1931: the group includes Niels Bohr, Marie Curie, Paul Ehrenfest,

Huygens' clock depended on a simple formula for the time taken for one oscillation of the pendulum:

$$T = 2\pi\sqrt{l/g},$$

where l is the length of the pendulum. This requires the weight to be concentrated at the end of the pendulum, as one can see from any grandfather clock. The period can then be adjusted by moving the weight. Once Huygens' principle was accepted, clock-makers applying it produced timekeepers of unprecedented accuracy – to say nothing of their usefulness to science.

Huygens remained in Paris for fifteen years, from 1666 to 1681, a period in which he worked on fundamental theories about the cause of gravity and the wave theory of light. His work continued after returning to Holland, but was continually frustrated by ill health. He died in 1693, after a lifetime of exceptional achievement in many different branches of science. Why then is his reputation today overshadowed by that of so many contemporaries: Descartes, Boyle, Hooke, Wren, Halley and, above all, Newton?

The real problem with Huygens was that he was a dilettante; with his background, he could afford to be. This led him to neglect that familiar rule of the academic life, 'Publish or perish' – which was just as true four hundred years ago. An example makes the point clear: Huygens, aged twenty-one, applied a principle, first established by Torricelli, to prove the 'general theorem that for a floating body in equilibrium the distance between the centres of gravity of the body and of its submerged portion is a minimum'. The proof was written up under the title 'De Iis Quae Liquido Supernatant' (About things which float in a liquid) in 1650; publication, however, had to wait for the *Oeuvres complètes* in 1908.[38]

Again, in the 1650s, Huygens looked at the theory of collisions. a phenomenon familiar from the game of billiards. This had also interested Descartes, whose results were published in 1644 in his *Principia Philosophiae*. Huygens showed that Descartes' results were completely erroneous and based on rules which simple observation would not support. He even wrote up his own results in his 'De Motu Corporum ex Percussione' (The motion of bodies as a result of collision) – 'the first comprehensive account of perfectly elastic collision between bodies of any size and speed that was in agreement with the experimentally observed facts'.[39] Apart from seven propositions, published without their proofs in the

Journal des sçavans of 18 March 1669, this major work was only published in *Opuscula Postuma* in 1703.

The instances of Huygens' failure to publish are legion, and their effect on his posthumous reputation disastrous. Who knows today that in physics he was far superior to Descartes, and often anticipated Newton,[40] whose reputation he helped establish?

Newton was also lax about publishing, but then he had to get even with Robert Hooke and had Edmond Halley prodding him to make sure that he did so. At the same time Newton had a clear sense of the cosmic significance of his work. This Huygens never had: he was content to have his quite exceptional gifts admired by the people around him, whether in Paris or the Hague. At the end of the day his destiny was to be a big fish in a small pond. Like too many other Dutchmen, he had brought this upon himself, and the story of his life (in which, significantly, he never married) is one of missed opportunities.

The conversion of energy

Towards the end of the eighteenth century it was becoming clear that diverse scientific phenomena, such as combustion by fire or the generation of heat by friction, shared a common factor, the conversion of one form of energy into another. In the nineteenth century the different ways that this could take place were further explored, and notably by one man, J. P. Joule (1818–89), who, in spite of his name,[41] was as English as the city he came from, Manchester.

James Joule was born into a prosperous family, popular in the city and owners of its largest brewery. Considered to be a delicate child, he was educated at home by tutors. In 1834, Joule, aged fifteen, was sent by his father to study chemistry privately with John Dalton, then nearly seventy, but still highly regarded in Manchester for having established at the beginning of the century the atomic theory described in Chapter 6. Joule, who was one of Dalton's last pupils, always acknowledged his debt to him for forming a 'desire to increase . . . knowledge by original research'.[42]

Joule also developed a taste for experiment, with little regard for safety – like so many of his contemporaries. On one occasion, a servant girl, subjected to shocks from a powerful voltaic battery, became unconscious. In spite of such mishaps, Joule's father was highly supportive, even going to the length of having a special laboratory built for him after a move to a new and larger house.

Joule's life in science began at a time of radical change in the understanding of its different branches. The Cambridge philosopher of science William Whewell (1794–1866) had defined the Newtonian achievement, comprising rational mechanics, planetary astronomy and optics, as the 'finished sciences', leaving botany, physiology, zoology, geology, chemistry, heat, electricity and magnetism to be defined as the 'progressive sciences'.[43]

In the first forty-odd years of the nineteenth century, many of the progressive sciences, notably heat, electricity and magnetism, were, in Whewell's words, 'brought within the jurisdiction of mathematics' to comprise physics, a term recognised by the British Association for the Advancement of Science in 1834. This, the world of experimental as opposed to natural philosophy, was where Joule would make his name. At this early stage, most of the work had been done in France, by such men as the Marquis de Laplace (1749–1827) and the Baron de Fourier (1768–1830) – both men raised to the nobility in recognition of their (mainly theoretical) scientific achievements.

In the field in which he was to make his name, Joule set off on the right track as the result of the work of another Frenchman, Sadi Carnot (1796–1832), who had related heat to work in such a way that every unit increase in temperature was equivalent to a corresponding unit of work. For Carnot this was pure theory: Joule established his reputation by measuring, with utmost accuracy, the actual mechanical equivalent of heat and, in doing so, proving experimentally that Carnot's theory was correct.

Joule was not content with one line of experiment: a notable success came when he made use of the revised gas laws (established at the beginning of the nineteenth century) relating the pressure, temperature and volume of gases. The apparatus was simple. A hollow copper cylinder, a foot long and with a diameter about a third as much, was placed in a container of water, well insulated against loss of heat. Once Joule was assured that the temperature of the water and that of cylinder (which contained air at atmospheric pressure) were the same – the state of thermal equilibrium – he pumped air into the cylinder until the pressure reached 22 atmospheres. The mechanical work required to reach this point could be calculated mathematically, applying the elementary gas law relating pressure to volume established by Robert Boyle in 1662. The result of this operation, as required by the gas laws, was to increase the temperature of the water. With a new thermometer made by a local instrument maker, for which Joule claimed an accuracy of 0.005°F, a temperature increase of 0.285°F was measured. This enabled the mechanical equivalent of heat – a physical constant now expressed in units of joules per calorie – to be calculated.[44]

The basic apparatus designed by Joule could be refined in any number of ways. In particular, working with two cylinders containing air at different pressures, but capable of being connected by turning a stopcock, produced new results, confirming and extending those of the original experiment.

Joule, having used the compression of air as a source of heat, then switched to a purely mechanical system designed for the same purpose. His apparatus was an adaptation of that already used by a fellow scientist in Manchester to investigate the friction of liquids. The core consisted of a calorimeter in the form of a copper drum filled with water, in which a paddle driven by a vertical shaft rotated. With baffles in the drum preventing the water rotating with the paddle, the underlying principle was that the energy supplied to the driving shaft should be converted into heat, to be measured by the increased temperature of the water. The driving shaft derived its power from a system of falling weights: the mechanical energy could then be calculated from their mass, and the distance and speed of the fall.

Although the operation of this apparatus was more transparent, the first results produced were not as good as those derived from the compressed air experiments. The best results came when Joule, following a suggestion of George Stokes (1819–1903), a man noted for his study of hydrodynamics, substituted mercury for water. By 1847, there was very little difference between the results from both methods presented by Joule to the annual meeting of the British Association in Oxford; however, those present had some difficulty in accepting the counter-intuitive idea that a liquid was heated merely by stirring it. On the same principle, water at the bottom of a waterfall should be warmer than that at the top, and Joule invited the sceptical to take appropriate measurements 'amid the romantic scenery of Wales and Scotland'.[45]

The compressed air experiment had one significant corollary, also presented to the Oxford meeting. According to the gas laws, as applied by Joule, the pressure of all gases should be zero at 480°F below the freezing point of water. William Thomson (1824–1907) – later Lord Kelvin – who was greatly impressed by Joule's contribution, proposed his absolute scale of temperature, based on absolute zero, for the first time in the following year.[46] In the coming years the collaboration between the two scientists would prove to be extremely productive.

This is a good stage to take stock of Joule's achievements, relating them at the same time to the prevailing concepts of energy. Joule was remarkably successful in devising experiments to reveal the equivalence between different kinds of energy. The experiments described above converted mechanical energy into heat, but Joule's earlier experiments, also relating to the conversion of energy, involved electricity and magnetism.

At a time when the steam engine dominated transport and industry, and the first prototypes of electric motors were being developed, experiments such as those carried out by Joule were of great practical interest. The key problem confronting both the experimenter and the engineer was the loss of energy in the conversion process, which, in turn, was a measure of its efficiency.

In this field, a fundamental principle of early-nineteenth-century science was that of *vis viva*, the store of energy waiting to be turned to practical use. This could take an number of forms: the head of water at a weir, fuel ready for combustion, the electrolyte in a battery, or even food waiting to be consumed. In practice, no machinery could apply all the energy supplied to it to the purposes for which it was designed. In a measured period of time, a hoist powered by a water-mill would never raise a weight to a given height, equivalent to that of the fall of the water driving the water-wheel. Friction generated by the operation of the mill would dissipate much of the power in unwanted heat. In practice any number of countermeasures are taken: machinery is lubricated, pipes are lagged, and so on.

All this was known to Joule. His problem was how to take into account the fact that no apparatus could work with 100% efficiency. The best steam engines in the Cornish mines had only 10% efficiency. The accuracy of his results (which have stood the test of time) is above all a tribute to the efficiency of his apparatus.

Significantly, towards the end of the 1840s – the period of Joule's pioneering experiments – Hermann von Helmholtz (1821–94), a German physicist and definitely a heavyweight in nineteenth-century science, published a paper, 'On the Conservation of Force',[47] which is now regarded as the 'first comprehensive and scientifically satisfactory statement of the principle of the conservation of energy'.[48]

With hindsight, Joule can be said to have provided the experimental underpinning for the conclusions drawn by von Helmholtz, who was very much a theorist. The latter's paper, although citing Joule, does not do him justice, suggesting that his experimental work was unimpressive and only occasionally relevant. In fact Joule was only just beginning to be recognised in England. In the end, however, his experimental results were acknowledged as fundamental to a correct understanding of energy.

Finally, the growth of industry and technology in the nineteenth century led to unprecedented diversity and complexity in possible energy conversions. In an electric power station, the combustion of coal would produce heat to convert water into high-pressure steam that in turn would drive the turbines generating electricity according to the principles established by Faraday. The different stages can be separated from each other, so that the steam for the turbines can in today's world be produced by a nuclear reactor (something which Faraday could never have envisaged), or the turbines themselves could be water-driven, which is what defines the hydroelectric power station.

However it is generated, electricity is an all-purpose source of power, whether for motors, heating, lighting, or whatever – the list is still being extended. The character of the chain of successive conversions and the loss of power at every link in the chain determine the efficiency of the system in relation to any particular use. The practical consequences shape any modern economy.[49]

6

Chemistry: matter and its transformations

Premodern chemistry

CHEMISTRY AS the science of matter in all its forms and transformations goes back to prehistory. Humans, unwittingly, have always been both proactive chemists and observers of spontaneous chemical processes, such as combustion and fermentation, in everyday life. Cooking, firing pottery, making metal alloys, mixing paints and dyes, concocting medicines and poisons, producing adhesives, and blending floral essences into perfumes are all chemical processes.

Although, in the course of several millennia, humankind became increasingly skilful in all these matters, the knowledge applied was the product over long periods of time of trial and error. Such principles as there were (and in some fields they were well entrenched) rested on concepts of the basic constituents of matter, which, for some two to three centuries, we have known to be totally erroneous. The science defined by these principles is known as alchemy, and, until about the mid-eighteenth century, all chemistry was alchemy – that is, it was totally unscientific by modern standards.

Paradoxically, for almost the whole of the seventeenth and eighteenth centuries, Newtonian physics and alchemy existed side by side. Newton himself was a dedicated alchemist, carrying out any number of failed experiments from the laboratory he had set up in Trinity College, Cambridge. Newton's contemporary, Robert Boyle (1627–91), tried to put the shop in order, and in *The*

Sceptical Chymist (published in 1661) attacked both alchemical principles and Aristotle's four-element theory, based on earth, water, air and fire. Although he proposed, as an alternative, other simple and primitive elements, he came nowhere near analytical chemistry, as it began to evolve in the eighteenth century.

As a physicist, Boyle achieved notable results (related in Chapter 5) with gases and developed a versatile new air pump. But he never came close to a correct understanding of the gaseous state of matter, which in the eighteenth century, with men such as Joseph Priestley (1733–1804) and Antoine Lavoisier (1743–94), unlocked the door to modern chemistry as we know it.

Before looking at this new world, two points are worth making about chemistry, or alchemy, at any stage of history. The first is that the phenomena that are their concern are pre-eminently earth-bound. The world, as experienced by humans, is characterised by the solid, liquid and gaseous states of matter, and the capacity of one to transform into another, with or without human intervention. Such transformations belong to the wide class of chemical reactions, which largely define the subject. (The developing field of astro-chemistry is still, as shown in Chapter 8, very restricted in its scope).

The second point is that chemistry, even at its most esoteric, is inherently useful to humankind. This remains true when its ends are destructive, as they are in the development of explosives. The result has often been that chemists turn their skills to profitable enterprise. Ludwig Mond (1839–1909), who started life in Germany as a very competent chemist, ended up in England as one of the founders of Imperial Chemical Industries, whose success was based on the new methods he developed for producing ammonia and soda – two chemicals with wide industrial applications. None of this would have been possible without the eighteenth-century revolution in chemistry. This then is where the story begins.

Lavoisier: father of chemistry

Antoine Lavoisier (1743–94), rightly regarded as the father of modern chemistry, is important not only for his own original discoveries but also for providing the best possible framework for those of his contemporaries. In particular he was the first to establish the true nature of fire, although the experimental results that enabled him to do so were not his own. Lavoisier, with insights that others lacked, realised the true significance of chemical research carried out in the second half of the eighteenth century, not only in France, but also in England and Sweden. The essential focus of this research was air, and how it responded to various processes, notably heating. Air and fire were two of the four elements, since antiquity accepted (as explained in Chapter 2) as the fundamental constituents of all matter on earth.

In the eighteenth century, chemistry was the focus of a revolution in science. The century got off to a false start when the phlogiston theory of combustion, developed by the German chemist Georg Stahl (1660–1734), came to be the accepted explanation of all reactions involving heat and fire. Phlogiston was a vital essence thought to be present in all combustible substances, which was lost as they burn. This loss then explained not only the transformation of the substance itself, but that of the air around it. In a fire, wood lost phlogiston as it burnt to cinders, and air gained it, with notable changes in its properties – but not so that it ceased to be air. In the course of the eighteenth century, chemists were ingenious in incorporating phlogiston into the results of their experiments, even though their cumulative effect made the truth of Stahl's theory ever more improbable. Lavoisier, by showing that phlogiston was not needed to explain the results of the experiments, initiated the breakthrough into chemistry as we know it today.

Lavoisier, although born into a household that was prosperous and well connected, grew up in a family which stressed both ambition and caution. Lavoisier would prove to have too much of the former, and too little of the latter. As the only son to live beyond childhood it was natural that he should train as a lawyer, the accepted way to power and influence, but his interests were always in science. From a very early age his strategy was to look for offices and employment in which the opportunity to practise science would combine with material rewards sufficient to cover the costs of doing so. For someone dedicated to experiment, these could be considerable. In eighteenth-century France, however, Lavoisier's favoured strategy could only succeed if full account was taken of political reality. In the first forty-odd years of Lavoisier's life this meant not only accepting the dictates of a centralised state, in which the king was the ultimate source of power and patronage, but also advancing its interests.

Lavoisier, although always committed to improving public life in France, was still subject to the limitations of government, such as they were in prerevolutionary France, and these were critical in determining the course of his life. So much so, that when the revolution came, Lavoisier paid for his adherence to the *ancien régime* with his life.

In 1760, Lavoisier, aged seventeen, after a brilliant school career, enrolled in the University of Paris to study mathematics and philosophy. After a year, however, he switched to a three-year law course, but without abandoning his scientific interests. He attended public lectures, took private courses, and carried out fieldwork. Given the man he was, one of his ambitions, even at this early stage, was to be elected to the Académie des Sciences. Considering that the Academy had only fifty-four members, divided between six different sciences, with new elections dependent on a vacancy occurring, Lavoisier was certainly ambitious for a man in his twenties. None the less, he worked out a two-pronged strategy that brought him success when he was only twenty-five.

First, with royal permission, Lavoisier's entry in a competition for the best way of lighting the streets of Paris was awarded a special gold medal by the Academy. Second, with his sights set on chemistry – one of the six recognised sciences – Lavoisier presented two papers on the analysis of gypsum.

Scientifically these were boldly innovative. There were two recognised ways of chemical analysis, the wet and the dry. The latter, now known as destructive analysis by heat, was standard in eighteenth-century chemistry. Lavoisier applied it also to discover that gypsum, when heated, lost a quarter of its weight in a vapour that proved to be pure water. The process could then be reversed, with the water added back, to restore the original state. This, the wet way, now known as solvent analysis, showed that gypsum was 'a true neutral salt that becomes a solid by fixing water and forming crystals'.[1] This property made gypsum[2] the ideal material for casts for setting broken bones: this is the original plaster of Paris, which has only recently given away to fibreglass in orthopaedic surgery.

Lavoisier wrote up his experiments with gypsum in two papers read to the Academy in 1765 and 1766. Shortly after the second paper there was a vacancy for an adjunct (the lowest rank) in chemistry, but Lavoisier, although strongly supported and placed on the short list, was not elected. Never a man to give up, Lavoisier continued along the experimental path, by researching techniques for determining the specific weights of liquids. This means finding out the weight of a liquid in comparison with that of water in standard conditions: he worked with hydrometers, instruments made in two forms, variable immersion and constant immersion. (The former is familiar in the form used by garages to measure the acidity of car batteries.)

This work supplemented the results that Lavoisier had already obtained with chemical balances and, as always, his extremely high standards of accuracy produced important new results. These related particularly to measuring acidity in the light of Lavoisier's hypothesis that it was caused by a single element. Once again papers were read before the Academy in the early months of 1768. There was again a vacancy for an adjunct chemist, and this time Lavoisier achieved a majority of votes. Appointment was, however, the prerogative of the king, who, although nominating another candidate – on the grounds of seniority and service to the state – also allowed Lavoisier in as a supernumerary adjunct. On 1 June 1768 he was formally installed as a member.

Anticipating not only his election but also the prospective costs of continuing his experiments, Lavoisier was to take a step early in 1768 that he would later pay for with his life. His grandmother had died in January and, following advice from a family friend, he invested a considerable legacy in a share in the royal Tax Farm. Under the French monarchy this would ensure him a substantial revenue from commissions for collecting tax – an activity supported with all the power of the state. Given that popular discontent in France was largely caused by harsh and unjust taxes, Lavoisier had, unwittingly, allowed himself to become a considerable hostage to fortune. As a member of the Academy, with a strong record of service to the state, all this – in a time without tabloid newspapers – did not count for much in Lavoisier's circles.

For twenty years and more, the young academician would continue his researches, combining them with a succession of science-related public appointments. Lavoisier was progressive in that almost everything he did in the public sphere was focused on using the French national resources, often in a state of extreme neglect, more efficiently. His work extended far outside Paris, so he was often travelling, and in a countryside where long-held special interests could block any reform, he still achieved a great deal. At the same time Lavoisier continued with experiment and publication. The direction he was moving in was to establish the properties of air as fundamental to chemistry, a subject which, when he had first looked at it in the early 1760s, he found to be 'composed of absolutely incoherent ideas and unproven suppositions . . . with no method of instruction, and . . . untouched by the logic of science'.[3] The phenomenon at the centre of his research was simple, although it had been little noted. Iron and copper exposed to air change into powdery substances, rust and verdigris, at the same time increasing in weight. By this process air was fixed in a solid metallic compound.

Lavoisier also found that production of sulphuric acid, made by mixing burnt sulphur with water, involved a similar weight gain. The same result could be achieved with phosphorus. The only explanation was that in the process of burning, air was once again fixed in the element. Believing this 'air' to be a universal constituent of acids, he coined the term 'oxygène', from the Greek *oxys*, meaning 'acidic'. In this he was mistaken: we now know that not all acids (hydrochloric, for example) contain oxygen. Oxygen was first discovered by Joseph Priestley (1733–1804), as Lavoisier acknowledged, but this was just another case of British discovery and French explanation.[4] Lavoisier was then able to summarise his results in general terms:

> Air exists in two modes in nature. Sometimes it appears as a highly attenuated, highly dilated, and highly elastic fluid, such as the one we breathe. At other times it is fixed in substances and combines intimately with them, losing all its previous properties. Air in this state is no longer fluid but rather becomes solid, and it can only regain its fluidity if the substance with which it is combined is destroyed.[5]

This analysis still fails to get to terms with the complexity of air as it exists in nature. The atmosphere around us provides us with a mixture of oxygen and nitrogen, from which animal organisms derive the oxygen necessary for sustaining life. It also contains carbon dioxide, essential for plant life (in the process of photosynthesis described on page 169), but with the supply continuously replenished by animal respiration.

Carbon dioxide (CO_2), produced by reacting sulphuric acid with chalk, was described by Joseph Black (1728–99) in 1754, so that this 'fixed air' became the first of the gases to be examined chemically.[6] Black produced it by heating limestone, and showed that it supported neither life nor combustion. The existence of gases other than air was thus demonstrated for the first time, but it would take many years before the full implications were appreciated – a process in which Lavoisier would play a key role.

In 1772 Paris learnt that Priestley in England had succeeded in fixing this new 'air' in water. By a somewhat complicated process of reasoning, this led Lavoisier to use the Academy's giant focusing lens to see what happened to a diamond when heated in an evacuated chamber. The fact that a reaction took place, but without combustion (which is what he expected), led him to search for a general explanation as to why 'a flame, but not concentrated sunlight, will ignite inflammable substances'. Before his explanation could become complete, there was much ground to cover.

In 1766, the English chemist, Henry Cavendish (1731–1810), investigating the 'factitious airs' which made up the earth's atmosphere, isolated hydrogen. He named it 'inflammable air', identifying it with phlogiston because of its power to react in almost spontaneous combustion. In 1783, Lavoisier, working with Pierre Simon de Laplace (1749–1827) – later to be well known in mathematical physics – turned his experiments to 'inflammable air' (produced by reacting iron with sulphuric acid). These culminated in a demonstration that inflammable air and vital air, burnt together, form water. (A year later, in 1784, Cavendish achieved the same result by using an electric spark to explode hydrogen.) For the first time water was proved to be not an element but a compound, a discovery that would open the way to a 'revolutionary new set of chemical theories'.[7] In the same year, 1783, the Montgolfier brothers, as related in Chapter 3, successfully launched the first hot-air balloon. Lavoisier realised immediately that hydrogen could fill balloons with something much lighter than hot air – an insight with fateful consequences for travel by airship in the twentieth century.

In November 1774, Priestley visited Paris and told Lavoisier of the remarkable properties of the 'air' released by heating, intensely, the red precipitate of mercury (a so-called calx produced by heating, moderately, metallic mercury). Lavoisier, investigating 'air' produced in this way, found that it supported both respiration and combustion much better than ordinary atmospheric air. The conclusion he came to was that 'the air we breathe contains only one quarter true air', the rest, which is non-respirable, proved to be nitrogen, identified by Priestley as one of the two constituents of ammonia,[8] the other being hydrogen. Ammonia is itself a gas, which Priestley discovered by heating its natural salts in a retort (a process which led to the discovery of a number of other gases – notably nitrous oxide, commonly known as 'laughing gas'). The path of experiment then led Lavoisier to separate atmospheric air into its salubrious and mephitic parts. The former category consisted simply of oxygen, while the latter subdivided into mephitic air produced by respiration (that is, carbon dioxide), and the non-respirable portion of the atmosphere (named 'mofette' by Lavoisier[9]).

Lavoisier also used results from experiments conducted by Priestley and presented to the Royal Society in London to explain the role of oxygen in respiration. To support his conclusion that respiration is combustion in the lungs, he noted that both blood and the oxides of metals such as mercury, lead and iron, are coloured red. Finally, on 12 November 1777, he presented to the Academy his *Memoir on the General Nature of Combustion* (which did, however, note that further experiments were still necessary). None the less the paper put forward 'a hypothesis that explains in a highly satisfactory manner all the phenomena of combustion, calcination, and even, in part, those that accompany the respiration of animals'.[10]

Lavoisier then proceeded to dispense with phlogiston, putting his own alternative explanation:

> The matter of fire or light is a very subtle and very elastic fluid that surrounds all parts of the planet we live on, which penetrates with greater ease all the bodies of which it is composed, and which tends, when free, to distribute itself uniformly in everything . . . this fluid dissolves a great many bodies, . . . it combines with them the same way that water combines with salts and acids combine with metals . . . the bodies so combined and dissolved in fluid fire lose some of the properties they had before the combination and acquire new ones that make them more like the matter of fire.[11]

With the benefit of hindsight, this is all very unsatisfactory: Lavoisier was out of the frying-pan into the fire. Equating fire with light was certainly an advance, but Lavoisier plainly failed to view fire from the perspective of a violent reaction between oxygen and some combustible material. Lavoisier may have banished phlogiston, but caloric, which he put in its place, raised more problems than it solved.

Fire appears as both cause and effect of the reactions to which it belongs. Lavoisier failed, however, to realise that it was essentially a by-product, made observable by incandescence, as described on page 91. This can hardly be held against him, since the basic physics (described in Chapter 7) was unknown until the second half of the nineteenth century.

When revolution came to France, Lavoisier continued to be a dominant in the world of science and, as Chapter 3 shows, played an important part in establishing a standard international system of weights and measures. None the less, he had been a tax farmer under the *ancien régime*, and as such he was, with all others in the same category, condemned to the guillotine in 1794. The loss to science of a man at the height of his powers was incalculable. As his contemporary the mathematician Joseph Louis Lagrange (1736–1813) said, 'It took only a moment to sever his head, and probably one hundred years will not suffice to produce another like it.'[12] With this brutal departure, it is now time to see what *la perfide Albion*, safe from revolution, had to contribute to chemistry.

Humphry Davy

Sir Humphry Davy
Detested gravy.
He lived in the odium
Of having discovered Sodium.[13]

When Sir Humphry was chosen as the subject of one of E. C. Bentley's clerihews, his name was well known in Britain, where schools taught that the invention of the miner's safety lamp made him one of humankind's great benefactors. All this was some time in the first half of the twentieth century, when coal-mining was still important in the British economy.[14] At the beginning of the nineteenth century, he was even more famous, being the first scientist since Newton to be honoured with a knighthood. If today, he is largely forgotten, his discovery of sodium remains an important milestone in the history of chemistry. This was but one of his achievements. Before looking at what this all meant, it is best to look at the man himself.

Humphry Davy was born in Penzance, at the furthest end of Cornwall, in 1778. The county was known for its shipwrecks, tin-mines and mild winter climate – all factors that would play a role in Davy's early life. Although apprenticed as a youth to a surgeon apothecary (with the prospect of a good career), Davy's interests were always wide-ranging. He loved being outside, particularly for fishing and shooting, but he also studied philosophy and history, and his talents as a poet (which he never lost) later earned him the friendship of Wordsworth and Coleridge.

Davy's interest in science was stimulated by a shipwrecked French surgeon, who encouraged him to experiment and also introduced him to Lavoisier's *Eléments de chimie*.

At much the same time, James Watt, whose invention of an

efficient steam pump had transformed Cornish mining, sent his tubercular son Gregory to lodge with the Davy family, to benefit from the Cornish climate. Thomas Wedgwood, son of Josiah, the famous potter, also came to Penzance for his health. In such company, Davy found not only an audience for his own scientific ideas, but also gained an entrée to the Lunar Society of Birmingham, whose members combined wealth and influence with a devotion to scientific discovery.

In 1798, Thomas Beddoes, noted for relating chemistry to medicine (in which he advocated the wide use of opium), was another visitor to Cornwall, where Gregory Watt acquainted him with experiments on heat and light carried out by Davy. Beddoes was so impressed that he arranged for the results, later repudiated by Davy, to be published. Beddoes went much further. Helped by funding from the Wedgwoods, he established the Pneumatic Institution (a scientific think-tank *avant la lettre*) in his own house in Bristol and appointed Davy as his assistant.

In Bristol Davy experimented, sometimes dangerously, with various gases, discovering that nitrous oxide (N_2O), popularly known as 'laughing gas', could be used as an anaesthetic. This work, combined with researches into carbon dioxide, which nearly ended his life, led him to conclude that 'chemical properties clearly did not depend in any simple way on material compositions' – a counter-intuitive principle that no prospective chemist should ignore. At the age of twenty-one Davy was, in the year 1799, already a man to be reckoned with. In this same year the world of science was to witness two important but unrelated events, which in combination would determine the future direction of Davy's life.

The electrochemical breakthrough

The first of these was the invention of Alessandro Volta's (1745–1827) pile, as related in Chapter 4. Although the commercial future of electricity never occurred to Davy and his contemporaries, the scientific possibilities of Volta's pile became immediately apparent. When Volta's paper reporting his discovery was received by the Royal Society in 1799, the two referees found that the electric current could be used to decompose water.[15] Davy, working with a battery of 110 double plates provided by Beddoes, found that it would not work with pure water as the electrolyte. On the other hand, nitric acid was extremely effective.

Davy also found that the copper electrode could be replaced with charcoal. (Graphite, another form of carbon,[16] is still used in electrodes.) Davy, noting that the zinc electrode oxidised in the process of use, concluded that the current was produced by a chemical reaction: contrary to Volta's view, mere contact was not sufficient. This was a key insight, since it opened the way to reversing the process, that is, using a current to cause a reaction.

The result was the electrolytic cell. The basic model with two electrodes separated by an electrolyte was the same, but a current from an outside source was then to be passed through the cell to see what reactions would occur. The possibilities were immense and went further than Davy could possibly have conceived of. Any number of substances could be chosen both for the electrodes and for the electrolyte. Useful reactions proved to need considerable electric power, which explains the vast battery provided by Beddoes. Even this was not enough for Davy, and it was in London, not Bristol, that his major discoveries based on electrolysis would be made.

This bring us to the second key event of the year 1799, the founding of the Royal Institution by Count Rumford, whose contributions to science are described in Chapter 5. The object was to encourage and popularise the application of scientific principles. Ever restless, Rumford only stayed three years in London, moving in 1802 to Paris, where he married Lavoisier's widow (who was not nearly so happy with him as she had been with her first husband). Even so, he did bring Davy to London, and by doing so initiated one of the most remarkable eras in the history of science.

Davy became lecturer in chemistry at the Royal Institution in 1801, professor in 1802, Fellow of the Royal Society in 1803, and one of the two secretaries in 1807 – all before he was thirty. The Royal Institution provided him with a magnificent laboratory, and a lecture hall where not only scientists, but London society, came to see him demonstrate the work in progress. He was a born showman, and a society lady attending one of his lectures noted, 'those eyes were made for something besides poring over crucibles'.[17]

The climax to Davy's work during his early years at the Royal Institution came with his Bakerian lecture, delivered on 20 November 1806. There he described his demonstration of how an electric current passed through pure water produces nothing but hydrogen and oxygen. To achieve this result he worked with an electrolytic cell made of agate and gold and with pure water from a silver still.

His battery had 100 double plates of copper and zinc, and the experiment ran for 24 hours. Count Rumford had certainly left the Royal Institution well endowed.

The decomposition of water was decisive in proving that electrolysis could proceed without requiring or producing any acid or alkali. Davy, taking over from Lavoisier, had established chemistry as much an English as a French science. The Institut de Paris awarded him a prize for the best work on electricity, which, on the instructions of Napoleon, was open to the citizens of any nation[18] – remarkably broad-minded seeing what the British had done to Napoleon at Trafalgar.

At the end of his Bakerian lecture, Davy revealed how the chemical properties of metals could depend upon their electrical state. In his electrolytic cell positively charged silver had proved to be reactive, and so converted to silver oxide, and negatively charged zinc, a metal higher in the reactivity series, inert.[19] According to his final summary,

> Amongst the substances that combine chemically, all those, the electrical energies of which are well known, exhibit opposite states . . . supposing perfect freedom of motion in their particles or elementary matter, they ought, according to the principles laid down, to attract each other in consequence of their electrical powers.[20]

In other words, molecules were bound together by electrical forces. This was, effectively, the beginning of electrochemistry – a field with immense potential, which Davy himself began to realise almost immediately.

In 1807 Davy was invited once again to give the Bakerian lecture, an unusual honour – which would also be granted in 1808. In 1807 he told of his discovery of a previously unknown element, potassium, which he had isolated by means of electrolysis. His starting point was caustic potash, an alkali then prepared from burned plants. This he placed in a tube, with a platinum wire sealed into its closed end and with the open end in a bath of mercury. The platinum and the mercury then constituted the electrodes, and the caustic potash the electrolyte. When connected to the battery, with the platinum as the negative and the mercury as the positive electrode, the mercury oxidised – the expected reaction – and a small quantity of some unknown substance (which Davy called an 'alkaligen') formed itself round the platinum. An experiment based on the same principles produced an entirely new substance from soda. Both experiments were described in the Bakerian lecture. The key question was whether the resulting products should be called metals. Davy's own thoughts, as noted in his paper 'New Phenomena of Chemical Changes', are worth quoting:

> The bases of potash and soda agree with metals in opacity, lustre, malleability, conducting powers as to heat and electricity, and in their qualities of chemical combination.

The problem was that both, as metals, were extremely light, with specific gravities way below that of water – a property shared by no other metal. Davy, pointing out that other known (but much heavier) pure metals varied greatly in weight, finally saw it as proper to call the new substances 'potassium' and 'sodium', names in a form only appropriate for metals (so allowing E. C. Bentley to compose the clerihew on page 153).

In any case, both newly discovered elements were very highly reactive in comparison with any known metals (which explains in part why it was so hard to decompose them). Bring potassium or sodium in contact with any other substance, including particularly liquids and gases, and a reaction, often violent, is almost certain to occur. No wonder that they never occur in pure form in nature. Davy himself, describing the way sodium formed at the negative electrode, told how 'the globules often burnt at the moment of their formation, and sometimes violently exploded and separated into smaller globules, which flew with great velocity through the air in a state of vivid combustion, producing a beautiful effect of continued jets of fire'.[21]

The discovery of potassium and sodium is a landmark in the history of science. Both are essential elements for all living organisms. In animals (including humans) electrically charged atoms, or ions, of both elements, play a key role as transmitters of impulses in the nervous system; the positive potassium ion is also essential to protein synthesis in plants. Sodium bonds with chlorine (which was named by Davy) to form salt, an extremely stable compound of two highly reactive and unstable elements. Salt, or Na^+Cl^-, is regarded as the prototype of the ionic bonds.

Davy, having started with potassium and sodium, continued with electrolysis to isolate calcium, barium, strontium and magnesium, although here his contemporary, the Swedish chemist, Jöns Berzelius (1779–1848) – also a pioneer in the use and understanding of electricity – was ahead of him. Berzelius visited Davy in London, but, although they used much the same techniques, they can hardly be called collaborators, as is clear from Davy's own continental travels, which started in March 1813. Davy only grudgingly accepted the achievements of foreign scientists, however eminent.

Working with electrolysis led Davy, in 1810, to discover the electric arc, by connecting the terminals of a giant voltaic cell to charcoal electrodes. The arc crossed the gap arising when the electrodes were separated. One who witnessed this phenomenon reported 'a most brilliant ascending arch of light, broad and conical in form in the middle. When any substance was introduced into this arch, it instantly became ignited; platina melted as readily in it as in the flame of a common candle; quartz, the sapphire, magnesia, lime, all entered into fusion... The light, which was so intense as to resemble that of the sun, produced a discharge through heated air of nearly three inches in length, and of a dazzling splendour'.[22]

The electric arc, still essential to welding, is a continuous electric discharge across the space between two electrodes: to start with the gap must be small, but the heat generated by thermal ionisation[23] creates a conducting medium, across the gap, allowing it to be made much wider. This facility is built into any arcing system, so that the arc, once created, bridges the widest possible gap that will sustain it. The arc generates not only heat but also light, whose properties are determined by the chemical composition of the electrodes, which are not necessarily carbon. (Until Thomas Edison's (1847–1941) invention of the incandescent light bulb in 1879, arcs provided the only possible electric lighting.)

Davy, after being knighted by the Prince Regent on 8 April 1812, married a rich widow, Jane Apreece, three days later. This did not prove to be a happy move, because Jane's haughty disposition and social pretensions endeared neither her nor her husband to the company they kept. She no doubt had much to complain of: even on their honeymoon, spent largely in Scotland, Davy brought with him a chest of apparatus, so that he could continue chemical research. Once married, he largely abandoned his work at the Royal Institution, and in October 1813 he set out for France on the first of his continental journeys.

Davy was accompanied not only by his wife but also by a young man, Michael Faraday (1791–1867), who had been taken on as his amanuensis in March 1813. Judged in the light of Faraday's later scientific achievements, the company was extraordinarily distinguished. It was not happy, however: Lady Davy treated Faraday like a menial servant, and he was delighted when she succumbed to seasickness on the cross-channel voyage. Although it is easy to see Faraday as Davy's great apprentice, he himself saw Davy as 'a model to teach him what he should avoid'.[24] None the less the three stayed together, and once ashore they headed for Paris, where all three behaved like typical Brits abroad. Already on disembarking, Faraday had found, as they were searched by French officials, that 'he could hardly help laughing at the ridiculous nature of their precautions'.[25]

In Paris they received, among many other distinguished scientists, Joseph Louis Gay-Lussac (1778–1850) and André Marie Ampère (1775–1836) – almost exact contemporaries of Davy. Davy, invited to hear Gay-Lussac lecture at the Ecole Polytechnique, noted the experiments carried out, which with the help of diagrams explained a new substance discovered by the lecturer. Davy immediately rushed a paper to the Royal Society, identifying a new element and calling it 'iodine' (because of its affinities with chlorine). Gay-Lussac, whatever he thought of being pre-empted in this way, kept his cool: Davy accused him of picking his brains, but accepted that he stood 'at the head of living chemists in France'. Considering that the Institut elected Davy as a Corresponding Member on 13 December, this was – to quote a well-known figure from the twenty-first century – inappropriate.

This was not all. In front of a distinguished French audience, Davy demolished Lavoisier's theory about acids – according to which oxygen was always an essential component – by showing that chlorine could replace oxygen. Faraday was much less arrogant, leading one French scientist, J. B. Dumas, to remark at the end of the day 'we admired Davy, we loved Faraday'[26] – a view which many in England would share. As for Davy, Napoleon is reported to have remarked that he held all the members of the Institut in low esteem. Davy himself, on leaving Paris with his wife and Faraday, spent the first night at Fontainebleau, where he wrote a poem predicting the fall of Napoleon. Here, as so often, he was right: the battle of Waterloo was only eighteen months away.

In Italy, where the grand tour continued, Davy was somewhat happier. In Florence, using the giant burning-glass of the Grand Duke of Tuscany to focus the heat of the sun on a diamond, he produced pure carbon dioxide, so confirming that diamonds were a form of carbon. (Although Lavoisier had conducted the same experiment in 1772,[27] Davy's result was still significant.) In Rome he experimented with iodine and chlorine and their compounds with oxygen.[28] Finally, he visited Volta, the man who had made everything possible in the first place, in Milan. Volta was dressed in his finest, Davy in his scruffy travelling clothes – British to the end.

Davy returned to England on St. George's Day 1815, just two months before Waterloo. Following the defeat of Napoleon, as predicted in his poem, he wrote to the Prime Minister, Lord Liverpool, urging that the French be treated with severity in any subsequent peace treaty. This came from a man who but two years beforehand had had every honour bestowed upon him in Paris: talk about '*la perfide Albion*'. To his mother he wrote, 'We have had a very agreeable and instructive journey, and Lady Davy agrees with me in thinking that England is the only country to live in.'

In the last ten years of his life, Davy was president of the Royal Society. There, in spite of his feelings about the French, he did his best to make the Royal Society more like the Académie des Sciences, a self-governing institution with election restricted to men with some recognised scientific achievement. Davy accepted that other scientific institutions in France, notably the Ecole Polytechnique, were better than anything in Britain – although by this time the British universities were beginning to reform themselves. Although, while Davy was president, a majority of the members of the Council of the Royal Society had published a scientific paper, radical reform had to wait until after his death.

Davy died in February 1829, barely fifty years old. Even so many of his greatest achievements date back to the 1800s, before he achieved wealth and a position in society. That he was a man spoiled by success is only half the truth: his constant devotion to nature, poetry and the sport of fishing reveal a mystic rather than a materialist. As a scientist and a reformer of scientific institutions, he was almost always right – too much so to make him at all lovable. He may have been as great a man as Volta, Berzelius, Gay-Lussac and Faraday – to name only a few of his great scientific contemporaries – but in purely human terms they all had something he lacked.

Dalton's law for compounding chemicals

John Dalton (1766–1844), to all appearances a classic wimp, is rightly regarded as one of the founders of modern chemistry. His background explains both aspects of his life. Dalton's parents were

Quakers in west Cumberland and brought their son up in a part of England where the Society of Friends was exceptionally well represented. This was the result of the industrial revolution, focused on Manchester – the hub of north-west England – in which Quakers and other religious dissenters had played a prominent role.

The result, for Dalton, was that he grew up in an area where education was largely in the hands of the family of believers into which he was born and of which he would remain a member for his entire life. The dissenting ethos (which owed much to the Quakers) defined a world, centred on Manchester, in which Dalton would always feel at home. With this background it is not surprising that Dalton always earned his living as a teacher; his first appointment came when, at the age of twelve, he took over as teacher in the local school. When he was fifteen, he moved to Kendal (forty miles from his home) to become an assistant in a school generously endowed by Quakers.

The school had a good scientific library, containing not only books but also scientific apparatus – including a telescope, a microscope and an air pump. More important to Dalton, at least in his own view, was the patronage of another Quaker, John Gough, the blind natural philosopher and friend of Wordsworth, who also lived in Kendal.

With Gough as tutor, Dalton's interests focused on science, particularly meteorology, and in 1787 he began to keep daily records – a practice he maintained until his death fifty-seven years later. This was the beginning of his interest in the gases to be found in the earth's atmosphere. Of the elemental gases, Cavendish had already discovered hydrogen, Priestley oxygen, and Lavoisier nitrogen. Even earlier, in 1754, 'fixed air' (which would prove to be a compound of carbon and oxygen) had been identified as different from the ordinary air we breathe.

In 1785 the principal of the Kendal School retired, and Dalton, helped by his older brother, took over. This does not mean that Dalton was content as a schoolmaster – far from it. He observed that 'very few people of middling genius, or capacity for other business',[29] become schoolmasters. Dalton's first attempts to find something better failed, but he did become known, even outside Kendal, from the public lectures on mechanics, optics, pneumatics and astronomy, given with the aid of his school's scientific apparatus.

In 1792, after visiting London for the first time, Dalton was appointed as the first professor of mathematics and natural philosophy at the Manchester Academy, recently founded by prominent local dissenters. This appointment brought him to just the right place, at least at the end of the eighteenth century. For as Disraeli later observed, 'What Art was to the ancient world, Science is to the modern; the distinctive faculty. In the minds of men, the useful has succeeded to the beautiful . . . rightly understood, Manchester is as great a human exploit as Athens.'[30]

In the event Dalton liked Manchester better than the Academy, where he resigned his position in 1800. Instead he opened his own Mathematical Academy, which was an immediate success. The fees paid, and the freedom to organise his own life, meant that he could pursue the serious business of chemistry much more effectively. He had already begun to publish, and his first paper, submitted to the Manchester Literary and Philosophical Society (to which he had been elected in 1794), proved to be surprisingly influential.

Dalton's chosen subject was colour-blindness, an affliction that he shared with his older brother (although it was not then known to be hereditary). His theory about its causes, argued in meticulous detail, later proved to be wrong – as was confirmed when his own eyes, on his instructions, were dissected after his death. None the less, the condition was long known as 'Daltonism', although Dalton himself would never write about it again.

Dalton's most important work began only after he left the Academy, when the Literary and Philosophical Society provided a home for his apparatus and experiments, to say nothing of an up-to-date library and a journal which would publish many of his most important results. What then did Dalton achieve in the remarkable opening years of the nineteenth century?

In the new century Dalton's work with mixed gases – a development of his interest in meteorology – was critical. Towards the end of the old century, his study of the evaporation of water in the atmosphere, presented in his first book, *Meteorological Observations and Essays*,[31] had led him to the conclusion that no chemical reaction was involved. The book also contained, in somewhat inchoate form, the proposition that in a mixture of gases every gas acts as an independent entity – 'Dalton's law of partial pressures'.

The result is that such a mixture operates effectively as a reservoir for all the gases contained in it, so that chemical processes can always access the particular gases they require and discharge the substances they then produce. A simple experiment shows how to observe this process at work: light a candle and, after allowing it to burn for a minute, cover it with a large glass. Within a few seconds, the flame will be extinguished. The process of combustion, a chemical reaction dependent on oxygen, having consumed the entire quantity contained in the glass, can no longer continue: so long as it lasted, carbon dioxide, a product of the reaction, replaced the oxygen to mix with whatever other gases were present in the atmosphere. The process would in no way have affected the chemistry of these other gases (although it might have affected both their temperature and pressure).

This was only half the story. By Dalton's day it was beginning to be clear that gases could combine in a reaction to make compounds. In particular, nitrogen could combine in two different ways with oxygen, one producing nitric oxide and the other nitrous oxide – the laughing gas familiar from Davy's early experiments. It could also combine with hydrogen to produce ammonia.[32] There were also two different ways carbon (although not a gas) could combine with oxygen, and two more with hydrogen. In these cases the solubility of the gases in water was the focal point of Dalton's experiments. And then, of course, water was itself a compound of hydrogen and oxygen – the only one known, until the discovery of hydrogen peroxide[33] in 1815. (Astrophysicists have now discovered a third such compound, hydroxyl,[34] in interstellar dust.) Dalton's experiments in the opening years of the nineteenth century confirmed that 'compounds were formed from the combination of constant amounts of their constituents'.[35]

Taking the case of the two oxides of nitrogen, in which the quantity of nitrogen in one is twice that in the other, there is no question of a gradual transition from one to the other. There is no way of progressively 'adding' nitrogen to nitric oxide so as to transform it into nitrous oxide: these two compounds of nitrogen are formed by quite different reactions. In the general case of two elements A and B having known compounds $A+B$ and $A+2B$, further compounds may well be possible not only in the forms $A+3B$, $A+4B$, and so on, but also in such intermediate forms as $2A+3B$. Additional elements C, D, and so on, occur in more complex compounds. For most elements the list of possible compounds is quite short; the great exception in carbon, whose capacity to combine in long complex chains underlies (as related later in this chapter) the whole of organic chemistry and biochemistry.

Dalton's discovery led him to weigh the different portions of each element occurring in a compound. Since this procedure can be carried out with any quantity, the only constant results it can produce are the ratios of the weights of the different portions to each other. This was Dalton's system of equivalent weights.

Now Dalton was concerned with a process of transduction, by which he pursued equivalent weights conceptually to the point at which the smallest unit still retaining all the properties of the element occurs. This led to Dalton's theory of the atom, although the concept of such a small fundamental unit was already current among his contemporaries. The problem, when working with portions compounded in molecules built out of such atoms, was how to be certain about the proportions of the elements occurring at this fundamental level.

Dalton's treatment of water is exemplary. The relative weight of the portions of oxygen and hydrogen was 7 to 1. Dalton, assuming a one-to-one ratio at the atomic level, calculated a molecular weight for water, not having any means of knowing that the water molecule in fact contains two hydrogen atoms. His weighing was also inaccurate, since the relative weights in a water molecule are 8 to 1, so that with two hydrogen atoms, an oxygen atom must in fact weigh 16 times as much as a hydrogen atom. Dalton, knowing no better, commonly assumed the simplest possible combination – e.g. with one hydrogen atom – not only with water, but also with ammonia (NH_3) and other compounds. The problem always defeated him, as he was to note in 1814:

> After having the atomic principles in contemplation for ten years, I find myself still at a loss, occasionally, to discriminate between the combinations which contain two atoms of a given body from those which contain only one atom.[36]

Dalton was, however, on the right track. In the course of time his mistaken proportions could and would be corrected. In particular, his support for the so-called integral weights hypothesis was extremely hesitant. According to this hypothesis, stated by Dalton's contemporary, William Prout (1785–1850), in 1815, the atomic weights of all the elements are multiples of that of hydrogen.[37] On the other hand, in 1827 – rather late in the day – Dalton did accept that 'the greatest desideratum at the present time is the exact relative weight of 100 cubic inches of the element hydrogen'.[38]

Dalton should also have paid more attention to the Swedish chemist, Berzelius (introduced on page 167), who recognised him as 'one of the most ingenious physicians of our age'. Berzelius, a much more accurate experimenter, obtained results which both extended and modified those of Dalton. In particular, Berzelius accepted the law stated by Gay-Lussac in 1808 that when gases combine chemically, there is always a simple numerical relationship between the volume of those consumed and that of those produced. This then opened the way to finding in 1814 the correct ratio, 2 : 1, of hydrogen to oxygen atoms in water, which, if known to Dalton, would have saved him much trouble.

On the other hand, both Berzelius and Dalton ignored the important hypothesis stated by the Italian chemist Amedeo Avogadro (1776–1856) that equal volumes of all gases contain equal numbers of molecules when at the same temperature and pressure. Quite simply, Avogadro, coming from Turin, was not well placed to make an impact on the world of science. Nearly a half century later, his compatriot, Stanislao Cannizzaro (1826–1910), convinced the world of the truth and importance of Avogadro's hypothesis. By this time the Russian chemist Dimitry Mendeleyev was well on the way to establishing the periodic table of the elements.

Chemical notation

The modern world is now so conditioned to chemical notation that there is little concern as to what lies behind it, whether in terms of its meaning or of its historical origins. A headline such as 'CO_2 emissions threaten global warming' begs any number of questions about the true nature of CO_2. The word emission suggests that it occurs as the result of some sort of chemical process, and the notation suggests that this involves combining one measure of carbon (C) with two of oxygen (O). This principle is also enshrined in the familiar name 'carbon dioxide'. But what are these measures? Familiar dimensions, such as mass and length, provide no more than the first step towards the right answer. This is intuitively obvious, given that carbon is a solid and oxygen is a gas. At least since Lavoisier's day it has been clear to chemists that everyday compounds, whether gaseous (e.g. CO_2), liquid (e.g. water, H_2O) or solid (e.g. common salt, NaCl), combine elements with quite different characteristics. The problem, historically, is that the compounds came first, although, particularly in the case of gases, they are not always recognised as distinct substances.

Returning to the headline, the picture evoked, of a factory chimney discharging industrial waste in the form of gas into the atmosphere, could also have been that of a scenario from before the year 1754, when 'fixed air' – today's CO_2 – was first identified as a substance different from ordinary air. Even in these early days Thomas Newcomen's steam engines were already polluting the atmosphere. One hallmark of the revolution in chemistry is the transition from a vocabulary that in 1750 was chaotic to one that in 1850 was completely systematic. In 1750, the chemical lexicon, based on local language, varied from one country to another, and was hardly consistent in any one of them; in 1850 it was based on universally accepted abstract symbols. On the one hand, the transition was simply the result of lexical reform, the subject matter of this section; on the other, it required Mendeleyev's discovery and organisation of the periodic table of elements, explained later in this chapter.

The movement for lexical reform started in the Académie des Sciences, and, although it was originally confined to French, the principles governing it could be, and in due course were, applied to other languages – notably English. Lavoisier, who led the reform, set the ball rolling with a paper entitled 'The Need to Reform and Improve Chemical Nomenclature',[39] presented to the Academy in 1787. Lavoisier was not an entirely disinterested reformer: he intended to restate chemistry in terms which implicitly accepted the correctness of his new theories. His textbook, *Eléments de chimie*, published in 1789, was largely written to confirm this position. The question is, what did Lavoisier have to work with? And how did he then devise a suitable lexicon?

The basic material available was simply state-of-the-art chemistry, as Lavoisier regarded it in the mid-1780s – incorporating, therefore, even those parts of his own work still open (rightly as we now know) to dispute.

Lavoisier's list of simple substances was divided into four parts. Part 1 contained 'simple substances belonging to all the kingdoms of nature, which may be considered as the elements of bodies'. The status of oxygen, azote (nitrogen) and hydrogen, the three elemental gases known to Lavoisier, is still unquestioned, although time would add another seven to this category – one, chlorine, within a generation. Light was included because Lavoisier saw it as the fundamental principle of vegetable chemistry (observed as photosynthesis), and caloric as the basis of heat and expansion.

By Lavoisier's day, sulphur and phosphorus were well established as non-metallic elements, which is the defining property of part 2. Charcoal, as representative of carbon, is in the same class. The three radicals had not been correctly identified. In the 1770s, the muriatic and fluoric radicals, derived from chlorine and fluorine, were discovered by the Swedish chemist, Carl Scheele (1742–86), who passed on his experimental results to Lavoisier. In 1808 the boracic radical would lead to the independent discovery in both France and England of the element boron (never found free in nature).

Part 3 contained seventeen metals, of which six, bismuth, cobalt, manganese, molybdenum, nickel and tungsten, were discovered in the middle years of the eighteenth century (three by Scheele), while the rest had been known much longer, most since antiquity. Lavoisier's comprehensive description, 'oxydable and acidifiable simple metallic bodies', derived from his own experiments, attributed properties hardly relevant to metals, such as gold and platinum, low on the scale of reactivity.

Part 4 – 'salifiable simple earthy substances' according to Lavoisier's heading – contained five substances commonly occurring in nature, known since antiquity and useful for any number of practical purposes. Although they are all compounds, a particular element, not isolated in Lavoisier's day, characterises each one. These elements, calcium, magnesium, barium, aluminium and silicon, would all be isolated within a generation of Lavoisier's death.

Taking a synoptic view of the four categories, twenty-three of the substances listed are actual elements, while eight foreshadow elements on the threshold of discovery when Lavoisier died. Surprisingly, perhaps, Lavoisier failed to include potash and soda, the source of the first two elements, potassium and sodium, to be isolated by Davy – as related on page 156. Leaving aside light and caloric, Lavoisier's table provided a remarkably solid basis for developing chemical nomenclature and notation.

Lavoisier's work was carried on by two younger contemporaries, Claude Berthollet (1749–1822), who had collaborated with him, and the Swedish chemist Jöns Berzelius (1779–1848). Berthollet's realisation of the essential connection between the way a reaction takes place and the mass of the reagents opened the way to chemical formulae as we now know them. Berzelius, who himself discovered three elements (including silicon), lived to see the elemental base of all the substances in Lavoisier's table established; although he never reached seventy, of the fifty-four elements known when he died, thirty had been discovered during his lifetime.

Berzelius, a year younger than Davy, is regarded by many as his equal, but his real focus (shared in correspondence with Davy) was on the work of John Dalton. Like Davy he achieved distinction early in life, becoming, at the age of twenty-eight, professor of medicine in Stockholm, but sadly – again like Davy – his last years were troubled by professional discord. Unfortunately he never found a good biographer, and although he corresponded in English and French, his annotated letters are to be found only in a Swedish edition.[40] The record, therefore, consists largely of his published scientific work. This reflects 'his massive contribution . . . unique in the history of chemistry. His systematic mind saw the need for a structure in which chemistry could grow with the precision and the articulation of a living organism. The basic principle of his design was atomic composition.'[41]

Berzelius's new chemical symbolism was intended to replace that developed by Dalton. Where Dalton used circular signs, Berzelius used letters. Although he was not the first to do so, his system, as it developed in the course of some twenty years, proved superior to any its rivals and is still that in use today (so that expressions like CO_2 even appear in tabloid headlines). These were 'destined solely to facilitate the expression of chemical proportions',[42] but they were also much easier to write or print.

Photosynthesis and the life of plants

If the phenomenon of light, generally regarded, belongs to physics, it still governs a number of important processes in chemistry. The best known is that of photosynthesis, in which the light of the sun is essential to the growth of green vegetation, which in turn stores the energy absorbed from the sun. The actual process, or rather its results, was observed by the earliest representatives of humankind: it was, after all, essential for the food chain. An understanding of the process had to wait, however, until the end of the eighteenth century.

The first breakthrough came in 1771, when Joseph Priestley noted that green plants emitted oxygen. Eight years later, in 1779, Jan Ingenhousz (1730–99) (a Dutchman who had come to live in England), established that sunlight was essential to their growth, and in the same year the Swiss Jean Senebier (1742–1809) published in his *Action de la lumière sur la végétation*, the general principle: this is that a plant, by absorbing carbon dioxide (CO_2) from the atmosphere, and water (H_2O) from the ground, produces glucose (the basis of all carbohydrates), at the same time releasing oxygen (O_2).

Chlorophyll, a green compound to be found in the leaves of plants,[43] is essential to the process, which can be stated formally as follows:[44]

$$6CO_2 + 6H_2O \underset{\text{chlorophyll}}{\overset{\text{sunlight}}{\longrightarrow}} C_6H_{12}O_6 \text{ (glucose)} + 6O_2$$

The importance of the process is that it produces an astonishingly wide range of organic compounds – essentially the whole of plant life – which by chemical reaction can release energy. In living species this is the process of metabolism by which life is maintained. Equally plants can be used as fuel for combustion, either in the form of their natural growth or after being subjected in a decadent form to external forces (mostly geophysical) over long periods of time. Typically the former process produces wood, and the latter, coal and oil – the so-called fossil fuels. As already noted on page 152, it was one of Lavoisier's great insights to see that the two processes were essentially the same. Since early times both have been extended by human invention, from cooking vegetable matter to burning charcoal – to say nothing of modern industries such as oil refining.

Photography: the inorganic chemistry of light

According to a phenomenon already observed in the sixteenth century, certain naturally occurring silver salts become dark on exposure to light. In the first half of the nineteenth century it was discovered that these salts were those of chlorine, bromine and iodine – all elements discovered in the period 1810–26.[45] Louis Daguerre (1789–1851) in France and William Henry Fox Talbot (1800–77) in England, acting independently, used their understanding of the basic phenomenon to invent photography – that is, a chemical process by which a fleeting image could be recorded in permanent form.

The process itself belongs to technology rather than science. Myriads of minute silver salt crystals are uniformly distributed over a flat surface covered by an emulsion. An image focused on the surface then causes each separate crystal to darken in proportion to the light incident upon it. The result is a photographic negative: the bright parts of the image became dark, and the dark parts, bright. The problem, still familiar to amateur photographers, is to stop the process and fix the image when it has reached the most satisfactory stage of development.

The solution came by realising that, on exposure to light, the ions composing the salt (e.g. positive silver and negative chlorine) separate and, by electron transfer, transform to the basic elements, a process completed by a developing solution. The untransformed salt crystals (i.e. those not exposed to light) can then be eliminated, leaving only a residue of black metallic silver constituting the photographic negative. If the surface containing the emulsion is transparent, the process can be repeated with emulsified paper, reversing the colours so as to produce a positive image.

Basically this is all photography adds up to, at least before the recent invention of the digital camera (although the colour photography developed by Gabriel Lippmann (1845–1921) and Lord Rayleigh (1842–1919) should perhaps be noted). In the early days the usefulness of photography to science was limited by poor resolution and long exposure times, but better optical systems and creative chemistry in the emulsions (which continued to be based on silver) cured these defects to the point that from about 1850 onwards photography became an essential adjunct to astronomy (particularly after Lippmann's coelostat enabled a telescope to follow the movement of the stars). The first ever photographs of a solar eclipse, taken in 1851, represent a key breakthrough. Following astronomy, almost all optical instruments made use of photography in one way or another, and as we shall see later in this chapter this use extended beyond light waves to other forms of radiation such as X-rays. This is particularly true of spectroscopy (the subject of the following section).

Spectroscopy: the rediscovery of light

In optics, until the end of the eighteenth century, there was little advance on Newton's discovery that a prism resolved the light of the sun into the series of colours that make up the spectrum. In the middle of the century, a young Scotsman, Thomas Melville (1726–53), had noted the presence of a brilliant yellow light in the flame of burning alcohol, when other substances, notably salt, were added. This, the distinctive line of sodium, then aroused little interest, perhaps because Melville, dying at the age of twenty-seven, had too little time to establish a scientific reputation.

This was not the case with William Wollaston (1766–1828), a wealthy man born into a family of scientists, who in 1802 observed seven dark lines in Newton's spectrum when this was obtained from a beam of light passed through a narrow slit only one-twentieth of an inch wide. This was just the beginning. In the years 1814–17, Joseph von Fraunhofer (1787–1826), a Bavarian master glass-maker, following up Wollaston and using his own improved lenses, developed the prism spectrometer. This, an optical instrument of unprecedented precision, made possible a whole new science of spectroscopy, which, as it developed and expanded in the following two centuries, would transform the universe of science.

Fraunhofer, using a telescope to observe spectra, saw not only Wollaston's seven dark lines, but hundreds of others. He counted 600, which he recorded on a map, in which the most prominent were given the letters A, B, C,... starting at the red end of the spectrum – the system still used today. This was a remarkable discovery, but Fraunhofer, an instrument-maker rather than a scientist, could never explain its significance – perhaps because he also died too young. The breakthrough came with two other German scientists, Robert Bunsen (1811–99) and Gustav Kirchhoff (1824–87) – two of the best-known names in nineteenth-century science. Although in early nineteenth-century Germany culture and language were more or less uniform, there was no German state. Instead, the common culture area was ruled by a great number of princes, whose courts largely determined not only the character of law, politics and administration but also the support given to science. Matters of state reflected a general concern for the impact abroad of the French Revolution, which had led to the fall of Europe's most powerful kingdom – not a happy augury for German princes – and introduced any number of unsettling new principles into politics, religion and education. The result in Germany was that both radical and reactionary ideas flourished, subject at local level to the politics of the court, which were generally conservative.

This climate favoured science. More than any other Germans, scientists were cosmopolitan, and at home, not only where German was spoken, but in France, England and even, occasionally, Sweden and Russia. France, throughout the revolutionary period, continued to lead in scientific discovery, and many a German prince, anxious not to be left behind, patronised and encouraged science (hoping at the same time to avoid the dangerous ideas that could accompany it). This was a great gain to both the universities and the scientists who taught there. The princes were not disinterested sponsors: the King of Bavaria, when he appointed the American Count Rumford to introduce the potato and supervise the royal arsenal, set a precedent for combining science with practical politics and economic policy. In the nineteenth century, the German princes were keenly aware of the practical advantages of sponsoring science.

Such was the background to the life and work of Bunsen and Kirchhoff. Both came from comfortable families of academics and court officials, established in old university towns – Göttingen in the Kingdom of Hanover in the case of Bunsen, and Königsberg in the Kingdom of Prussia in the case of Kirchhoff. In middle life their paths would cross, first in Breslau (in Saxony) and then in Heidelberg (in Hesse), where, as close friends, they collaborated over a long period. Both would live to see the unification of Germany, in 1870, orchestrated by Prince Bismarck, the Iron Chancellor of Prussia.

Bunsen, although a sometimes impetuous schoolboy (who once overturned his desk when the master made a joke at his expense) achieved, aged seventeen, a distinguished *abitur* – or high-school diploma – and went to follow a broad science syllabus at his home university. He shone as a student, but he also learnt glass-blowing (useful for a chemist) and made a chemical balance good for weights from 10 milligrams to 200 grams – a remarkable range. This also came in useful in a life of scientific experiment.

In 1830 Göttingen became a centre of radicalism at the time of the French July revolution, with professors refusing the loyal oath to the King of Hanover. Bunsen, already busy with his doctoral thesis, maintained a low profile, and his prudence was rewarded by a government travelling scholarship. This led to two years of travel, in which Bunsen visited almost every continental centre of scientific research. He spent the longest time, eight months, in Paris, and then, in 1833, in St. Etienne, much further south, he saw a railway for the first time, and travelled by train – predicting, correctly, that trains would also come to Germany (as they did two years later). Ironically, Austria, because of the revolutionary events in Hanover, first denied him entry, but he still made it, finally, to Innsbruck and Vienna. (In contrast to the German princes, the Habsburg emperors were seldom patrons of science.)

Once home, he completed his doctorate with distinction, and while busy working to qualify as *Privatdozent* in chemistry, accepted, in 1836, a well-paid job in Kassel (in Hesse) with plenty of time for his own research. He was still, apparently, impetuous: he was first nearly blinded by a laboratory explosion, and then, while researching the chemistry of poisons, spent several days near death after experimenting with potassium cyanide. The chapter of accidents continued throughout his life, and when he was fifty-seven only his left hand, held before his eyes, prevented his being blinded by another explosion. Finally, when he was already sixty, all his papers, photographs and drawings were lost in a fire. Undiscouraged, he built up his collection again before the publication, in 1875, of his comprehensive *Spektralanalytische Untersuchungen*.

From the beginning, Bunsen's exceptional talents were recognised. Although his work at Kassel was important economically for Hesse, Bunsen, never out for gain (unlike many distinguished contemporary scientists – Liebig, Mond, Siemens – among his compatriots), was noted for saying 'work is fine, acquisition, contemptible'.[46] Aged twenty-eight, his talents brought him a professorship at Marburg where he invented a cheap and efficient zinc–carbon battery,[47] which he then used to produce electric arcs between metal electrodes: during his evening lectures he beamed the brilliant light of the arcs on the neighbouring Elizabethkirche.

Reactionary politics in Hesse, following the revolutions that swept Europe in 1848, led Bunsen to move to Breslau (in Saxony) in 1851. He stayed only a year, moving on to Heidelberg in 1852, attracted by a high salary and the promise of a new laboratory – so foreshadowing the familiar career structure of twentieth-century academia.

In Breslau, Bunsen, with the help of his zinc–carbon battery, was the first to use electrolysis to produce magnesium on a large scale, and in Heidelberg he added chromium and aluminium. There, also, with the gas supply in his new laboratory, Bunsen adapted a gas-burner brought from England by Henry Roscoe, so as to produce a burner in which the gas–air mixture could be controlled to produce different sorts of flame. This became the world-famous Bunsen burner, part of the essential equipment of any laboratory (and also the basis of today's gas cookers). For Bunsen himself, it made possible experiments in spectroscopy, by which he, together with his friend Gustav Kirchhoff, would transform both physics and chemistry.

Bunsen met Kirchhoff for the first time in Breslau and in 1854 arranged his appointment to a chair in Heidelberg. So began a most fruitful and amicable collaboration between two men of exceptional ability. Kirchhoff, although no less gifted than Bunsen, differed from him in both manner and appearance. Bunsen was a self-confident extrovert, with a massive and somewhat rough-hewn physique, unconcerned about the risks his experiments involved. Kirchhoff, diffident and uncertain, a theorist rather than an experimentalist, was small and somewhat effeminate – the perfect foil to Bunsen. It was said later that 'Bunsen's greatest discovery was Kirchhoff'.

As a schoolboy and student Kirchhoff shone as brightly as Bunsen. At the age of twenty-two, he published his doctoral thesis, which contained the first version of the fundamental laws governing electric currents and conducting systems – now known simply as Kirchhoff's laws. This success led the Physikalische Gesellschaft in Berlin (the capital of Prussia) to offer a grant for a year's study in Paris. Kirchhoff never made it, frightened by the revolutionary political situation in France. Instead he spent the time in Berlin, where he continued to build his reputation to the point that in 1850 at the age of twenty-six he could accept the chair of experimental physics in Breslau.

Sadly, as so easily happens in academic life, he clashed with another professor in the department, took sick leave on doctor's advice, and went back home to Königsberg. There he met Hermann von Helmholtz (1821–94), recently famous for his work on the conservation of energy, who immediately befriended and admired him. But at Breslau Kirchhoff had met Bunsen, who together with Helmholtz orchestrated his appointment to Heidelberg in 1854. (Bunsen had wisely advised Kirchhoff, a colossal self-doubter, not to give way to untimely modesty in making his formal application.)

At Heidelberg, Kirchhoff equalled Bunsen as a lecturer (with two of his students, Gabriel Lippmann (1845–1921) and Heike Kamerlingh-Onnes (1853–1926), later becoming Nobel prize-winners). Kirchhoff also showed that electric currents in narrow wires propagate as waves with the speed of light, a key result in the later experimental work with radio waves (explained in Chapter 7) of another of his students, Heinrich Hertz (1857–94).

Bunsen and Kirchhoff, almost inseparable, were familiar in the streets of Heidelberg, in deep scientific conversation as they walked together. (In proportion they must have resembled Helmut Kohl and François Mitterand). On one of the walks, a sunset seen from the wooded heights above Heidelberg led them to look at spectral analysis. Although others were already active in this field, Bunsen and Kirchhoff, by consolidating and extending their results, established spectroscopy at the heart of both physics and chemistry.

After Fraunhofer's discovery of dark spectral lines in sunlight, the next key finding was that other light sources, besides the sun, had distinctive lines. The Norwegian Anders Ångström (1814–74) discovered that a spark between two metal electrodes contained the spectral lines of both the metal and the gas medium, which, in contrast to those of sunlight, were light rather than dark. Further advance was then blocked by failure to understand a distinctive yellow line occurring in almost all spectra. In 1857, William Swan (1818–94) showed that this was always a sign of a sodium compound, present even when common salt[48] was but one part in 2,500,000 of the substance producing the spectrum. Bunsen and Kirchhoff then showed that it corresponded to Fraunhofer's D-line, and, given the ubiquitous traces of salt in the atmosphere – the result of oceans covering two thirds of the earth's surface – the practical problem was how to eliminate it from the laboratory.

One problem remained unsolved: why were the lines of the sun's spectrum dark, whereas those of spectra produced in the laboratory were light? According to Kirchhoff, 'this was either a nonsense or something very important'.[49] In 1859 he examined the sun's spectrum through a yellow sodium flame, to find that instead of masking the dark sodium line (as he had expected) it accentuated it. He then obtained the same result in the laboratory by substituting an intense white incandescent light for the sun. Kirchhoff's explanation (which proved to be correct) was that light of a given wavelength would absorb incident light of the same wavelength, in a phenomenon comparable to resonance. For this reason the dark lines became known as absorption lines (in contrast to the bright emission lines).[50]

This is the fundamental principle that for 'rays of the same wavelength at the same temperature the relation between the emission and absorption power is the same with all bodies'.[51] This led to Kirchhoff's concept of black bodies absorbing all light incident upon them – later to become a key factor in heat-radiation research (as described in Chapter 7).

This also explained the Fraunhofer lines: the sun's spectrum, with its dark lines, is nothing other than the inverse of the spectrum produced by the sun's atmosphere. Spectral analysis of the sun's atmosphere could then proceed by taking the dark lines to correspond to the bright lines produced by any substance when heated in a flame. This culminated, in 1860, in the publication of Kirchhoff's *Chemische Analyse durch Spektralbeobachtunge*,[52] based largely on experiments carried out by Bunsen.

In spectroscopy the practical problem was to produce, first, in purest possible form, different salts for each metal, and then, by use of electric arcs or different flames, to make them incandescent – for otherwise there would be no light to analyse. It was essential that the character of different flames, with their vast differences in temperature, would not affect the exact location of the separate spectral lines of the metals investigated. Bunsen and Kirchhoff showed how bright lines, according to their location in the spectrum, indicated the presence of particular metals. This was an extremely accurate analytical tool, especially for small quantities, although some results were foreshadowed by work done by others.[53]

One experiment detected sodium vapour diluted to one part in 20,000,000,[54] and although no other metal had such a prominent spectrum, very small quantities of lithium, strontium, calcium, potassium and barium could be observed in the same way (an economically significant result when it came to prospecting for metals whose sources were widely dispersed).[55]

Bunsen and Kirchhoff also used spectroscopy to detect new elements. In particular, they expected the discovery of a fourth alkali metal next to potassium, sodium and lithium – showing only two lines in their spectroscope. They described the characteristic lines of the then undiscovered caesium in the blue part of the spectrum, and went on to do the same for rubidium with its characteristic dark red lines. Finding actual specimens of these two elements was very difficult: 44,200 kilograms of salt solution had to be processed to produce 7.272 grams of caesium chloride and 9.237 grams of rubidium chloride.

Finally, the correspondence between absorption lines observed in sunlight and iron emission lines observed in the laboratory was so exact that it could not be the result of chance. The only possible explanation was that iron was present in the sun's atmosphere. This line of reasoning led to the further discovery on the sun of thirteen known metallic elements, together with hydrogen.[56]

Kirchhoff's absorption research also showed that the sun has a very hot light core, covered by a cooler layer containing the vapour of the elements discovered spectroscopically. Similar methods could be applied to analyse the composition of stars, which, earlier in the century, the French philosopher Auguste Comte had cited as an example of things that were inherently unknowable. In historical perspective, Bunsen's and Kirchhoff's disproof of Comte's claim introduced the new science of astrophysics.

The pioneering work of Bunsen and Kirchhoff was soon followed by others, using high-quality custom-built apparatus. First came William Crookes (1832–1919), who in 1861 discovered thallium after noting a bright green line in the spectrum; two years later indium was discovered in Germany; and at the end of the day, spectroscopy was to play a part in the discovery of twenty-one out of the twenty-seven elements discovered between 1860 and 1910. (Some of these, notably helium, were first discovered on the sun before being found on earth.)

Bunsen and Kirchhoff were never able to discover the theory underlying the exact location (in terms of wavelength) of the distinctive spectral lines characteristic of any element: this, when it came, belonged to physics rather than chemistry and was one of the fundamental achievements of the great Danish physicist Niels Bohr (1885–1962). By this time, spectral analysis had extended far beyond the outer limits of the visible spectrum, although as early as 1800 William Herschel (1738–1822) had found that the heat spectrum extended into the infrared, while a year later Johann Ritter (1776–1810) showed that the darkening of silver halides (as in photography) continued into the ultraviolet. In this way three spectral zones were established, but the full implications only began to become clear at the end of the nineteenth century – a story told in Chapter 7.

Mendeleyev's periodic table

For more than a hundred years the periodic table has been an icon for chemistry, a status – to judge from the title of Primo Levi's *The Periodic Table*[57] – extending far outside the discipline. For those who know how to read the table, which means almost any serious chemist, the key properties of the different elements (of which something over ninety occur in nature), systematically arranged in the order of their atomic weights, are immediately apparent. None the less, before its discovery in 1869 by Dimitry Ivanovich Mendeleyev (1834–1907), the inchoate world of chemistry was only beginning to discover the order established by his table of elements (of which the first draft is to be seen in Figure 6.1).

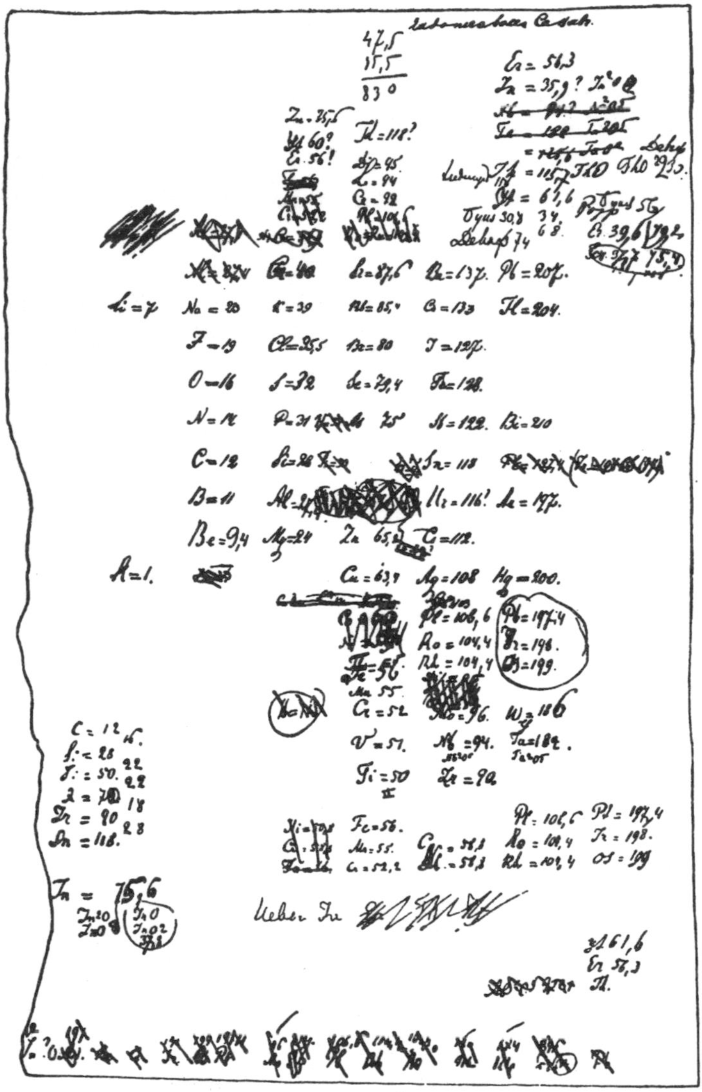

Figure 6.1 Mendeleyev's first draft of the periodic table of the elements.

The life of Mendeleyev gives a distinctive Russian twist to the familiar story of the precocious schoolboy, who, after running through a gamut of academic distinctions, makes it to full professor at the age of thirty. His appearance, with gleaming eyes, long hair and an even longer beard, is that of a character from Tolstoy. This would fit in with his being born in Siberia. The city of Tobolsk, however, was far from being the back of beyond – at least, culturally. Mendeleyev was the youngest of seventeen children. His father was the rector of the local gymnasium – equivalent to an English grammar school. The city was also home to a number of Decembrist exiles, so named because of the part they had played in the unsuccessful *coup d'état* of December 1825, which was staged as part of the succession crisis of the death of Tsar Alexander I. This provided the city with an intelligentsia, whose presence brought contacts with metropolitan culture much appreciated within the Mendeleyev family.

Unfortunately the rector became blind and lost his job when his youngest son was only a year old. The family was rescued by his wife, an intelligent and energetic woman, who took over and restored to prosperity a small family glass-works, mainly producing for pharmacists, in Aremzyanskoye, twenty miles outside Tobolsk. The youngest Mendeleyev therefore grew up with molten glass, learning the trade from the craftsmen employed. In 1840, however, the family returned to Tobolsk so that he could go to school.

Although only six years old, Mendeleyev soon proved to be a brilliant pupil in mathematics, physics and geography, although he hated Latin. In 1847 his father died, and his mother set off for Moscow with the younger children, travelling the whole way by coach.[58] The object was to enrol the youngest, still only thirteen years old, in the university. He was rejected, because his gymnasium diploma from Tobolsk was not recognised. This meant another move, to the capital city, St. Petersburg, where Mendeleyev was admitted to the Main Teacher Training College to study mathematics and the natural sciences. He still could not escape misfortune: in 1850, when he was only sixteen, the deaths of both his mother and his older sister left him on his own.

In St. Petersburg Mendeleyev did, however, have the advantage of being taught by a brilliant chemist, Professor Alexander Voskresensky, at a time when, following the work of Dalton and Berzelius, great advances were being made in the world of chemistry. In this favourable climate, Mendeleyev completed major research into isomorphism, the process in which similar elements replace each other in some chemical combinations without changing their crystalline form.

This was significant for Mendeleyev's future research, since 'the similarity of behaviour of atoms of different elements was . . . one of the most important characteristics on the basis of which elements may be grouped in their natural order'.[59] After taking his degree and winning a gold medal at the Main Teacher Training College in 1855, Mendeleyev published his 'Isomorphism in Connection with Other Relations between Crystalline Forms and Chemical Constitutions' in 1856.

Like any young academic, Mendeleyev looked for a job, preferring a location with a good climate because of ill-health. In Russia this meant the Black Sea coast: his first choice was Odessa, but a bureaucratic error led to another man being appointed there, so he ended up instead in Simferopol in the Crimea. The climate was just as good, but, because of war with the French and English (remembered by the latter for the Charge of the Light Brigade and Florence Nightingale), the university had closed Mendeleyev's department. So as winter approached he took off for Odessa anyway, wearing a short fur coat, bearskin boots and a tall fur hat.

This turned out to be a good move: Mendeleyev found a job at the local gymnasium that combined well with research at the Novorossiisk University. He wrote up his subject, specific volume, in a thesis presented to the University of St. Petersburg in 1856, which led to his being appointed reader in chemistry the following year.[60] He was only twenty-three.

Mendeleyev, clearly recognised as a high-flyer, soon gained permission to study abroad. His chosen destination was Heidelberg, already known for Bunsen and Kirchhoff's research into spectral analysis. However, in his two years in Germany (1857–59) he chose not to work in their shadow, preferring to research molecular cohesion in his own small laboratory. There he researched 'constant' gases, which, according to the prevailing wisdom, could never become liquid. Mendeleyev's work on liquefaction and absolute boiling points showed that there were no such gases – a key result, long confirmed by the state of the art in low-temperature physics.

Only two years after returning to St. Petersburg Mendeleyev published *Organic Chemistry* (1861), a revolutionary textbook. This work enunciated the key principle that 'every living phenomenon is the result not of some peculiar force or peculiar reason, but of the general laws of nature. There is not one living process, taken separately, that may be attributed to a peculiar power.'[61] This result may have been anticipated by Charles Darwin (whose *Origin of Species* was published in 1859), and Louis Pasteur was working towards it in France during the 1860s,[62] but even so Mendeleyev was ahead of his times. (Not surprisingly his outspoken materialism would later commend him to Soviet scientists.) During the 1860s Mendeleyev also worked on solutions, reaching the paradoxical conclusion that their properties (density, conductivity, etc.) 'change in leaps against the background of the general steady change in the proportions of the components of the solutions'.[63] This followed from tests made with 283 different substances, with the changes always taking place at 'specific points', whose study became critical in physical chemistry.

In 1862 Mendeleyev, seldom lucky in family life, made an unhappy marriage to Feozva Leshcheva, who never showed the interest and support that a Russian of his generation expected from a wife. In 1876 he got a divorce, having met his true love, Anna Popova. His marriage to her in 1880 transformed his home into a salon open to the scientists of the day. Under Russian law, however, it was bigamous, since less than seven years had elapsed since the divorce. The Tsar, informed of this lapse, observed, 'Mendeleyev may have two wives, but Russia has only one Mendeleyev.'[64]

In 1867 Mendeleyev, still only thirty-three, succeeded Voskresensky to the top chair at St. Petersburg, where his lectures, in which he would often 'digress into mechanics, physics, astronomy cosmogony, meteorology, geology, the physiology of animals and plants, agronomy and also into different branches of technology, including air navigation and artillery',[65] were greeted with rapture.

To meet the needs of his students, Mendeleyev started working on a general manual for chemistry. Trying to establish a basic principle for systematising chemical knowledge by comparing atomic weights, Mendeleyev concluded that 'the properties of the elements are in periodic dependence on their atomic weights'.[66] This fundamental insight came to him in a dream on 17 February 1869; he had taken a brief nap while working on his book, and 'when he awoke, he set out the chart, in virtually its final form'.[67]

Mendeleyev's table of elements was presented to the new Russian Chemical Society on 18 March 1869. The fundamental law was then stated in his *Foundations of Chemistry*, published later in the year and described by Mendeleyev as 'his favourite child', and *An Outline of the System of the Elements* was circulated outside, as a pamphlet, in 1870.

Mendeleyev knew of 61 separate elements: these could be ordered according to their increasing atomic weights. This was clearly the starting point for any systematic classification. Elements could also be classified according to their properties, so that the alkali metals (e.g. sodium and potassium) belonged together, as did the alkaline earth elements (e.g. magnesium and calcium) and, at the other extreme, the non-metallic halogens (e.g. chlorine and iodine).

The periodic table (see Appendix A) then shows how, with the increase in atomic weight, elements at first acquire entirely new, changing properties, and then how these properties recur in a new period, in a new line and row of the elements and in the same sequence as in the preceding row. Thus the law of periodicity may be expressed as follows: the properties of the elements, and thus the properties of simple or compound bodies of these elements, are dependent in a periodic way on the magnitude of the atomic weight of the elements.

In compiling the periodic table Mendeleyev faced four different problems (although the last remained hidden from him). First, nine of the atomic weights in the records were incorrect, so that that of beryllium was recorded as 13.7 instead of 9, and that of calcium as 20 instead of 40. Mendeleyev made the necessary corrections, as he did also with gold, platinum, osmium, iridium, yttrium, indium and erbium. In all these cases the fault lay with inaccurate work done by others earlier in the century.

Second, if there was any regularity in the rate of increase, then there were at least six gaps in the table. These in fact represented as yet undiscovered elements. That three of the gaps related closely to boron, aluminium and silicon in the periodic system led Mendeleyev to name the missing elements eka-boron, eka-aluminium and eka-silicon – *eka* being the Sanskrit for 'one'.

In 1875 a French chemist, Lecoq de Boisbaudran (1838–1912), in a spectral analysis of zinc blende from a mine in the Pyrenees, noted the presence of an unknown element, whose properties corresponded to those of eka-aluminium predicted by Mendeleyev. De Boisbaudran, patriotically minded, named the newly discovered element 'gallium'. Its discovery led to something of a dispute between the Académie des Sciences in Paris and the Russian

Physical-Chemical Society. Mendeleyev immediately saw his own prediction confirmed, but while de Boisbaudran's atomic weight, 68,[68] was acceptable, his measure of density, at 4.7, was inconsistent with the value, predicted according to the periodic system, of 5.9. Both sides stuck to their guns, but in the end Mendeleyev was proved right (although, ironically, the undisputed number for atomic weight was to prove to be too low).

Mendeleyev had an easier ride with eka-boron and eka-silicon: the former, isolated by a Swedish chemist in 1879, was appropriately named 'scandium'; the latter was named 'germanium' after being isolated by Clemens Winkler (1838–1904) in 1886. Internationally this was an important breakthrough for the periodic system, since, for the second time, Mendeleyev's predicted values for atomic weight and density had independently proved correct – as was immediately recognised and acknowledged by the finders in both cases. Moreover, Winkler's germanium would dissolve in water but not in acids, had an oxidization formula GeO_2, and produced a chlorine compound, $GeCl_4$, with a boiling point of 83°C and a specific weight of 1.887 – in near perfect accord with Mendeleyev's predictions.

Mendeleyev's third problem was that if the classification in the columns was to be consistent with what his system required, then tellurium and iodine should change places, as should cobalt and nickel, even though this would reverse the order based on atomic weight.[69] Mendeleyev made the necessary changes, but the full explanation as to why they were correct had to await the discovery in 1913, six years after his death, of atomic numbers by Henry Moseley (1887–1915). (The relation between atomic numbers and atomic weights, together with the related phenomenon of isotopes, are explained in Chapter 7.)

The fourth problem was somewhat esoteric: although never suspected by Mendeleyev, there were gaps at the end of every row in the table, so that, moving to the next row from fluorine, chlorine, bromine and iodine in column 7 led directly to sodium, potassium, rubidium and caesium in column 1. (Note how the respective suffixes, 'ine' and 'ium' reflect the common properties of the elements in the two columns; this is the whole point to the periodic table). Mendeleyev might have noted in all these cases a relatively large increment, of the order of 4 or 5 (where otherwise there is a maximum of about 3) in atomic weight, but, if so, he did not realise its significance.

The noble elements

Then, in 1893, Sir William Ramsay (1852–1916) looked at the question put by Lord Rayleigh (1842–1919) as to why nitrogen separated from air always proves to have an atomic weight higher than that of nitrogen produced in the laboratory. To find an answer, Ramsay separated both the oxygen and nitrogen from a sample of air using established laboratory methods. He found a residue, which could only be some other gas, equal to about 1% of the original volume. This was argon, and two years later, Ramsay isolated another similar gas by boiling a mineral called clevite, which spectral analysis proved to be helium, first discovered nearly thirty years previously in the sun's atmosphere.

On the basis of atomic weight, argon had a place between chlorine and potassium in the periodic table, while helium should fall between hydrogen and lithium – the two elements with lowest recorded atomic weights. The need to accommodate these new elements could be met by introducing a new column – column 8 – but helium and argon would then occupy only two of the six places in it, at the end of rows 1 and 3, leaving rows 2, 4, 5 and 6 still to be filled.

Ramsay was plainly on a winning streak. In 1898, working with liquefied air (which technology had only recently made available) he once again eliminated the oxygen and nitrogen, and examined the residue spectroscopically. In this way the distinctive lines of three previously unknown elements, neon, krypton and xenon, appeared. The gaps at the end of rows 2, 4 and 5 were filled. Finally, in 1907, Ramsay showed that radon, discovered in 1900, should be placed at the end of row 6: column 8 of Mendeleyev's table was therefore complete in the year of his death.[70]

How could it be that not one but six elements had escaped the notice of chemists until the end of the nineteenth century, when the great rush to find 'new' elements, led notably by Davy, had started at its beginning? Chemistry was transformed in this century, which closed appropriately with the discovery of the 'new' elements in column 8. But why were they not discovered earlier?

The answer to this question is to be found in the basic phenomenon of all chemistry, the reaction between different substances, whether elements or compounds. Strong, sometimes violent, reactions are characteristic of the elements in columns 1 and 7 – just think of all those laboratory explosions. The reactions are often such that extremely unstable elements combine to form the most stable compounds: here common salt, a so-called ionic compound (Na^+Cl^-) of sodium (column 1) and chlorine (column 7) is exemplary. The column to which an element belongs defines its reactive potential, but what if an element will not react with any other? In the first place, there is then no chemical phenomenon, whether occurring in nature or the laboratory, that will betray its existence. No-one can smell or taste it, or, if is a gas, feel it, and it will have no power to corrode, contaminate, or even decay, for all these everyday processes depend on chemical reactions. No wonder argon was given that name, which means 'lazy' in Greek. Argon was discovered only when it occurred to Ramsay to use chemical reactions to separate the other gases, oxygen and nitrogen, occurring in air. He could not touch argon in this way, and it was only towards the end of the year 2000 that the first chemical compound, ever, was produced from argon.

Appropriately the six elements in column 8 are known as the inert gases: this is just what they are. They are also known as the noble elements, because of their disdain to combine with the lesser elements in columns 1 to 7. They could only be detected when, in incandescent state, they produced distinctive spectra: for all except helium this state was created artificially in the laboratory. Helium, as already noted on page 177, was observed first in the sun's atmosphere, where its incandescence is the result not of a chemical but of a nuclear reaction – a vast difference the significance of which is explained in Chapter 7.

Mendeleyev first compiled the periodic table in 1869. With the emergence of nuclear physics in the twentieth century, with its capacity to produce new elements with atomic weights approaching 300, it is difficult to state a precise number for those discovered since that year – somewhere around forty is the best possible estimate. No matter: whatever the number, twenty-one had been forecast by Mendeleyev. This is but one measure of the achievement of one of the most remarkable men in the history of science.

Organic chemistry

For more than a hundred years, in the popular understanding of science, no distinction has been more clear-cut than that between organic and inorganic chemistry. Two hundred years ago, when the achievements of Lavoisier, Dalton, Priestley, and many others were beginning to make chemistry a hard science and give it its modern form, this particular distinction would have been meaningless. At the beginning of the nineteenth century, organic chemistry, according to one of the pioneers in the field, Friedrich Wöhler (1800–82), was 'like a dark forest with few or no pathways'.[71] In 1828 he showed that urea extracted from a dog's urine was identical to ammonium cyanate produced in the laboratory – a major step in bringing light into the dark forest. This was a sensational discovery, since it showed that no special 'vital' principle governed chemical processes taking place within a living organism.[72] The end of *vitalism* can be said to open the way to biochemistry, although its potential for development was little realised at the time.

It was not so much Wöhler as his friend and colleague Justus von Liebig (1803–74), who 'revealed the source of richness . . . of organic chemistry, that the simple elements of Carbon, Oxygen, Hydrogen and Nitrogen could combine together in myriads of different ways to produce millions of different compounds'.[73]

During his youth, Liebig could witness, at first hand, the state of the art in proto-industrial chemistry. His father, who sold paints and other household wares in a *drogerie* in Darmstadt – the capital city of a German Grand-Duchy – produced much of his stock-in-trade in his own workshop. With eight children, the family was far from prosperous, and Liebig had to spend much of his time helping in his father's business. In his spare time, however, Liebig was able to study books on chemistry in the library of the Grand-Duke Ludwig of Hesse-Darmstadt. By good fortune Karl Wilhelm Kastner (1783–1857), a professor of chemistry, was one of his father's clients. Although in later life Liebig was derisive about Kastner's competence as a chemist, the professor still took him on as an assistant and then recommended him to the Grand-Duke.

With noble patronage Liebig was able to spend some months in Paris, working with Gay-Lussac, one of the greatest scientists of the day. There he also met the great German naturalist, Alexander von Humboldt, who recommended him to the Grand-Duke for an academic appointment. This led Liebig to the University of Giessen in 1824, and in 1825 he became effectively head of a new department of chemistry with a well-endowed laboratory. There he focused his research on organic chemistry and built up one of the best teaching and research schools in Germany. The University of Giessen is now named after him. Talk about success.

The great Swedish chemist Berzelius, a generation older than

Liebig, had always seen 'the discovery of the rational constitution' as the main problem in organic chemistry, insisting at the same time that 'organic bodies obey the same general laws as... inorganic combinations'.[74] The key was to be found in the idea of the radical, which since the time of Lavoisier was defined as the 'stable part of a substance that retains its identity through a series of reactions even though it was known to be a compound'.[75] (This recalls the 'molecule integrante' proposed by Haüy for the analysis of crystals – a process described later in this chapter.)

By 1830 not only Berzelius and Liebig, but also Bunsen at Göttingen, were all sold on radicals. The problem was that for too many well-known substances different radicals could be found, and the choice of different radicals could lead to conflicting results. There were endless disagreements about how molecules combined, partly because of confusion about atomic, equivalent and molecular weights. (This was a generation before Mendeleyev sorted out such matters.) By the mid-1830s the term radical came to indicate a hydrocarbon group or chain, so that the fundamental nucleus becomes an unsaturated hydrocarbon, C_8H_{12}. (Strictly a hydrocarbon is any chemical compound containing only hydrogen and carbon, but the number of such compounds is very great.)

In the mid-1840s some order was brought by the 'homologous' series introduced by the French chemist Charles Gerhardt (1816–56), who had earlier studied with Liebig. This is the so-called 'ladder of combustion': the formula $C_nH_{2n}O_2$ for primary alcohols,[76] is an example, with each successive value of *n* defining one member of this class.

In 1847, a young man with the unlikely name of Friedrich August Kekulé von Stradonitz, having come to Giessen to study architecture, fell under the spell of Liebig, who advised him to study chemistry in Paris. Once there he became a friend of Gerhardt. As chemists, though not as friends, the two were soon to part company. Kekulé forsook Gerhardt's principle based on one key type of atom for one in which 'no atom was more important than another in a constitutional formula'. Kekulé's method of classification depended on how many other atoms or groups a given atom combined with, so that, for example, carbon is classed as 4, both oxygen and sulphur as 2. In 1859 Kekulé introduced the principle of catenation, according to which carbon atoms were linked in a chain.

According to some sources the idea came to Kekulé in the summer of 1854, when he fell asleep on the top of a London omnibus; but, however that may be,

> In the cases of substances which contain several atoms of carbon, it must be assumed that at least some of the atoms are in the same way held in the compound by the affinity of carbon, and that the carbon atoms attach themselves to one another, whereby a part of the affinity of the one is naturally engaged with an equal part of the affinity of the other . . . For example a group of two carbon atoms, C_2 . . . will form a compound of six atoms of monatomic elements, or generally with so many atoms that the sum of the chemical units of these is equal to six.[77]

An elementary example of this is carbon and hydrogen combining to form ethane (C_2H_6). The rule stated by Kekulé is fundamental, and led finally to his defining organic chemistry as 'the chemistry of carbon compounds'.[78] At the same time it was also applied to the general principle of valency[79] (developed by Edward Frankland (1825–99) in England) which governs all structural chemistry, by stating precise numerical rules according to which atoms bind together to form molecules. This led to the familiar glyptic or 'croquet ball' models of molecules. The standard three-dimensional version, illustrated in Figure 6.2, which shows a molecule of phosphorus pentoxide (P_4O_{10}) and to be found in any introductory textbook, was only introduced by the Dutch chemist Jacobus van't Hoff (1852–1911) in 1874.[80]

Van't Hoff also proposed that the four carbon bonds are directed to the corners of a tetrahedron, which then determines the arrangement of the atoms in the molecule of any carbon compound. More generally, according to the Russian physicist Alexandr Butlerov (1828–86), who had popularised the phrase 'chemical structure',

> Only one rational formula is possible for each compound, and when the general laws governing the dependence of chemical properties on chemical structure have been derived, this formula will express all of these properties.[81]

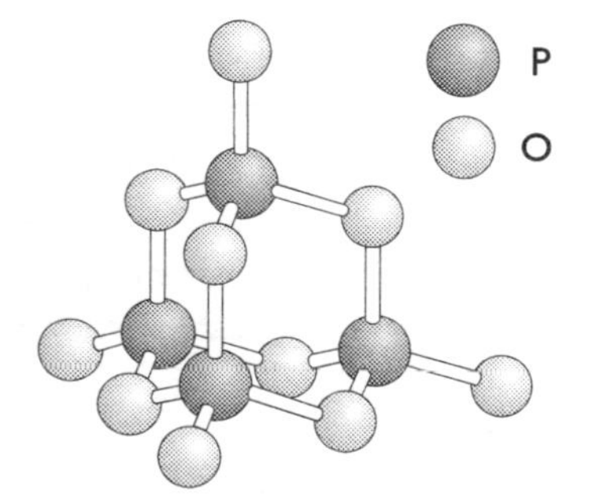

Figure 6.2 Phosphorus pentoxide.

This was an oversimplification, and by the 1920s it was becoming clear that the question of structure was extremely problematical. None the less it represented a triumph in nineteenth-century chemistry.

Kekulé is best known for his discovery of the benzene ring – at least in the world of chemistry. Outside, Kekulé's renown is more problematic, if only because the problem of benzene is not very transparent. Benzene was first identified, in compressed oil gas, by Faraday in 1825 (when he was still working as a chemist under the shadow of Davy). This result was published (although the name 'benzene' was coined by Eilhard Mitscherlich at Berlin in 1834). In 1850, the German chemist, August von Hoffman (1818–92) showed that benzene belonged to the chemical family of aromatics, of which it is now the prototype.

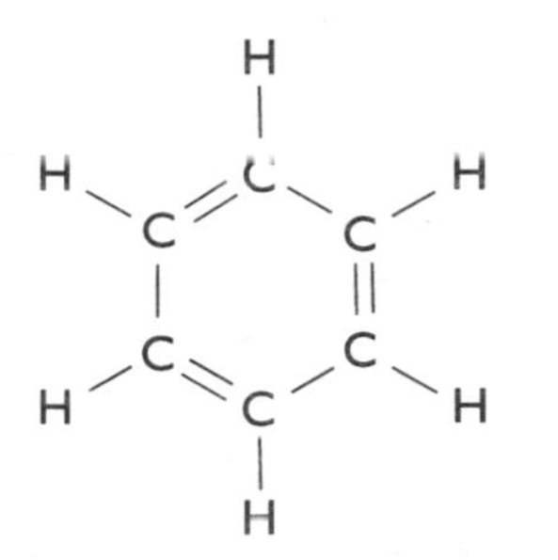

Figure 6.3 Benzene

The problem was how to reconcile its chemical formula, C_6H_6, into the bonding principles established by Kekulé for the carbon atom. Once again, tradition has it that the solution came to Kekulé in a dream, this time after he had fallen asleep by his fireside. The answer was to form a ring of six carbon atoms, each of which was bonded to a hydrogen atom, as shown in Figure 6.3, which also illustrates the principle of valency. After he had dreamt of this solution, Kekulé asked, 'What else could a chemist have done with two valencies left over?'[82] Kekulé thought he had discovered 'an inexhaustible treasure trove', and at a memorial lecture in London, two years after his death in 1896, it was stated that Kekulé's benzene theory was

> the most brilliant piece of scientific production to be found in the whole of organic chemistry... three-fourths of modern organic chemistry is, directly or indirectly, the product of his theory.

Organic chemistry, as practised in the nineteenth century, took little account of the fact that the whole chemistry of plant and animal life (in geological time the source of many organic compounds, such as all fossil fuels) was also based on molecules containing carbon atoms. This is the realm of biochemistry, a science that did not develop significantly before the twentieth century,[83] although some phenomena, such as photosynthesis, had been examined much earlier. In the second half of the twentieth century, molecular biology led to any number of remarkable discoveries, and with the complete decoding of the human genome, these are set to continue in the present century. This defines a science quite beyond anything that Kekulé, the great dreamer of nineteenth-century chemistry, ever conceived of. It is also beyond the scope of this book.

The magic of crystals

Crystals, one of the most common forms taken by solid matter, have played a part in human culture since prehistoric times, with magical powers often being ascribed to them. Their appearance, although far from uniform, always discloses something of their distinctive properties. These are remarkably wide ranging, and relate to optics, electricity, chemistry, geology and even the life sciences. Crystals' close relationship to minerals, the basic substance of rocks, defines such characteristic properties as hardness, lustre, colour, cleavage and fracture. The problem is the sheer diversity of crystalline minerals, so that classification has always been an important part of crystallography. Some crystals, such as quartz, are extremely common. Even so, this, the most abundant of all minerals in the earth's surface, occurs not only as common rock crystal but also as amethyst, chalcedony, agate and jasper. It is no wonder then that it has always attracted humankind, and continues to do so in its modern applications in timepieces, electronics, optical instruments and abrasives – to say nothing of jewellery.

In the written record of science, going back to antiquity, crystals are seen to have interested any number of great men. Democritus, Plato and Aristotle – each with his own theories on this subject – head the list, but it goes on to include Newton, Hooke, Boyle, Huygens, van Leeuwenhoek, Lavoisier, Linnaeus, and many others: the list could continue indefinitely. There are also the lesser-known actors such as, notably, the Abbé René Just Haüy (1743–1822), who gave Linnaeus the honour of being the father of crystallography, a title he could well have claimed for himself.

In recent history the subject was transformed by one single event, the discovery, in 1912, by the German physicist, Max von Laue (1879–1960), that X-rays were diffracted by crystals. Not only did this demonstrate that X-rays (whose discovery by Röntgen is described in Chapter 7) belonged to a specific band of wavelengths in the electromagnetic spectrum, but it also established X-ray crystallography as one of the most useful methods of modern science. This linked one of the most ancient branches of science, crystallography, to one of the most modern, electromagnetic wave theory.

Returning, first, to science as it was before this breakthrough, what, at the end of the day, is the defining characteristic of crystals? The answer was given by Henry Baker (1698–1774), who, after sketching dozens of crystalline substances observed with his microscope, noted how 'we see every Species working on a different plan, producing Cubes, Rhombs, Pyramids, Pentagons, Hexagons, Octagons, or some other curious figures, peculiar to itself'.[84] All these objects are geometrical solids, with plane faces, intersecting in straight lines. What is more, as noted by Aristotle,[85] these curious shapes – which in any given crystal take only one form – must have the capacity to fill space.

This fundamental defining property of any crystal leads immediately to the geometry of three dimensions, a constant and often controversial leitmotiv of the whole history of crystallography. The basic principle is simple: the unit cell in any crystal system must be a parallelepiped, that is a solid with six faces, all of which are parallelograms, so that they divide up into three groups of two, with the two faces in each group being parallel. Subject to this limitation, there are seven possible crystal systems, of which the simplest is the cubic.

A system can be defined by the lengths of the sides converging at a single vertex, and the angles between them. Designating the former as a, b and c, and the latter as α, β and γ, with α being the angle between b and c, β, that between c and a, and γ, that between a and b, the seven systems are as follows:

(1) cubic: $a = b = c$ and $\alpha = \beta = \gamma = 90^\circ$;

(2) tetragonal: $a = b \neq c$ and $\alpha = \beta = \gamma = 90^\circ$;

(3) orthorhombic:[86] $a \neq b \neq c$ and $\alpha = \beta = \gamma = 90^\circ$;

(4) hexagonal: $a = b = c$ and $\alpha = \beta = \gamma \neq 90^\circ$;

(5) trigonal: $a = b \neq c$ and $\alpha = \beta = \gamma \neq 90^\circ$;

(6) monoclinic: $a \neq b \neq c$ and $\alpha = \gamma = 90^\circ \neq \beta$;

(7) triclinic: $a = b = c$ and $\alpha \neq \beta \neq \gamma$.

The way this works out in practice is shown in Figure 6.4, which makes it clear how every one of the seven systems can be reached by starting with the first, a perfect cube. If this is thought of as being made of some elastic material – say India rubber – all possible crystalline forms can then be obtained by squashing it in different directions, but always in such a way that not only opposite faces but also the sides of each face remain parallel. Given this restriction, only the seven forms given above are possible. With the simplest form, the cube, all faces are perfect squares and all angles are 90°. With the most complex, triclinic, all adjacent faces are different, as are the angles between them, none of which is 90°. Any uniform substance, will, if crystalline, contain only one form of crystal. (With aggregates, needless to say, any number of different forms are possible.)

The development of crystal systems to their present canonical form continued until well into the nineteenth century, by which time

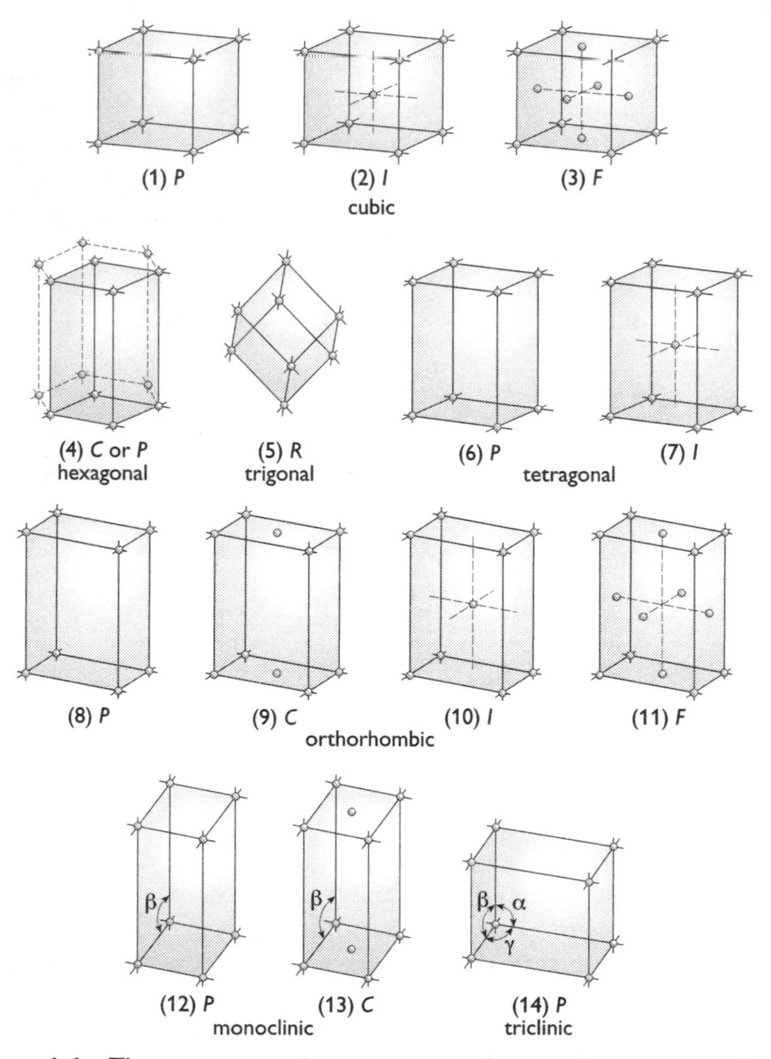

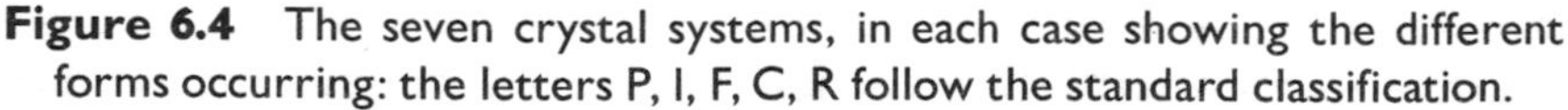
Figure 6.4 The seven crystal systems, in each case showing the different forms occurring: the letters P, I, F, C, R follow the standard classification.

it had become clear that all known crystals (which are counted in thousands) belong to just one of them, and that this defines many distinctive properties of the individual crystal. It was only at the end of the eighteenth century that Lavoisier and others established that the constant repetition of definite geometrical figures was the essential defining principle, even though van Leeuwenhoek's microscopic investigations should have made this clear much earlier.

Huygens, even earlier, had established the fundamental crystallographic law that the constancy of interfacial angles comes from regular repetition. (In his *Treatise on Light*, published in 1690, he had noted how the same crystals found in France, Corsica and Iceland were completely identical). Huygens' measurements of interfacial angles were extremely accurate, and his examination of the Iceland crystal led to the first ever 'accurate mathematical analysis of the form of any crystalline substance', establishing also the importance of the crystal axis. These results were neglected for more than a century, simply because 'the faith of scientists in mathematics as their most potent tool and as the language of scientific expression was not yet so firmly entrenched that they were prepared to accept the fact that the wording could be extremely complex'.[87] Where Kepler, after pages of complicated mathematics, was able to state his three laws of planetary motion in simple form, Huygens, in presenting his results with crystals, was unable to do likewise.

As the geometry of crystals became clearer, there was still constant debate as to whether they were linked to specific chemical forms, such as acids, alkalis or salts. Every such link had its own advocates, but in the end counter-examples were always found to disprove the case. On the other hand, from about 1720, it was clear that all metals were crystalline.[88]

Another key question related to the forces that bound crystals together. The diverse properties of crystals made it difficult to establish a principle governing all cases. Many crystals are subject to cleavage, but only in certain directions, which explains why mica, for instance, can be reduced to very thin sheets (with countless experimental uses.) What force is it that binds mica in one direction but not in others? Any number of theories were put forward to answer this and other questions derived from particular properties – mechanical, optical, or whatever – of crystals. The great men of the day were at odds, so that Newton emphasised cohesive force where Huygens rejected this possibility. It is now known that 'the planes of cleavage . . . are those across which the forces between the atoms are weakest' and that this 'may be due to the type of atomic bond or to a greater spacing of atoms or ions in the crystal at right angles to the plane of cleavage'.[89] This, however, is the science of the twentieth century. There were still many bridges to cross to get to this point, although from the end of the eighteenth century it was generally accepted that the way molecules were superimposed within a crystal explained the different forms.[90]

With countless instances of known crystals, by the end of the eighteenth century it had become clear that systematic classification was essential to real progress in understanding them. The history of science constantly shows how 'the development of comprehensive theoretical systems seems to be possible only after a preliminary classification of kinds has been achieved'.[91] When it came to crystals, the problem was similarity of form among diverse substances combined with variation in shape in the same substance: how then were both the similarity and the diversity to be explained?

Classification characterises the mind-cast of the naturalist rather than that of the physicist, for where the former 'looks for the regularities of given forms' the latter 'seeks the form of given regularities'.[92] It is not surprising, then, that Carolus Linnaeus (1707–78), famous for his system of botanical classification, also turned his hand to the 'minute and accurate description of the shapes of many crystals and explained in what ways certain configurations were related to or differed from others',[93] nor that this led to his being called the founder of the science of crystallography.

The person responsible for this tribute was the Abbé René Just Haüy (1743–1882), to whom it might equally apply, were it not for the fact that almost all his fundamental theories in the end proved to be mistaken. Haüy came from a humble but devout provincial family that moved to Paris when he was seven years old. Although he trained for the priesthood, he also studied science, becoming a Master of Arts in 1761 long before becoming a priest in 1770.

As a scientist he started in botany, but a series of lectures in mineralogy at the Jardin du Roi turned his interest to crystallography. Inspiration came to him, so it is told, after he had dropped on to the floor a group of calcite minerals that had crystallised in the shape of hexagonal prisms. The fragments were rhombohedra, in precisely the shape of Iceland crystal, and following Archimedes (translated into French) he cried out 'tout est trouvé'. What he claimed to have discovered was that rhombohedra must be the nuclei of all calcite crystals. This was actually rediscovery of a fact already quite well known, although not, apparently, to Haüy. In any case, he was set on the path of establishing the taxonomy of crystals according to an intricate scheme largely of his own devising.

Haüy's problem was that at a time when increasingly accurate measurements, using an instrument known as the goniometer, were being made of the angles between crystal faces, he stuck to an elementary geometrical theory, which the new measurements failed to support. Haüy discovered through cleavage of a variety of substances that the number of different nuclei could be reduced to six primitive forms (which in turn would break down into the crystal systems listed on page 191) but in the end it all came to next to nothing. He persisted in his work during the period of the French Revolution, in which life for a priest was very uncertain, and he is still regarded as one of the first modern crystallographers. At the same time, the way in which his theories were faulted by others produced important results not only for crystallography but also for chemistry and optics. The three-dimensional geometry of crystals, depending on whichever of the seven crystal systems defines any given case, tells nothing of the number and ordering of molecules, atoms and ions, in each unit cell. However, the complete regularity of crystal structure, essential to Haüy's theory, means that whatever the pattern, it is repeated, like some three-dimensional wallpaper, in an indefinite series of lattices. The characteristic of any such series will in turn depend upon the direction, defined by an axis of the crystal, in which the lattices defining it are superimposed upon each other. This is critical for cleavage, which occurs where the bonding between the superimposed lattices is weak.

In the first half of the nineteenth century, a German chemist, Eilhardt Mitscherlich (1794–1863), was the first clearly to recognise a phenomenon known as a solid solution. This apparent contradiction in terms describes a compound solid in which ions of one chemical element may be substituted for another. This happens with certain alloys, as of gold and copper, but it also occurs with some so-called double salts. In any case the process only occurs in crystals: it is then a simple form of isomorphism, where chemically related substances take similar forms. Sometimes, the one element substitutes more or less at random for the other in the lattices, but in others, this occurs in a regular pattern, in which a series of superlattices is created. None of this fits in with Haüy's theoretical structure; nor does it fit the related phenomenon, also studied by Mitscherlich, of polymorphism. Haüy knew of Mitscherlich's work, but by never coming to terms with it ensured the eclipse of his own theory.

At least until 1912, the most interesting properties of crystals were optical. Some, such as double refraction, first noted in Iceland crystal by Erasmus Bartholin (1625–98) in 1669, had long been known, although – in spite of the interest of men such as Huygens and Newton – never explained. The optical properties all depend upon locating optical axes, of which there may be only one, in the crystal itself. The defining property of an axis of a crystal is symmetry, which means that if the crystal rotates about the axis, through a given angle, it will coincide with its previous position. A cube, for example, rotating about a line joining the centre points of two opposite faces, will achieve this result where the angle is 90°: this is the case of fourfold symmetry, since coincidence occurs four times in a complete rotation of 360°. In fact only two-, three-, four- and sixfold symmetry (for which the angles of rotation are 180°, 120°, 90° and 60°, respectively) are possible if, as required, space is to be filled. The cubic system is unique in having the first three types of symmetry, and for each of them, the axes of rotation intersect at right angles. Such a system is isometric, and is the only in which double refraction does not occur. The cubic system, to which many crystals belong, alone has this property.

In crystals belonging to the other systems, either one or two of the axes are *optical*: an optical axis has the unique property of defining a direction in which light transmitted through the crystal is not subject to double refraction. Crystals belonging to tetragonal, hexagonal and trigonal systems have only one such axis, and are *uniaxial*: quartz and tourmaline are examples. Those belonging to rhombic, monoclinic and triclinic systems are *biaxial*: mica is the most familiar crystal with this property.[94]

In the uniaxial case, a light ray not parallel to the optical axis, on entering the crystal divides into two rays that are polarised with the light waves at right angles to each other. The *ordinary* ray, so-called because it obeys the normal laws of refraction described in Chapter 2, vibrates in a direction perpendicular to both the direction of propagation and the optical axis. The *extraordinary* ray, not subject to the normal laws of refraction, vibrates in a plane containing it and the optical axis, with the direction of vibration depending on the angle between the two. The biaxial case, although more complicated, is subject to similar rules relating to each of the two optical axes. The problem for Haüy was that his theory did not allow for the distinct optical status of either one or two of a crystal's axes.

Double refraction required Etienne Malus's (1775–1812) discovery of the polarisation of light in 1809, before it could be properly understood, which explains why Huygens', Newton's and Haüy's explanations all failed. Even so Malus failed to account for the difference between uniaxial and biaxial crystals, which had to await the discovery, by Jean-Baptiste Biot (1774–1862) of biaxiality in 1812. It is odd, in this context, that polarisation, which is fundamental to the understanding of light, took so long to be discovered, when it can so easily be demonstrated. This is because light incident on a plane mirror at an angle of some 57° becomes polarised on reflection, so that double reflection in two planes at right angles to each other will extinguish it. This means that the glare from a reflecting surface is partially polarised in the vertical plane (depending on how close the angle of reflection is to 57°): Polaroid sun-glasses, based on a double refracting material that polarises light at right angles, then filter out some of the glare.

In 1756 the German physicist Franz Aepinus (1724–1802), experimenting with crystal tourmaline, discovered that opposite faces acquired polarised electric charges as the result of heating. This, the phenomenon of pyroelectricity, was a further demonstration of the importance of the crystal axis. This result was also known to Haüy, but his theory of crystal structure could not accommodate it. Haüy was greatly respected in his own lifetime, and Napoleon commissioned him to write a physics textbook to be used in all French *lycées*: his discovery, at a relatively early stage, that crystals always break to produce rhomboidal fragments led to the formulation of the basic laws of crystallography. Haüy, however, made too much of a good thing and insisted on maintaining theoretical constructs, which, however simple and elegant, failed to hold up in the face of accurate observation of the crystalline phenomena described above.

Christian Weiss (1780–1856), having attained his doctorate at Leipzig when he was only twenty, became interested in mineralogy, and as a young man free to travel ended up in 1807 at the feet of Haüy in Paris. (He was already busy translating Haüy's *Traité de mineralogie* into German.) Haüy proved to be a tyrant, intolerant of any criticism, so that Weiss, on insisting on views which contradicted his own, was banished from his circle. No matter, Weiss returned to Leipzig as professor, and, under his direction, nineteenth-century crystallography 'became concerned with establishing the mathematical relationships evidenced in the crystalline end products of natural processes'.[95] There he took the first steps to establishing the crystal systems as they are used today, but he was hesitant about accepting the possibility of symmetry systems in which the axes were not mutually perpendicular. Weiss's contemporary Friedrich Mohs (1773–1839), by grasping this particular nettle, established the present triclinic and monoclinic crystal systems. According to Mohs, the main concern of crystallography was simply figured space, so that geometry should determine everything. The mathematics of symmetry became the main preoccupation of crystallographers. Chemistry could not be entirely disregarded, because Mitscherlich had shown that different substances could crystallise in the same form and the same substance in different forms.

Following Mohs' principles of classification, three types of symmetry combine to assign any crystal to one of the thirty-two possible classes or point groups, belonging to six crystal systems. Optical and electric properties of any crystal are then determined according to the class and system to which it belongs. Internal symmetry is constrained by the rule that any particle component is a node in a three-dimensional space lattice, so that all particles of a given type, say silicon atoms, have identical lattice environments. (This is subject to substitutions in the cases of iso- and polymorphism already described). In the course of the nineteenth century, fourteen different spaced lattices proved to be possible, each subject to two additional symmetry operations. Even so, Weiss's and Mohs' geometry told nothing about the shapes of the elements that were the basic constituents. The geometrical problem was finding the number of symmetrical ways of arranging points in space so the each one's environment was identical to those of the others, but not necessarily similarly ordered in the space lattice. Correlation between actual crystals with space groups led to a procedure with so many alternatives that a final solution to this problem defeated the nineteenth-century crystallographers. This had to wait until 1912, the year of one of the most far-reaching events in the history of scientific experiment – the discovery of X-ray diffraction by the German physicist Max von Laue (1879–1960).

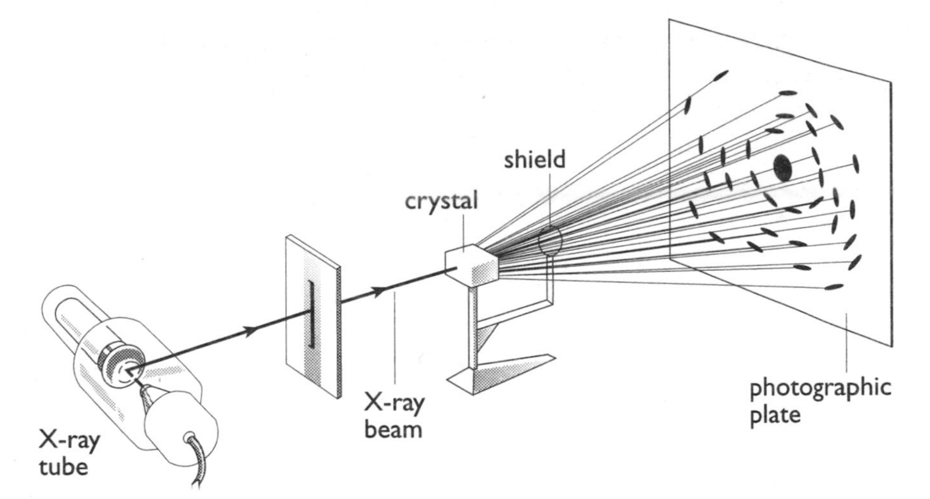

Figure 6.5 X-ray diffraction apparatus.

Although von Laue got most of the credit (and a Nobel prize two years later) the actual experiment was carried out by a research assistant and a doctoral candidate, who, by passing a narrow X-ray beam through a crystal of copper sulphate, recorded an image on a photographic plate. A similar image was obtained after substituting zinc blende. The basic apparatus is illustrated in Figure 6.5. Von Laue then demonstrated that only the diffraction of very short electromagnetic waves by a regular arrangement of atoms or molecules in the crystal could account for this phenomenon. This could only mean that X-rays were part of the electromagnetic spectrum in the range 10^{-11} to 10^{-9} centimeters, because only these limits could be accommodated by the dimensions of the known particle structure of the crystals. Physicists, such as, notably, J. J. Thomson, had used X-rays in the laboratory ever since their discovery by Röntgen in 1895 (as related in Chapter 7), but before von Laue's experiment their true nature was unknown. Oddly enough, von Laue himself almost immediately turned to other matters.[96]

Even so, the importance of X-ray diffraction as a research tool was realised almost immediately. Its usefulness was much increased in 1913, when the American physicist William Coolidge (1873–1975) found a new method for producing X-rays, by directing accelerated electrons on to a metal target under a high vacuum. In 1916 he added another improvement: a hot tungsten cathode in place of a cold aluminium cathode. (The Coolidge tube is the prototype of all modern X-ray apparatus: its inventor, dying at 102, had nearly sixty years to enjoy its success.) Above all, a father and son team, William

(1862–1942) and Lawrence Bragg (1890–1971) (who would share the Nobel prize for physics in 1915, a year after von Laue), using X-ray diffraction with their own improved 'Bragg' spectrometer determined the arrangement of atoms in common salt, pyrite, fluorite and calcite. This was only the beginning. At the same time Lawrence stated the eponymous Bragg law, which related the wavelength of the X-rays to the separation distance of the layers of atoms forming the crystal planes and the angle of incidence of the X-rays. Crucially, the separation distance of the crystal planes is of the same microscopic order of length as the wavelength of the incident X-rays; if it were otherwise, there would be no diffraction. To a large extent, the Braggs' results confirmed the already established theory of the space lattice and groups described above, but even so X-ray crystallography, as an experimental technique, opened up whole new branches of science, such as solid-state physics (which relates to the electrical properties of semiconductors[97]) and, at a later stage, molecular biology.

One man, more than any other, realised the full potential of this new field. This was Linus Pauling (1901–94), who went on to win two Nobel prizes – for chemistry in 1954 and peace in 1962.

Oregon, in 1901, was still something of a frontier state, and there Pauling was born, the first of the three children of an unsuccessful druggist, who would die in 1910. Although born in the depths of the countryside, Pauling grew up in Portland, the state's largest city. At Washington High School, he was remembered as being bright and very sure of himself: he left, however, without a diploma, after refusing to take a required course in civics.

This did not prevent him from gaining a place at the Oregon Agricultural College, where his fellow students were immediately taken by the ease with which he passed through first-year courses. He also published (in 1920) his first paper: there was little premonition of his future achievements in its title, 'The Manufacture of Cement in Oregon'. Once at college Pauling's ability was obviously noted by the professors, who in his third year offered him for $100 a month a job as instructor for the second-year course in quantitative analysis that he had just taken himself. This was a lifesaver, for Pauling had to support his gravely ill mother.

Although Pauling had thought of becoming a chemical engineer after graduation, in the course of his studies he set his sights on becoming a professional scientist. His special interests were in chemical bonding (the way in which atoms are bound into molecules, and molecules to each other) and the properties of crystals, such as hardness, cleavage and colour. Although Pauling's first choice for graduate study was the University of California at Berkeley, chance had it that he should go to the California Institute of Technology at Pasadena – long known, familiarly, as Caltech – the leading American research institute for chemical bonding. He arrived there, twenty-one years old, in September 1922.

Pauling's doctoral research would be on the structure of crystals, using the techniques of X-ray diffraction as developed by the Braggs in England. Pauling recorded the value of this technique, as he saw it at the end of his life, in these words: 'Our present understanding of the nature of the world of atoms, molecules, minerals and human beings can be attributed in large part to crystallography.'[98] And no-one made better use of it than Pauling. But what then was he looking for?

Given Pauling's long life in science, this question has any number of answers. A key event was the discovery of quantum mechanics, mainly by physicists in Germany in 1925. A year later Pauling himself was in Germany, where he spent much of his time at the Institute of Theoretical Physics in Munich. Just as he arrived, in 1926, Erwin Schrödinger had published his key paper of wave mechanics, one particular aspect of quantum mechanics. Pauling reacted by investigating the chemical applications.

Before these developments, Bohr's theory of the elliptical orbits of electrons in atoms (described in Chapter 7), with their capacity to be used in chemical bonding, was the last word in physical chemistry at the atomic level. This theory did not, however, go far enough: it clearly had potential for development, but the way forward opened up in 1925 when a generation of young physicists, de Broglie, Heisenberg, Schrödinger, Dirac and Born, established quantum mechanics.

Pauling applied quantum mechanics to research into the structure and properties of substances, particularly crystals – at least in the early days. He was particularly successful in investigating the two types of possible bonding, ionic and covalent, in silicate crystals. In the latter case, in which outer orbital electrons are shared between different atoms in a molecule, his methods led to a correct understanding of many basic substances, such as benzene (whose significance is explained on page 189) and graphite.[99] He adhered consistently to the principle, stated by Eddington in relation to physics, that science is a quest for structure rather than substance.[100]

Pauling illustrated the point by comparing diamonds with graphite, both consisting only of carbon atoms. The palpable difference between the two is explained by the so-called quadrivalence of diamonds, in which every atom is bound to the four others closest to it in a regular pattern that defines the crystal (which belongs to a simple cubic or isometric system[101]). It is characteristic of all hard substances that there are bonds connecting all the atoms in a crystal into one giant molecule. This is the case with quartz (SiO_2), held together by strong silicon–oxygen bonds. Alternative arrangements produce mica, which splits into sheets, and asbestos, which splits into fibres.[102] X-ray crystallography proved to be the ideal means of investigating such structures with unprecedented precision. Its scope as an investigative tool has transformed subjects, such as biochemistry, only remotely connected to crystallography. The complete mapping of the human genome, the most recent triumph of molecular biology, accords perfectly with the principles and practice of Pauling – one of the most creative minds of the twentieth century.

7

The new age of physics

Rays and particles

In the year 1875, the world of science, both in chemistry and physics, seemed to rest on firm foundations. Mendeleyev's periodic table of the elements had been published in 1871, and Maxwell's *Treatise on Electricity and Magnetism* in 1873. Mendeleyev (as related in Chapter 6) established a scheme of fundamental particles, the atoms of some ninety-odd elements, out of which all matter was constituted. Maxwell showed how the fundamental relation between electricity and magnetism, revealed above all by Faraday's experiments (described in Chapter 5), could be expressed in a theory of complete generality, predicated on a universe in which energy in many different forms was transmitted by electromagnetic waves.

Mendeleyev's table contained a number of gaps, and his legacy to chemistry was the challenge to fill them – a process not completed until well into the twentieth century. Problems still remained as to how different atoms combined to form molecules, but the atom itself was still conceived of as indivisible, the meaning of the Greek word *atomos*. At the same time, Maxwell's wave theory, whatever its experimental background, did not immediately point the way to new experiment: its scientific potential was not transparent.

On the practical side, in 1858 the invention of the mercury air pump had enabled Julius Plücker, in Germany, to evacuate a glass cylinder of all but one-thousandth part of the air contained in it, and so create a vacuum more perfect than any previously attained. With this apparatus he became the first to observe cathode rays, but without realising what they were. The technological breakthrough was radical, for 'the enormous stream of discoveries at the end of the nineteenth century that gave us such insights as the discovery of X-rays, working with radioactivity, and all that, is entirely due to the fact that the technologists developed decent vacuum pumps'.[1] Plücker's results became significant when, in 1878, William Crookes (1832–1919) replicated his experiments, but with a vacuum 75,000 times better.

In the last quarter of the nineteenth century, the firm foundations of physics would be shaken as never before. Since then it has always been in turmoil, with continuing interaction between experimenters working with ever more elaborate apparatus and theorists proposing a world of almost unimaginable complexity. In this new era, one theme above all others has dominated physics: this, simply stated, is the relation between elementary particles and electromagnetic waves.

I open this chapter with a short history of the way apparatus developed so as to allow this theme to be explored. This will establish a material background for theory, which consistently related to phenomena occurring on a microscopic scale, often with extremely short duration, so that direct observation was next to impossible. The problem then facing the experimenters was to contrive side-effects, or epiphenomena, which could be observed. This process often required new apparatus, which the experimenters had to design. In the time elapsed since 1875, this has meant a vast increase in scale not only of the apparatus used but also of the research establishment.

In 1887, Heinrich Hertz, with apparatus of his own design, created electromagnetic waves of a predetermined frequency. This was the foundation of all subsequent radio transmission, although much new technology (in a process initiated by Guglielmo Marconi) was needed before complex signals (as required by sound radio and television) could be sent. At the same time, the ability to receive and analyse signals at radiofrequencies would also prove critical for pure science.[2]

In 1897, long before television, J. J. Thomson's experiments with specially designed cathode ray tubes led to the discovery of the electron: then in 1899, helped by the prototype cloud chamber developed by his student, C. T. R. Wilson, he was able to estimate both the charge and mass of the electron. The integrity of the atom was lost forever.

Although Thomson's discovery of the electron in 1897 made it clear for the first time that the atom was no longer indivisible, the chance discovery of natural radioactivity by Henry Becquerel a year earlier had revealed another source of particles, which equally threatened the integrity of the atom. Becquerel's discovery required no special apparatus: it was the result of leaving a small sample of uranium, a copper key and a photographic plate in a drawer. Pierre and Marie Curie continued to research radioactivity, to discover polonium and radium, but still failed to discover the true nature of the phenomenon.

This was achieved by Rutherford, who discovered not one type of radiation but two, named α and β according to the nature of the particles emitted in the process of decay. The former proved to be doubly ionised helium atoms, the latter simply the same electrons as Thomson had just discovered. Significantly both particles carried an electric charge, positive for α and negative for β, which meant that their trajectories could be directed and amplified in an electromagnetic field (such as happens inside any television tube). This made both particles promising material for further experiment, provided always that a sufficiently powerful radioactive source was available. This meant, for many years, the radium or polonium discovered by the Curies: even with industrial production (mainly for medical use) the costs were high, so that research budgets were a critical factor in the success of experiments.

The working principle is that radiation from a radioactive source in an enclosed chamber is directed to a target (generally a thin metal foil) which scatters the particles: the directions in which the particles then shoot out is observed by means of a scintillator. This records each separate event in the form of a flash of light, which may last no more than a ten-thousandth of a second. In the early days of particle research (when Rutherford preferred simple apparatus held together by string and sealing-wax) the experimenter had simply to record the scintillations as they occurred. Because they were so faint, all this took place in near complete darkness.

Rutherford's painstaking work foreshadowed a problem which confronts even today's experimenters, that is, simply detecting some intelligible sign of nuclear events. New apparatus is 'an indolent monster without the equipment for analyzing what it produces. Nearly every advance . . . poses new detector problems.'[3]

Although today's scintillation counters are much more sophisticated (and spare experimenters the trials suffered by Rutherford and his colleagues), the need to count subatomic events remains crucial, as does their most common form, bombarding a selected target with particles. Somewhere around 1930 the limitations of the string and sealing-wax apparatus favoured by Rutherford became apparent. The device used by Chadwick in 1932 to discover the neutron was one of the last of this kind.

Acceleration and bombardment

There were essentially two problems: first, the particles emitted by radioactive decay moved relatively slowly; and, second, there were only the two kinds, α and β. In the right circumstances the problem could be circumvented by producing a stream of protons from a canal ray tube: this is similar to a cathode ray tube, but is filled with low-pressure gas, which ionises when a potential difference is created across the electrodes. If, then, holes are bored near the cathode the positive ions (which will be protons in the case of hydrogen) escape as 'canal rays'.

There were still critical limitations to what bombardment could achieve, which would be largely surmounted if a means of accelerating particles could be found: in 1931 Ernest Lawrence's cyclotron in California was the critical breakthrough. A year later, John Cockcroft and Ernest Walton, in Cambridge, developed an accelerator which they used to bombard lithium with protons – produced by a canal ray tube – across a potential of 710 kilovolts, so bringing about the first artificial disintegration of an atomic nucleus.

This is just the beginning of a very big subject, so first a warning from Abraham Pais's *Inward Bound*: 'A rapid glance at the growth of accelerators does not remotely suffice to convey the complexities of the developments which have led to the modern high energy laboratories.'[4]

In this can of worms, the choice is between circular and linear accelerators, and the competition between the two has now been fought out on a massive scale. The advantage of the circular accelerator is that with every complete circuit the particles can be given a further boost, simply by increasing the power of the electric and magnetic fields in which they move. The principle, analogous to that of the sling as an efficient means of launching a projectile, is easier to state than apply. For one thing, the direction of the fields must continuously correct the natural tendency of the particles to shoot off at a tangent (which is the way a sling operates). At the end of the ride, when the particles have reached the required velocity, they are diverted into a straight section, with the target at its far end.

One way of dealing with this problem is to increase the dimensions of the circuit and thereby reduce its curvature. This is the solution pioneered by CERN, which has constructed a circular tunnel, 26 kilometres long, deep underground near Geneva, to contain the largest of its ten accelerators. Until November 2000, this was the Large Electron Positron (LEP) collider, but this is now being replaced by the Large Hadron Collider (LHC), due to open in 2005. Experimental research is carried out in four underground stations, each with its entrance above ground: the particle track is then straight, with facilities for inserting the actual recording apparatus, from which the relevant subatomic event, such as electron–positron collisions, can be observed. Once the apparatus has been set up for a given experiment, the results will be read off from a computer screen. The costs are astronomical – so much so that in the mid-1990s a Texas project, with a tunnel more than twice as long as CERN's, was simply abandoned.

Another problem is that as a particle's velocity approaches that of light in a vacuum its momentum and energy increase: this is one consequence of Einstein's special theory of relativity.[5] It is a critical limiting factor at so-called relativistic velocities, which approach the speed of light. The power then required is gigantic: the LEP collider was only allowed to take electricity from the Swiss power net in summer, when other demand was low. The LHC will doubtless be subject to the same restrictions.

If the balance of advantage is with the circular accelerators, the linear model also has its proponents, notably in Stanford, in California's Silicon Valley, where a 2 mile long tunnel houses the SLAC accelerator. This consists of some 80,000 copper modules, each consisting of a cylinder, some 4 centimetres long with a 10 centimetre internal diameter, with a copper disc with a small hole at the centre defining the two extremes, and providing access to the discs on either side. SLAC operates by sending alternate bunches of electrons and positrons down the whole length of the tunnel, separated, in both cases, by a distance equivalent to three modules. The motive force is an oscillating electric field, which keeps in step with the alternating bunches of particles, so that both positive and negative are sent in the same direction. The velocity of the particles then approaches that of light in a vacuum, a result achieved by relating the dimensions of each module to the microwave frequency of the electric oscillations. The collisions achieved would not be possible with a circular accelerator, so this is hardly a case of true competition. Both types have their advantages, according to the experimental results being looked for.

The nuclear reactor

Although accelerators, starting with the cyclotron, are indispensable in nuclear research, the reactor has extended its range considerably, even though its main use has been in nuclear power stations. No other scientific instrument ever had such a dramatic origin. It is associated with one man, the Italian physicist Enrico Fermi (1901–54), and the day, 2 December 1942, when his atomic pile first went critical is a landmark in the history of science. Today's research reactors, such as the Euratom High Flux Reactor in the Netherlands, are a useful source of different particles needed for research and industrial use, particularly in medicine.[6]

The ultimate instrument in particle and high-energy research is the Russian-invented *tokamak*, a toroidal device for producing controlled nuclear fusion by confining and heating a gaseous plasma. Today's European Fusion Development Agency works with one tokamak, known as the Joint European Torus (JET), which is located at Culham, just south of Oxford.

The history behind this location goes back to the 1970s, when many European Union countries were competing for this prestigous scientific object. In the end Britain and Germany were the front runners. At a critical moment a Lufthansa aeroplane was highjacked and forced to land at Mogadishu, the capital of Somalia. The German chancellor, Helmut Schmidt, was desperate to rescue the passengers, but only the British Army had a special team trained for such purposes. This was called in to help the Germans, with such success that all the passengers were rescued.

Schmidt was so grateful that he offered the British prime minister, Jim Callaghan, to name a reward. The response was to ask Schmidt to give up Germany's claim to the tokamak, which explains how JET opened for business in Oxfordshire in 1978.[7] But how does the monster work?

To begin with, the inside of the torus is a perfect vacuum. Into this vast space a minute quantity of gas – less than 100 grams – is introduced. A large electric current is then passed through the inner poloidal field coils, which, by the standard operation of a transformer, generates a current in the gas in the vacuum chamber.

This, on a principle similar to that of an ordinary lamp filament, becomes extremely hot and at a stellar temperature of some 40,000,000°K the gas becomes plasma. At the same time, the magnetic field generated by thirty-two large D-shaped coils prevents even the smallest contact between the plasma and the walls of the vacuum chamber: otherwise the immense heat would melt the walls. The outer poloidal field coils then regulate the actual configuration of the plasma cloud within the torus.

The transformer action requires the plasma current to operate in pulses, following the basic principle established by Faraday in 1831. A single pulse can last up to a minute, but the interval between pulses must be at least twenty minutes. The Culham site is next to the Didcot Power Station, and two of its generators are needed to supply the power required by JET, which is power-hungry on the same scale as CERN. What, then, is the payoff?

According to the history of JET, as told on the internet,[8] it achieved fusion power in 1991 and, in 1997, 'World Records in Fusion Performance'. The recent European Fusion Development Agreement provides for the use of the JET facilities until at least the end of 2002. The actual fusion takes place between plasmas consisting of deuterium and tritium, the two heavy isotopes[9] of hydrogen. This is essentially the nuclear reaction, discovered by Bethe in 1938 (see Chapter 8), which, by converting the sun's hydrogen into helium, accounts for its heat. The peak recorded for 1997 was for fusion power of more than 16 megawatts, an amount greater than the input from the Didcot Power Station.

All this is essentially applied science, the reason that the EU was ready to finance JET in the first place. Practical applications define the next stage, known as ITER, the International Thermonuclear Experimental Reactor, which is almost certain to be located in Germany.

JET has shown that it can achieve a positive balance of power: it can send out more than it takes in. It is not, however, self-sustaining, which it would be if some of the 16 megawatts could be fed back into the system, to replace the power taken from Didcot, at the same leaving a surplus for commercial distribution. This is ITER's remit.

The mind boggles. It has been estimated that if ITER foreshadows thermonuclear power generation, a single station would be sufficient for all the power requirement of the EU. There would be no nuclear waste, nor any toxic emissions. This is the scenario for the coming mid-century. Local inhabitants would learn to live with solar temperatures maintained inside a giant doughnut, once they were assured that all the heat would dissipate in a split second in any breakdown of the system.

The skeleton outline of the development of the instruments used in modern physics is now complete, as least so far as space allows. Any contemporary physicist could point out critical omissions, but at this point the skeleton must be clothed with flesh and blood. The rest of this chapter then focuses on the people who developed and used the apparatus introduced above, giving at the same time the results they achieved. The picture would still be one-sided, so next to the great experimenters, such as Hertz, Thomson, Rutherford, Geiger, Chadwick and Fermi, many of whom were also considerable theorists, there will be a short look at those seldom to be found inside a laboratory, Planck, von Laue, Dirac, Heisenberg and Pauli, with a few, notably Bohr, who achieved distinction on both sides of the line. Finally, no history of twentieth-century physics could omit its transformation as a result of the Second World War, and the vast investment in nuclear weapons represented by the Manhattan Project and continued by national research laboratories funded and equipped on an unprecedented scale.

The cathode ray tube

In 1878, William Crookes constructed a near perfect vacuum tube containing two electrodes, one positive and the other negative. Once the current was switched on, a pale dim light illuminated the thin air and the glass walls became fluorescent. The fluorescence came from the cathode, the negative electrode. He discovered that the rays would bend in a magnetic field, which precluded their being pure light – but then they were neither gas, liquid nor solid. Perplexed, he named them simply 'radiant' matter. Without knowing it, he had opened a whole new realm in experimental physics.

In the 1890s many noted physicists were experimenting with Crookes tubes. Two among them, J. J. Thomson (1856–1940) in Cambridge and Wilhelm von Röntgen (1845–1923) in Würzburg, noted that fluorescent material outside the tube glowed as a result of the discharge within it. Thomson was too intent on studying the cathode rays to pursue the cause of this effect, but Röntgen decided to do so. On 8 November 1895, he discovered that black paper covering the tube was no barrier to whatever was causing the fluorescent effect on the screen he had placed in front of it. Placing his hand between the tube and the screen, he noted that this slightly reduced the overall glow but in dark shadow he could see the bones. This result, published in December 1895, stunned the world.

Röntgen, at a loss to explain his discovery, resorted to the unknown, X, to coin the name X-rays, current to this day. None the less he became, in 1901, one of the first Nobel prizewinners for physics. At the same time, the discovery of X-rays, whatever they might be, had an immediate and profound effect on the research of others – among them J. J. Thomson. (The discovery of the true nature of X-rays by Max von Laue (1879–1960) in 1912 is described in Chapter 6).

J. J. Thomson, a generation younger than Crookes, was the son of a Scottish bookseller. As a boy he had wanted to become a railway engineer, and with this in mind he enrolled at Owens College, Manchester, at the age of fourteen. There he became interested in mathematical physics and, aged twenty-one, he went as a scholar to Trinity College, Cambridge. He was second wrangler in 1880, and in 1884 he became Cavendish Professor and a Fellow of the Royal Society: at the age of twenty-seven he had succeeded to the chair that only five years before had been held by Maxwell. With one of the world's finest laboratories, he could choose almost any direction for his research.

He chose to look again at the question that had defeated Crookes. What was is it in the rays observed in a cathode tube that made them respond to a magnetic field? He worked with every possible design of tubes and constantly varied the parameters – the level of the vacuum, the material of the cathode, the electric potential and the strength of the magnetic field. In April 1897, he gave his final answer: 'On the hypothesis that the cathode rays are charged particles moving with high velocities ... the size of the carriers must be small compared with the dimensions of ordinary atoms or molecules.' And in 1899 he went on to add that 'electrification essentially involves the splitting up of the atom, a part of the mass of the atom getting free and becoming detached from the original atom'.[10] Thomson had discovered negative particles of electricity, which, stripped from the atom, provided the means for transmitting electric energy. The current carried by an ordinary wire consists of such particles moving with a velocity of 160,000 miles per second.[11]

Thomson first referred to the basic particles as 'corpuscles', but the name soon changed to 'electron', and the cathode ray is now known as an 'electron beam'. The discovery of the electron revolutionised the understanding of matter. For the whole of the nineteenth century, atoms – as first propounded by Dalton – were regarded as the indivisible units out of which all matter was constituted. Towards the end of the century, the periodic table established properties common to different classes of atoms, but, even so, they retained their own separate identities. Thomson's electrons were completely uniform, common to all atoms, stable against decay. Now, more than a hundred years later, they still have no known size or substructure.

Thomson's discovery went further: sophisticated mathematics revealed the ratio e/m of the electron's electric charge to its mass, so that by determining the charge, experimentally, the mass could be calculated.[12] Thomson's result was of the order of a two-thousandth of the mass of the hydrogen atom.[13]

With a mass of approximately 10^{-27} grams, the problem of actually observing an electron would appear to be insoluble, the more so given their high velocity. The resolution of an optical microscope (to which there was no alternative in the early 1900s) is limited to a factor measured in thousands – no use whatever for such microscopic particles. Yet if the electron was to be credible, a way had to be found of observing it.

Thomson, helped by tea-time conversations with his research students at the Cavendish Laboratory, grasped the nettle. The man who helped him out was C. T. R. Wilson (1869–1959), another Scotsman, who had started life as a meteorologist. Wilson had observed a coloured halo surrounding shadows cast by the sun shining through the Scottish mountain mist. From this starting point, he devised an apparatus in which moisture contained in air condensed around microscopic dust particles when the air was suddenly cooled by expansion. This effect, a result of so-called 'adiabatic' expansion, followed from Boyle's law (see page 133).

In 1899, Thomson, in measuring the charge of an electron (and derivatively its mass), had used an elementary cloud chamber produced by Wilson in which the stream of electrons ionised the vapour, to an extent that could be accurately recorded. This then made it possible to derive the charge of a single electron. (Another twelve years' work was needed to perfect the cloud chamber. In its final form, based on a mixture of air and ethanol vapour, charged particles, such as electrons passing through the supersaturated vapour, create a visible trail of ions which attract drops of moisture.

Although this apparatus only became available in 1911, it enabled a photographic image to be made of the trail of water droplets marking the high-velocity passage of a single electron.) The cloud chamber represents the first critical step in recording subatomic phenomena, and Wilson rightly earned the Nobel prize finally awarded to him in 1927.

Radio transmission

In 1871, the year after the new German state was formed with Berlin as its capital, the city university appointed Hermann von Helmholtz (1821–94) as director of its Institute of Physics. Von Helmholtz, with a distinguished research background in mathematics, physiology and physics, was a man of immense talent and drive. He was also a friend and colleague of Werner von Siemens, who had made a fortune by turning the wisdom of Faraday to the power generation of electricity. Germany under Prince Bismarck, its chancellor, provided the perfect climate for enterprise in almost any field, including science. It is not surprising, then, that von Helmholtz turned his attention to Maxwell's field theory.

The status of von Helmholtz's institute was such as to attract the ablest students, and they, rather than their mentor, established the breakthrough when it came to subjecting Maxwell's theory to experimental verification. Of these students, one, Heinrich Hertz (1857–94), immediately impressed von Helmholtz with his unusual talent. In years of friendship and collaboration, the latter's expectations were never disappointed. Hertz, for his part, never doubted his own ability. When he arrived in Berlin in 1878, he wrote to his parents, 'I grow increasingly aware, and in more ways than I expected, that I am at the centre of my own field; and whether it be folly or wisdom, it is a very pleasant feeling.'[14] Self-evidently, the Institute of Physics was a case of 'Deutschland über alles'.

In spite of all the hype at Berlin, Hertz, if he was to advance his career, could not stay there, and it was in the period 1884–89 spent as professor at the Technische Hochschule in Karlsruhe that he discovered radio waves and demonstrated experimentally the correctness of Maxwell's equations. He had already met the theoretical challenge of deriving the equations from fundamental laws, established by Ampère and Faraday, governing varying electric currents. In doing so, he discredited alternative theories, current in

Germany, but incompatible with Maxwell's equations. In his own words, 'we may infer without error that if the choice rests only between the usual system of electromagnetics and Maxwell's, the latter is certain to be preferred.'[15] If Hertz was conceited, he also had integrity.

When it came to testing Maxwell's theory experimentally, Hertz had first to generate electromagnetic energy and then detect it. By its nature, the energy will radiate from its source in waves that make no impact on the ordinary senses of sight or hearing. These are generated by the top half of the apparatus illustrated schematically in Figure 7.1. *A* is an induction coil, an elementary transformer, which will multiply the voltage supplied to its core by an ordinary battery. The basic principle is simply that of Faraday's induction ring described in Chapter 5. The multiplication factor is determined by the number of times the input and output wires are wound round the coil.

The input comes from a powerful battery (not illustrated in Figure 7.1). The output goes to two metal spheres 30 centimetres in diameter, C and C', joined by a wire with a small gap, pp', at its midpoint. If the induction coil is activated by connecting the battery, the immediate result is that the induced current charges the two spheres, one positively and the other negatively. This is

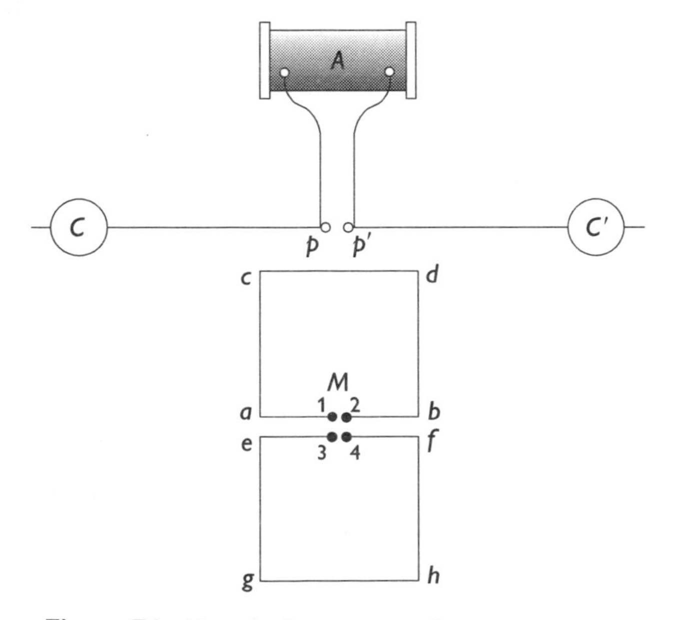

Figure 7.1 Hertz's electromagnetic wave apparatus.

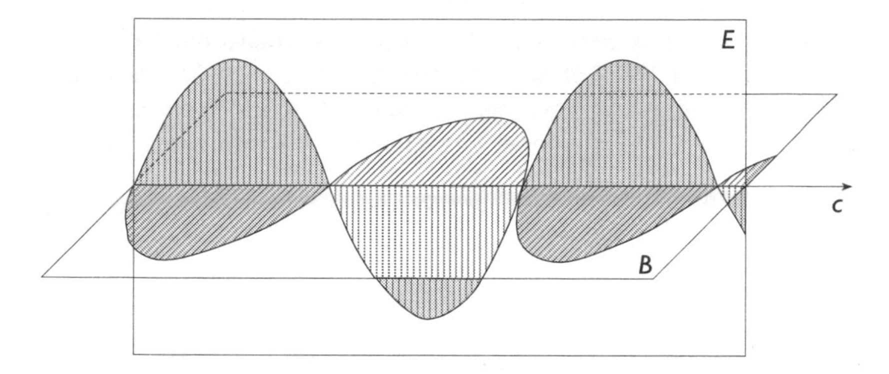

Figure 7.2 Electromagnetic field of light. A pencil of light moves in the direction *c*: the associated electric (magnetic) field vibrates in the plane marked *E* (*B*).

simply the principle of the Leyden jar (described in Chapter 4), and the layout devised by Hertz is known as a dipole.

In this process a point will be reached when the spheres are discharged by a current flowing, as a spark, across the gap – an effect already well known in the eighteenth century. (Remember Louis XV and the 700 monks.) The result is to distribute the charge evenly along the two halves of the dipole. The process, however, is not instantaneous, because it is resisted by counteracting forces[16] induced in the wire itself: it therefore takes a small, but still finite, time to complete. At this point the electric field has declined to zero, with its energy, in the process, being transferred to a magnetic field at right angles to it, as shown in Figure 7.2.

The magnetic field in turn collapses, generating – remember Faraday – a reverse electric current, so that C and C' are recharged, but with their polarity reversed. They will in turn discharge, creating a new magnetic field, which on collapsing will restore the original polarity. Back to square one, after a four-stage process. If there is only one original charge, the process will rapidly spend its force, since the electric potential across pp' will become too small to make a spark. The lost energy is then radiated from the system.

Hertz realised that if the induction coil was used to produce a high-frequency pulse (by continually connecting and disconnecting the battery), the four-stage process could be maintained indefinitely, to produce an electromagnetic field. The optimal frequency could be calculated by measuring the capacitance of the two spheres and the self-inductance of the wire.

For the first time in the history of science, apparatus had been designed and built to transmit an electromagnetic wave. (That these are also known as Hertzian waves is not surprising.) Hertz's problem was now to find a way of receiving or, better, detecting the wave: this explains the layout of the bottom half of Figure 7.1.

The square *abcd* is a simple wire loop with a spark gap 1–2. Electromagnetic waves encountering the loop transfer a proportion of their energy, causing an oscillating current to flow. The width of the spark gap can then be adjusted to the point where the flow is broken, thus providing a measure of the energy transferred – and, incidentally, the first means ever for detecting high-frequency voltages. At the same time Hertz noted that the maximum voltage at 1–2 corresponded to a null reading on the opposite side, *cd*, of the loop, proving that the current in the loop was not induced as in the Faraday ring experiment (described on page 115).

Hertz also showed that, by changing the lengths of the dipole and the loop, a state of maximum voltage could be reached. (This explains the presence of the second loop, *efgh*, which acted as a control.) This was a state of resonance, comparable to that familiar in sound – remember Wheatstone's magic lyre (described on page 116).

Finally Hertz was also concerned to measure the velocity of the propagating waves, which, if close to that of light, would confirm the validity of Maxwell's equations. It would have been sufficient to show that there was a simple numerical relationship between the wavelength and the lengths of the dipole and the loop. In fact both the dipole and the resonant loop peak when their length is half that of the wavelength.[17] This is all one needs to know, but Hertz, for some reason, never fully established the connection.

However important Hertz's results were for science, their practical consequences were world-shaking. Although Hertz's experiments were conducted within the confines of a single room, distance was no barrier to the propagation of electromagnetic waves, so that the distance separating the dipole CC' from the loop *abcd* could be measured in miles. Within ten years Rutherford in Cambridge and Marconi in Bologna were demonstrating this in practice, and well within a hundred years radio signals were being received from the most distant parts of the universe. Hertz lived to see none of this: he died, aged only thirty-six, in 1894. Von Helmholtz, heartbroken, wrote just after his death: 'There can surely be no thought of finding someone who could replace this unique man.'

Radioactivity

Henri Becquerel (1852–1908), born to a family well established in the French scientific world, inherited from his father, Alexandre-Edmond, an interest in phosphorescence. On 20 January 1896, the French Académie des Sciences learnt of Röntgen's discovery of X-rays. Becquerel, then aged thirty-nine, had achieved relatively little in the preceding five years, but this new phenomenon led him, and three others, to test the proposition that phosphorescence on its own could be a source of X-rays.

The three other scientists lost little time in producing evidence for the proposition – now long known to be false. Each of them had worked with different phosphorescent substances, and Becquerel did the same, but with no positive results. But then, working with a sample of uranium salts (a fifteen-year-old legacy from his father) he struck pay dirt – or so it seemed. As he reported to the Académie,

> One wraps a photographic plate . . . in two sheets of very black paper . . . so that the plate does not fog during the day's exposure to sunlight. A plate of phosphorescent substance is laid above the paper on the outside and the whole is exposed to the sun for several hours. When the photographic plate is subsequently developed, one observes the silhouette of the phosphorescent substance, appearing in black on the negative. If a coin, or a sheet of metal . . . is placed between the phosphorescent material and the paper, then the image of these objects can be seen to appear on the negative.[18]

So far so good, but Becquerel, a prudent man, decided on a corroborating experiment. This time he placed a thin copper cross on the black paper covering the photographic plate. February 1896 enjoyed little sunshine, but in the end Becquerel became impatient and developed the plate at a moment when William Crookes happened to be visiting his laboratory. Astonishingly the image of the copper cross shone out white against the black background.

Repeated experiments showed that only uranium produced this effect, and that to do so it did not need to be phosphorescent. By serendipity Becquerel had discovered radioactivity, although he did not use that name, nor explain its source.[19] The discovery aroused little immediate interest: there was much more interest in X-rays, which could be easily produced, while uranium was almost impossible to obtain. None the less, a young married couple, Pierre and Marie Curie, who worked quite outside Académie circles, found the subject very attractive, and Marie decided to make it her own. This was a turning point in one of the most remarkable lives in the history of science.

Maria Sklodowska, who first saw the light of day in Warsaw on 7 November 1867, was not born under a lucky star. The family belonged both to the intelligentsia of the city and the impoverished aristocracy of a countryside, which since 1797 had been brutally incorporated into the Russian Empire. Periodic uprisings always failed, to be followed by an ever more oppressive regime: after that of November 1830, the Russians closed Warsaw University, so depriving Maria's father of the chance of a degree. None the less he managed to make some kind of a career as a schoolmaster, in spite of constant harassment by the Russian officials in charge of the public schools.

Maria's mother did rather better. She became the head of a private school for girls, where the Russians, little interested in education for women, left her alone. When she had to give up this job, she taught herself shoemaking, and set up a little shop in the family apartment, where Maria shared her talent for manual work – later an important factor for success as an experimental physicist. Such activity was far from the ethos of a social class that had long regarded the work of the shopkeeper and the mechanic as degrading.

Tragedy then struck in a form all too common in nineteenth century Europe: Maria's mother succumbed to tuberculosis. This was before Robert Koch (1843–1910) discovered the tubercle bacillus in 1882 (although this did not lead to a cure). With middle class patients, doctors imposed a programme of cures in expensive foreign spas – which meant, in the Sklodowski family, that their modest means were further depleted, at the same time as they had to comfort a mother who was slowly dying. The scenario was all too familiar, and the experience was traumatic for Maria.

Three months after her mother's death, in 1878, Maria, aged eleven, enrolled at a girls' gymnasium, where predictably, she was unhappy. Physics, however, was well taught, and in 1883 Maria graduated, first in her class, and with a gold medal. This looked like the end of the road: Warsaw university, long reopened, did not admit women, so further education meant going to St. Petersburg or Paris – not an easy option given the family circumstances.

Maria set her sights on Paris, where she hoped to join her elder sister, Bronia. First, she had to work as a governess, ending up with four years on a country estate. There, in a dimly lit room, she spent her spare time studying mathematics and physics, working with textbooks in English, French and Russian. Maria was finally able to return to her father in Warsaw: there, also, the appointment of a cousin to be a director of the new Museum of Science and Industry gave her gave her access to a modern chemistry laboratory. At the same time, Bronia graduated in medicine at Paris, where her Polish exile husband was already in practice. They were ready to help Maria, and Bronia sent to Warsaw the course catalogue for the science faculty at the Sorbonne for the year 1891–92. Maria enrolled in November 1891, one of twenty-three women among 1825 students.

Although Maria could only afford a poorly heated attic room, Paris suited her perfectly. Her life there is best described in her own words:

> All my mind was centred on my studies. I divided my time between courses, experimental work, and study in the library. In the evening I worked in my room, sometimes very late into the night. All that I saw and learned that was new delighted me. It was like a new world opened to me, the world of science, which I was at last permitted to know at all liberty.

Maria made the best of her opportunities: when her studies ended in July 1894, she had been ranked first of her year in science and second in mathematics. In both subjects she had had the benefit of first-class teachers of international renown, like the young mathematician Henri Poincaré (1854–1912). After graduating she returned to Warsaw for the summer, and if it was expected that she would remain there, she had good reason for returning to Paris.

Pierre Curie (1859–1906), born into a family of inventors with extreme radical political views, received an important part of his education at home. Although Paris in the 1890s was very different from Warsaw, Pierre Curie and Maria Sklodowska were brought up in much the same ethos. It is not surprising then that when they met – shortly before Maria would return to Warsaw – they found that they had much in common, including their both being workaholics.

Pierre and Marie Curie married without fanfare on 26 July 1895: thus began one of the most remarkable partnerships in the history of science. Pierre, eight years older than Marie, already had substantial scientific achievements. With his elder brother, Jacques, he had investigated how the electrical properties of crystals changed with pressure – a phenomenon that came to be known as 'piezoelectricity' (and which was later turned to practical use in quartz clocks and other gizmos). The two brothers themselves invented a piezoelectric quartz balance, which proved essential for the delicate measurements required in the process of discovering radium.

In the autumn of 1897, two events gave a new turn to Marie's life: first, her daughter, Irène, was born on 9 September; second, she decided to research the phenomenon discovered by Becquerel in February 1896. As she noted later, 'the subject seemed to us very attractive and all the more so because the question was entirely new and nothing yet had been written upon it.'

Becquerel had noted that uranium turned air into a conductor of electricity, while others had found that a certain 'saturation point' defined a limit to its conductivity. The Curies decided to measure the energy given off by uranium, and also to test other elements. The worked with two metal discs, 8 centimetres in diameter, one above the other, the two being separated by 3 centimetres. Having covered the lower disc with uranium, they then charged it with a high-voltage battery and measured the rate at which the upper plate became charged. The experiment, which depended upon extremely accurate instruments, including the electrometer and quartz balance invented by Pierre, produced the figures needed for analysis.

Beginning with uranium, Marie went on to test thirteen other elements, including gold and copper. Then, on 17 February 1898, she turned to the mineral pitchblende, the ore from which uranium is extracted. This produced a much stronger current than uranium, a fact recorded without comment in Marie's notebook. Subsequent entries show, however, that her mind was not at rest. She tested several other uranium compounds, but they were all less active than pure uranium. Then, on 24 February, came another surprise: the mineral element thorium also proved to be more active than uranium, though less so than pitchblende.

Further experiments showed that, another mineral, natural chalcite, was more active than uranium. The only explanation seemed to be that they contained an undiscovered element more active than uranium, so that for the first time in the history of science radioactivity was proving to be a diagnostic for the discovery of new substances. The fact, also, that the energy level with uranium compounds varied according to the amount of uranium present meant that the phenomenon was a property of the atom.

The next step was to isolate the hypothetical new element. The Curies, starting with 100 grams of pulverised pitchblende, used chemical means to break it down into different products, at every stage concentrating on the most active of these. Finally they thought that they had a product pure enough to be examined spectroscopically, following the methods developed by Bunsen and Kirchhoff described in Chapter 6. They were disappointed: no new spectral lines appeared.

They asked the help of a chemist, Gustave Bémont, who, distilling small quantities of pitchblende, created products more active than pitchblende. By mid-summer 1898, Marie had something 300 times more active, while Pierre, working independently, had done better with 330 times. By this point the results suggested that pitchblende contained, not one, but two unknown radioactive elements, one accompanying bismuth in the breakdown process and the other accompanying barium, both elements normally present in pitchblende.

They persisted with bismuth, and when they attained a substance more than 400 times as radioactive as uranium, they prepared a statement which Becquerel read to the Académie des Sciences on 18 July:

> We . . . believe that the substance we have extracted from pitchblende contains a metal never before known, akin to bismuth in its analytic properties. If the existence of this metal is confirmed, we propose to call it polonium after the name of the country of origin of one of us.[20]

The title of the paper, 'On a New Radio-Active Substance Contained in Pitchblende', introduced the word *radioactive* to the world of science. After a long summer vacation, the Curies, working with a new load of pitchblende, succeeded in reducing the barium products to a substance 900 times more radioactive than uranium. This time spectroscopy did produce new, hitherto unknown, spectral lines, and on 20 December Pierre scribbled the name 'radium' in his notebook. They still failed, however, to discover its atomic weight: it took another three years to discover that the 'new' element was present in barium at a level of less than a millionth of one percent.

Marie Curie faced this challenge almost alone. Given the vast quantity of pitchblende needed for reduction to this level, she needed a vast amount of space; and this came in the form of what once had been a dissecting room for medical students. At the same time money given by Baron Edmond de Rothschild made it possible to import from Austria 10 tons of pitchblende residue, from which the uranium had already been extracted.

In the spring of 1899 Marie began working with 20 kilograms of material at a time, in a process that started with hours of boiling in cast-iron basins. It soon became clear that isolating radium from barium would be simpler than isolating polonium from bismuth, so this is where Marie concentrated her efforts. In both cases the products, at a sufficient level of concentration, became luminous, to the delight of both the Curies. (Little did they realise that they were also carcinogenic.) By this time others were beginning to be interested and to conduct related experiments. Finally, at the International Congress of Physics convened in Paris as part of the Universal Exposition of 1900, the Curies had a chance to present their results to the whole world of science. Their paper, 'The New Radioactive Substances', noted that 'the spontaneity of the radiation is an enigma, a subject of profound astonishment' and concluded by asking, 'What is the source of the energy coming from the Becquerel rays? Does it come from the radioactive bodies, or from outside them?'

It took another two years to produce a pure sample of radium, weighing one-tenth of a gram – little more than a grain of rice – but the paper presented in 1900 had put radioactivity at the centre of the stage. In 1903 Pierre, in a joint experiment, found that within an hour 1 gram of radium could raise the temperature of 1.3 grams of water from 0°C to 100°C.[21] In the same year the Nobel prize for physics was awarded to the Curies, together with Becquerel. (Marie was the first woman laureate; in 1935, her daughter, Irène, would be the second). In 1905 Pierre was finally elected to the Académie, and a special chair was created for him at the Sorbonne. Then tragedy struck once more: in 1906 Pierre died after been run over by a carriage.

Marie, inconsolable, continued with her work, being awarded the Nobel prize for chemistry in 1911 for her discovery of radium and polonium. In 1918 the Radium Institute in Paris was founded and Marie spent the last sixteen years of her life as director. She died of leukaemia in 1934, probably as a result of her long exposure to radioactivity. Ironically, radium, by this time, had become an important element in the treatment of cancer. Her scientific legacy is incalculable, but with the trials and sorrows that continued throughout her life, she paid a high price for her place in the Pantheon of Science.

The chance combination of circumstances is often the driving force of history. This is as true in science as in any other field. So it was in the year 1895 when chance brought Ernest Rutherford to Cambridge.

In England, the Great Exhibition of 1851 generated a large profit, to be devoted to education. In the course of time, part of the funds were applied to scholarships, offered to men from every part of the British Empire. In 1895, Cambridge set up for the first time a scheme for graduates from other universities to be admitted as research students to read for a higher degree. Also in 1895, New Zealand, entitled every other year to one scholarship, awarded it to Ernest Rutherford (1871–1937).[22] When told of his good fortune, Rutherford, who was digging potatoes on his parents' farm, which was mainly devoted to harvesting and processing flax, could only say, 'That's the last potato I'll dig.' He was right.

At this time Rutherford was already in his fifth year at the University of New Zealand in Christchurch. A brilliant all-round student, he had already qualified for an MA, with a double first in mathematics and physics, and was working for a B.Sc. Although nineteenth-century Christchurch, seen from Europe, must have been the back of beyond, Rutherford, and those who taught him, were remarkably up to date when it came to the latest developments in physics.

Rutherford himself, impressed by the recent discovery of radio waves, set up a Hertz oscillator in a 'miserable, cold, draughty, concrete-floored cellar', and then developed a magnetic detector which could detect radio waves at a distance of 60 feet and that after they had overcome a number of opaque obstacles. This was a line he would develop at Cambridge, where he impressed the whole academic community by detecting signals over distances measured in miles, not feet.

Once at Cambridge, Rutherford was an immediate success. J. J. Thomson, equally gregarious and outgoing, liked him, saw that he became a member of Trinity College, and before the end of year arranged for him to give a lecture, with experiments, to the Physical Society. Not only professors, but their wives, attended, to see Rutherford explain his apparatus for detecting radio waves. With the presence of what Rutherford called 'the usual vulgar herd', the atmosphere must have been reminiscent of Faraday's lectures at the Royal Institution half a century earlier. By the end of the term, Thomson had suggested that Rutherford send a report of his work to the Royal Society, and on 18 June 1896, he was invited to appear for the first time before the Royal Society (of which he would later become president). Plainly the grass was not growing under his feet.

Within a year or two of Rutherford's arrival in Cambridge, three new lines of advance opened up in physics: (1) the discovery of radioactive bodies; (2) the discovery of electrons; and (3) the ionisation of gases by X-rays. Although Rutherford's immediate interest was in the ionisation of gases, radioactivity would determine the road to his greatest discovery, that of the atomic nucleus, in 1911.

The moment that Rutherford became interested in uranium in 1897, he later described as the most important in his life. Starting with the effect discovered by Becquerel, a simple experiment showed that it comprised two different types of radiation. A uranium sample was covered with successive layers of aluminium foil. With the first three layers, the radiation steadily decreased, but at this point it appeared to become stable. It was only after several further layers were added that the steady decrease resumed. The only possible explanation was that there were two different kinds of radiation, which Rutherford named α and β. Experiments carried on in Germany and France, not with uranium, but with thorium would shortly reveal a third type, γ.

In 1898, Rutherford, aged twenty-seven, offered the chair of physics at McGill University in Montreal, accepted for two reasons: one simply was the pay, a matter always of interest to him; the second was the chance of his own laboratory in the MacDonald Physics Building, which had been endowed by a local millionaire. There he began to analyse the distinctive properties of the different kinds of radiation.

To start with, by examining the properties of gas surrounding a radioactive source, he discovered that its atoms acquired a strong positive electric charge – this is the process known as ionisation – from α-radiation, but only a weak charge from β-radiation. He then noted that the high level of energy dissipated could not be produced by a purely chemical reaction, with atoms recombining at molecular level. For the first time ever, this implied the possibility of atomic energy.

In 1899 Rutherford discovered that any substance in contact with ionised gas also became radioactive, but that this induced radioactivity declined very rapidly. In the event, all radioactivity proved to decrease with time, but at very different rates. A year later, in Europe, it was discovered that β-radiation, but not α-radiation, was deflected by a magnetic field – a property shared by the stream of electrons discovered by J. J. Thomson in 1897. (It took more than ten years to prove that the two phenomena were essentially one and the same.)

In 1901, Frederick Soddy (1877–1965), a young chemist, joined the McGill faculty, and with his critical mind forced Rutherford to consider the implications of the new discoveries. At an early stage, Soddy was able to show that thorium radiation produced an emanation lacking any chemical reactivity. This discovery, occurring within ten years of Sir William Ramsay's discovery of argon, led Soddy to exclaim, 'Rutherford, this is transmutation: the thorium is disintegrating and transmuting itself into argon gas.'

For the time being, however, those working in this new field were content to talk about thorium and thorium-X, or, where relevant, uranium and uranium-X. Returning from the Christmas vacation, 1901–2, Soddy discovered that the thorium he had been working with was more radioactive than ever, while the thorium-X was hardly radioactive at all.[23] This led him to take measurements in both cases. When plotted on a graph, these revealed for the first time the half-life of a radioactive element, that is, the period, long or short, over which its radioactivity decreases to half its original level. Soddy then discovered that, after the process of separation, both thorium and uranium produced only α-radiation, while uranium-X and thorium-X, only β-radiation.

Rutherford believed that 'experiment, directed by the disciplined imagination either of an individual, or, still better, of a group of individuals of varied mental outlook, is able to achieve results which far transcend the imagination alone of the greatest philosopher.' His genius as an experimenter is shown by the apparatus he designed for investigating α-radiation. First he had to wait for the Curies to send a sample of radium from France to Canada. With this as a sufficiently powerful radioactive source, a basic structure was made out of a number of small empty metal cases, joined together. This created a series of channels, open at both ends: radium, as a source of α-particles, was placed at one end of each channel, while half of the other end was blocked by a minute strip of metal. The next step was to create, within the channels, electric and magnetic fields of sufficient power to deflect the α-radiation. This was easier said than done. At a certain point the electric field would creates sparks from one channel wall to the other, which had to be prevented, and a magnet powerful enough for the magnetic field had to be acquired from the electrical engineers.

Everything worked out in the end and the apparatus then showed that not only could an electric and magnetic field block the flow of α-particles from the radium source but also that the α-particles were positively charged, so that they could be either hydrogen or helium ions. The results were published in the *Philosophical Magazine* for September and November 1902, under the title 'The Cause and Nature of Radioactivity'.

By using liquid air to cool down the radium emanations to the liquid state, Soddy, helped by Sir William Ramsay, showed experimentally that helium was the right answer, although it was not immediately clear that α-particles were helium atoms. Finally radioactivity was confirmed as a phenomenon of the individual atom, so half-life was a statistical effect – an expression of random occurrences. After all this, Rutherford and Soddy broke up – amicably – ending their last paper with these prophetic words:

> All these considerations point to the conclusion that the energy latent in the atom must be enormous compared with that rendered free in ordinary chemical change.[24]

(Rutherford and Soddy's results, by enabling rocks to be dated according to their half-life, were also important for geology. Moreover, Lord Kelvin had calculated, on the basis of heat-loss radiation, that the earth was 20–40 million years old; because of radioactivity this could be corrected to several hundreds of millions of years, a result compatible with recent geological research. This is the first appearance of the radioactive clock.)

In 1907 Rutherford accepted a chair at Manchester: the laboratory would be well equipped, and, at a time when physics was just getting started in America, the city was much closer to the important research centres – including the Cavendish Laboratory. Rutherford stayed in Manchester for twelve years – the happiest and most productive period years of his life. In particular he worked with exceptionally able colleagues: the first of these, Hans Geiger (1882–1945), was expert in finding means for calculating the rate of particle emission – giving his name to today's familiar Geiger counter.

The counting apparatus relied on the fact that a single ion, such as an α-particle, moving in a low-pressure gas, subject to an electric field, could produce several thousand more ions as a result of colliding with the gas molecules. With a sufficiently powerful field, low pressure and sensitive electrometer, this effect could be measured before the particle lost its power of ionisation. The hard core of the apparatus was simply a plain brass tube, about 60 centimetres long, with a wire running down its axis. With the tube sealed, and the pressure reduced to the lowest possible limit, a pencil beam of α-particles was allowed to enter through a minute hole, covered by a thin sheet of mica. The electrometer would then register every single particle.

Not only did the experiment provide the best possible evidence for the existence of atoms as material realities, it also allowed the electric charge of the hydrogen atom to be calculated, together with Avogadro's number, which counts the molecules in a cubic centimetre of gas at standard temperature and pressure. This, according to a hypothesis stated, but never proved, by the Italian chemist, Amedeo Avogadro (1776–1856), is standard for all gases. Rutherford, a physicist, had finally confirmed the correctness of the hypothesis, which had in fact been generally accepted since the 1880s.

Finally, a German glass instrument maker in Manchester produced a device which allowed Rutherford to separate the emanation from radium in a separate chamber, and submit it to spectroscopic analysis. This, the final and conclusive test, showed that the α-particle was undoubtedly a helium atom, whose mass and charge had already been calculated: with its high speed and large mass, it would then become the main tool that Rutherford would use to investigate the interior of the atom. The results would be breathtaking.

The atomic nucleus

In 1909 research continued, with its focus on the scattering of α-particles. This was a phenomenon that Geiger had observed in his brass tube: on impact with gas molecules, the particles were deflected in ways he could not explain. In particular, some deflections were far outside normal variations. A young research assistant, Ernest Marsden, having placed a thin sheet of gold foil in the apparatus, observed a number of α-particles, from the pencil beam focused upon it, bouncing back: these he was able to count. Rutherford, astounded, said that 'it was as though you had fired a fifteen-inch shell at a piece of tissue paper and it had bounced back and hit you.'[25] A similar effect had been noted with β-particles, but then they had less than a thousandth of the mass of an α-particle. The experiment, repeated with other metals, showed that the greater the atomic weight, the more often the effect occurred. With aluminium, the lightest of the metals used, it occurred with one of every 8000 particles. The effect occurred far too often to be explained statistically as the result of a 'multitude of small random deflections'.[26]

Rutherford then saw that the effect would be explained if the atom, whatever the metal in the foil barrier, was represented by a single point carrying a positive charge, so that a positively charged α-particle approaching it would be deflected on a hyperbolic path. This was elementary mathematics familiar to Rutherford, and the numerical relationship between the charge at the point and that of the particle gives results conforming with actual observation.

Rutherford saw what the experimental results implied:

> In order to explain these and other results, it is necessary to assume that the electrified particle is considered for a type of atom which consists of a central electrical charge concentrated at a point and surrounded by a uniform spherical distribution of opposite electricity equal in amount.[27]

This reported, in effect, the discovery of the atomic nucleus, although the word was only introduced for the first time in a later paper,[28] published in 1913. In particular, the results from Manchester suggested that the central charge of the atom was proportional to about half the atomic weight. Both this hypothesis and that of the nuclear atom succeeded not only because they fitted the experimental results but also because of their explanatory power. This extended to chemistry, where it elucidated the nature of the elements and the differences between them, and also to the physical phenomena of spectra, this latter providing the research focus for Rutherford's most gifted collaborator at Manchester, the Danish physicist Niels Bohr (1885–1962). The discovery of the atomic nucleus led Arthur Eddington to comment: 'In 1911 Rutherford introduced the greatest change in our idea of matter since the time of Democritus.'

There was no getting round the fact that Rutherford's atomic nucleus was positively charged. However, since the atom in its normal state is uncharged, the nucleus must be surrounded by orbiting electrons whose combined negative charge equals the positive charge on the nucleus. Quite simply, the two cancel each other out. But then the orbiting electrons, as they lose their energy by radiation, should collapse into the nucleus, in accordance with normal Newtonian principles. (It would be as if the moon slowed down, and as result spiralled towards the earth.) Obviously if this were to happen, the whole nuclear system would break down. The objection was a formidable one, and it was Bohr who showed how to overcome it. This achievement, and the results that followed from it, placed Bohr at the forefront of world physics, a position he never lost.

The Rutherford–Bohr atom

Bohr was not the only young physicist attracted to Manchester by Rutherford. Another, Harry Moseley, a refined and upper-class Englishman, who joined Rutherford from Oxford in 1910 (and shuddered at this colonial's lack of culture), also achieved remarkable results. Moseley's study of the alpha-line, the most prominent in any spectrum, revealed that its frequency always increased by the same amount for successive elements in the periodic table (which by 1910 was nearly complete). Bohr suggested that this corresponded to one unit charge of the atomic nucleus, so that the place of any element in the periodic table corresponded to the positive charge of its nucleus and therefore, in the neutral state, to the number of electrons in orbit around it. At the same time, Moseley, helped once again by spectroscopic analysis, confirmed this result by showing that it was the charge, not the mass, of an atom that determined the behaviour of the outer electrons. The result, now absolutely standard in both physics and chemistry, is that any element has two numbers, one, the so-called atomic number, defining the charge of its nucleus, and the other, the atomic weight,[29] its mass. To take a simple case, that of manganese, this can now be designated as $^{55}_{25}Mn$, to show that it is number 25 in the periodic table with atomic weight 55. In fact, $_{25}Mn$ is a pleonasm, the number and the letters necessarily imply each other.

Atomic weight still posed problems. The key problem, to explain the increments between successive elements in the periodic table, became more puzzling in cases such as cobalt ($_{27}Co$) followed by nickel ($_{28}Ni$) or tellurium ($_{52}Te$) followed by iodine ($_{53}I$), when there was no increase, but a decrease, in atomic weight. Here Soddy, who had been working with Rutherford since 1901, produced the right answer in 1911 – his last major scientific contribution. Quite simply, any element could occur in forms with different atomic weights, each separate case constituting a distinctive *isotope*. Moreover, nothing prevents isotopes from one element, measured according to their atomic weight, overlapping those of another. To take the relatively simple case of cobalt and nickel, the former occurs in nature with a single isotope, ^{59}Co, while the latter has five, ^{58}Ni, ^{60}Ni, ^{62}Ni, ^{61}Ni and ^{64}Ni, occurring, respectively, as 68%, 26%, 4%, 1% and 1% of any sample.

Chemically, there is nothing to distinguish the isotopes of a given element; physically their properties can, and do, vary widely. In particular, some natural isotopes are radioactive and, in certain cases, such as uranium, at the top end of the periodic table, all are. (One reason why uranium still occurs naturally is that the half-life of its most common isotope, ^{238}U, is of the order of a hundred million years.) What is more, experimental physics has developed artificial radioisotopes in numbers far exceeding those occurring in nature. (Indeed their exceptionally short half-lives, sometimes measured in fractions of a second, rule out natural occurrence, save as a very short phase in radioactive decay.)

The discovery of isotopes, combined with Rutherford's discoveries relating to α- and β-particles, opened the way to the displacement law which describes the effects of radioactivity. Since an α-particle is the most common isotope of helium (atomic number 2), ^{4}He, the emission of such a particle means that the atomic weight decreases by 4, and the atomic number by 2, so that the result is always an element two down in the periodic table. For example, the most common isotope of uranium (92), ^{238}U, on decaying by α-emission, becomes ^{234}Th, itself a much more unstable isotope of thorium (90), which means that the process of radioactive decay will continue almost instantaneously. In contrast, the loss of a β-particle (an electron) increases the atomic number by 1, while not affecting the atomic weight, so that for instance with carbon (6), the radioactive isotope ^{14}C (with a half-life of great interest to geologists and archaeologists) undergoes the transformation $^{14}C \rightarrow {}^{14}N$ – the most common (and non-radioactive) isotope of nitrogen (7). These, as Bohr realised, are all phenomena of the atomic nucleus.

If, for scientists such as Eddington, quoted above, Rutherford's discovery of the atomic nucleus was his greatest achievement, his success in splitting the atom, first reported in 1919, made many more headlines. The mathematics underlying his research had shown the enormous energy locked up in the nucleus and released by the expulsion of α-particles, but he did not immediately see any way of influencing radioactive changes. The way was opened in 1914 when Ernest Marsden explored experimentally the effect of α-particles hitting a hydrogen nucleus. A suitable target then produced 'H-particles' moving with 1.6 times the velocity of α-particles and with a range about 4 times greater – factors which gave a quite distinctive pattern in the detection apparatus. The problem was that the number of H-particles observed was far greater than expected: there was plainly some new factor at work, but Marsden failed to find it.

With hindsight, Marsden's experiment had already split the atom without his realising it. At first he thought that the unexplained H-particles came from the same source as the α-particles, but once Rutherford got on to the problem, in late 1917, his experiments ruled out this possibility: the source of the H-particles must be the air through which the α-particles passed. Now air is a mixture of oxygen and nitrogen, and Rutherford saw that if CO_2, a compound of carbon and oxygen, replaced the air, the results would show which of the three elements was the source of the H-particles. The new experiment showed that with CO_2 there were no H-particles, so nitrogen was their only possible source. Rutherford went on to devise a number of ingenious control experiments, changing both the metal compounds in the detection apparatus and the source of the α-particles from radium to polonium, but the results were the same. The α-particles were splitting the nitrogen atoms and knocking out hydrogen nuclei.[30] The effect was extremely rare, since, as Rutherford noted, 'for every one thousand million collisions . . . in only one case does the alpha-particle pass close enough to the nucleus to give rise to a swift H-atom'.

When Rutherford, in 1917, returned to the question first raised by Marsden's experiment, he was fully committed to war work for the international anti-submarine warfare committee. He made a classic apology for any possible neglect of this work: 'If, as I have reason to believe, I have disintegrated the nucleus of the atom, this is of greater significance than the war.'

The Second World War began within two years of Rutherford's death on 19 October 1937. The significance of his work in winning this war is one measure of how his achievements changed the course of history. Rutherford's success is best measured by an often-told story: on being asked, 'Do you always ride on the crest of a wave?' he replied, 'After all, I made the wave, didn't I?'

Christian, the first Bohr ancestor of the physicist Niels Bohr to be a Danish citizen, lived – like Hamlet – in Elsinore. His arrival there,[31] in 1776, marks the beginning of a very remarkable family. At first they were teachers, but Niels' father, also Christian, moved a step up the academic ladder, to become professor of physiology, in 1880, at the University of Copenhagen (where in 1907 he would be recommended for a Nobel prize). In 1881, he married Ellen Adler, who came from a wealthy Jewish family, active in public life. According to the official record of the marriage, any children born were to be brought up in the Mosaic faith, but when they came of age, they were, according to the register of births, without religious affiliation.

In practice the climate of Christian's household was tolerant and enlightened. Religion played a minor role, and although the children were baptised in the Lutheran Church[32] (in which their parents had been married) this meant little to them and Niels formally resigned from it in 1912.

In today's jargon, the household in which Niels, his sister Jenny and his brother Harald (later to become a distinguished mathematician) grew up, was supportive. (The support even extended to seeing Niels become a member of the Danish Olympic football team in 1904). Niels was born under a very lucky star. Never cynical, he was always affectionate, and in the vicissitudes of the twentieth century, in which he was involved as much anyone, he never lost his naive faith in the goodwill of his fellow men.

Inevitably Bohr, graduating from high school just before his eighteenth birthday, became a student at the University of Copenhagen in 1903, to see his father become Rector two years later. His distinction soon proved to be on the same level, and in 1907 the Royal Danish Academy of Sciences awarded him a gold medal for a prize essay. A young man in a hurry, he set his sights high and, after gaining his doctorate in 1911, set sail for England to work with J. J. Thomson at the Cavendish Laboratory.

This was not a good choice. Bohr did not start off well when, on first meeting the great man, he said, referring to a page in a book of JJ's which he had in his hand, 'This is wrong.' Bohr was in fact right, but at this stage in his life, JJ was not a man to allow himself to be corrected. He was always very friendly and charming, but not ready to be interested in the work of a young man like Bohr, who suffered also from a poor command of English. Bohr hung on for a few months, attending a few lectures and playing occasional football, but according to his final judgement, 'the whole thing was very interesting in Cambridge but it was absolutely useless'.[33]

If JJ lost the chance of being the patron of perhaps the most gifted man he would ever meet, the loss was entirely his. Bohr knew where he must go if he was not at home in Cambridge: the next stop was the physics laboratory at Manchester University. There Rutherford proved to be the father figure Bohr was looking for, and the collaboration between the two men was astonishingly productive.

Bohr arrived in Manchester in 1912. The previous year Rutherford had shown that the atom must have a positively charged nucleus surrounded by orbiting electrons. The problem was to find out how these orbits were determined, given that at some level the atomic system had to be stable. Bohr was immediately interested, but achieved little before returning to Copenhagen on 24 July, to marry, a week later, Margrethe Nørlund, after a two-year engagement. This was the beginning of a marriage that would last for more than fifty years. Once again, we see a man with an exceptional gift for human relationships.

In Copenhagen Bohr had hoped for a professorship, but he had to be content to work at a lower level, as *privatdocent* and *assistent*. The research begun in Manchester with Rutherford continued, and Bohr saw that the answer to the problem of the electrons must come from looking at spectra, particularly that of hydrogen. In the 1860s four lines in the spectrum had been observed, and their frequencies measured with extreme accuracy by the Swedish physicist Anders Ångström (1814–74). Then, in 1885, Johann Balmer (1825–98), a teacher at a girls' school in Basle, discovered an elementary mathematical formula,

$$\nu_{ab} = R\left(\frac{1}{a^2} - \frac{1}{b^2}\right),$$

that not only fitted Ångström's results but remained true as new lines in the hydrogen spectrum were discovered. In this formula a and b can be any whole numbers, as long as a is always greater than b. This gives an unlimited range of possible frequencies ν_{ab} corresponding to different values. Balmer discovered that, given the right value of the constant R,[34] the frequencies of the first four lines discovered by Ångström are given by $b = 2$ and $a = 3, 4, 5, 6$. All lines in the hydrogen spectrum revealed by later research proved, without exception, to correspond to whole-number values for a and b.

Bohr came across Balmer's formula in February 1913 – perhaps not for the first time – and by early March he had written a paper interpreting it.[35] This, the beginning of the quantum theory of atomic structure, was a landmark in the history of science. It established, for the particular case of hydrogen, the 'Rutherford–Bohr' atom, in which the single electron in the atom can move only on one or another of a discrete set of orbits, so that if its orbit changes – a very frequent occurrence – there must be, in today's jargon, a quantum leap from one orbit to the other.[36] On this analysis there must be an innermost orbit, defining the smallest possible distance between the electron and the nucleus, which relative to the size of either is very large indeed. This 'ground state', according to Bohr, had to be stable and represented the hydrogen atom at its lowest possible energy level.

All other states must then be unstable, so that the electron in any but the innermost orbit (that of the ground state) will always tend to move to an inner orbit, and in doing so will emit one light quantum corresponding to its loss of energy. Any possible orbit a corresponds to an energy state E_a, so that any transition from a to b will release one quantum of light with frequency ν_{ab} according to the formula $E_a - E_b = h\nu_{ab}$, in which h is Planck's constant (introduced on page 272). Such a transition is the only way a hydrogen atom emits (or absorbs) radiation. If all this sounds very abstruse, the theory still allows for elementary experimental confirmation, simply by activating a gas discharge tube containing hydrogen to produce light. A microscopic examination of the spectrum will then reveal the lines, whose position will conform with Balmer's formula. This could all be done in a school laboratory.

Hydrogen is, however, the simplest case. Bohr's theory also required ionised helium to have certain spectral lines, which were confirmed by experiment. Continuing up the periodic table the principle still holds good, but simple formulae, such as Balmer's, no longer govern its application.

The year 1913 was an extremely productive one for Bohr. His work with hydrogen and helium represents a decisive turning point in the history of physics, particularly when it is coupled with his demonstration that β-rays (consisting of nothing but electrons) originate in the atomic nucleus, while the outermost ring of electrons determines the atom's chemical properties.

In 1914, Bohr, still denied a professorship at home, moved back to Manchester as a lecturer. By this time war had come to the United Kingdom, Rutherford was away from Manchester and his young colleagues were serving on both sides of the front line. Geiger was wounded while fighting for Germany; a Turkish bullet killed Moseley at Gallipoli – probably the greatest loss to science in the First World War. Fortunately for Bohr, Copenhagen finally appointed him professor, and he started there in May 1916: later in the year Christian, the first of his six sons, was born. In neutral Denmark, Bohr was able to get on with his researches: the pace hardly slowed.

By the time he took up his professorship, Bohr's own results had forced him to confront quantum physics. This subjected him to great stress, because, as he himself wrote, 'all the time I am gradually changing my views about this terrible riddle which quantum theory is'.[37] At the heart of the riddle was, in the words of a colleague, 'the abyss, whose depth Bohr never ceased to emphasise, between the quantum theoretical mode of description and that of classical physics'.[38] In classical physics, a unique cause leads to a unique effect: quantum physics deals only in probabilities, a stumbling-block that Einstein would never overcome.

While Bohr was agonising about quantum theory, he was working hard to establish an institute for theoretical physics in Copenhagen. The Institute, officially opened in 1921, immediately became a centre to which physicists came from all over the world. In 1920, Bohr had already met two great contemporaries in Berlin, Albert Einstein and Max Planck: later in the year Einstein, and Rutherford also, both made their first visits to Copenhagen. In 1924, a young German physicist, Werner Heisenberg, made a first short visit, to return later, for two much longer periods of close collaboration in the period 1924–27. This was the beginning of an intimate relationship, which ended in 1941, in the heart-breaking circumstances portrayed in Michael Frayn's play *Copenhagen*[39] (and related later in this chapter).

In a lecture to a scientific audience in February 1920, Bohr told how 'through the recent developments in physics . . . a connection between physics and chemistry has been created which does not correspond to anything conceived of before'.[40] He was right: the advances made in the twenty-odd years from 1916 were revolutionary.

In 1916, the German physicist Walther Kossel (1888–1956) established the first successful links between the Rutherford–Bohr atom and the periodic table of elements. Kossel, noting the number 8 as the difference between the atomic numbers, 2, 10, 18, of the first three noble gases, helium, neon, argon, concluded that the electrons in such atoms orbited in 'closed shells' – the first containing only two, and the second and third, eight each. This shell model is now standard in school chemistry textbooks.[41] Its implications are far-reaching, and if the following paragraphs look complicated the mathematics never goes beyond elementary whole-number arithmetic as taught in primary school. (Any reader who made it that far has no excuse for skipping the next page or two, although it must be admitted that it took some of the ablest twentieth-century scientists years to unravel what the numbers meant. It would help to have a bookmark at the periodic table in Appendix A).

Kossel's configuration is highly stable, but he also noted how this was far from true if one was subtracted or added to the atomic numbers, to give, respectively, the series 1, 9, 17 and 3, 11, 19. The former begins with hydrogen, to continue with the first two halogens, fluorine and chlorine, while the latter contains the first three alkali elements, lithium, sodium, potassium – all, as shown in Chapter 6, highly reactive. What was important for Kossel was that the halogens easily picked up an electron, to turn into negative ions, while the alkalis easily lost one, to turn into positive ions.

Applying an elementary process of induction, the two previous paragraphs suggest that, while the first shell contains a maximum of two electrons, the number is eight for all subsequent shells. The periodic table immediately shows that this is not so: looking at the final elements in periods 3, 4 and 5 (see Appendix A), one sees that 18 elements separate argon (18) from krypton (36), and the same number, krypton from xenon (54). What then is the explanation?

At this stage a young Swiss physicist, Wolfgang Pauli (1900–58), joins the magic circle around Bohr. In 1924, he established an 'exclusion' principle, determining the maximum number of free electrons that can orbit in any of the shells of an atom. This is based on four 'quantum' numbers: the first three of these, n, k and m, define the configuration of an atom. Here n is the number of shells, which is never more than 7, while k, never greater than n, is the number assigned to each possible state within a shell, so that n_k, designates the kth state in the nth shell. The number of possible n_k states is m ($= 2k - 1$). The fourth quantum number, only discovered by Pauli in 1925, is either 1 or 2, and takes into account that any orbiting electron may have a complement. The exclusion principle can now be stated in Pauli's own words:

> In the atom there can never be two or more equivalent electrons for which . . . the values of all four quantum numbers coincide. If there is an electron in the atom for which these quantum numbers have definite values then the state is occupied.[42]

Working out how the quantum numbers apply in practice is like one of those induction puzzles in which, in a pattern of numbers, letters, shapes, or whatever, blank spaces have to be filled in.[43] In the orbiting electron puzzle, the going is easy for the first two shells. Obviously, if $n = 1$, then $k = 1$, and derivatively $m = 1$, so there is only one possible state, 1_1, which, with the fourth quantum number, allows for 2 possible elements: these are hydrogen and helium, numbers 1 and 2 in the periodic table.

Applying the same rule to the second shell gives one 2_1, and three 2_2 states, together making 4, which multiplied by 2 gives 8. These are all filled by elements 3 to 10 in the periodic table. Going on to the third shell, the same elementary arithmetic gives 8 as the total for the 3_1 and 3_2 states, but Pauli then allows for 10 $(2 \times (2 \times 3 - 1))$ 3_3 states. This may be, but there are no 3_3 states in the third shell, so that instead of 18 states[44] (the maximum allowed by Pauli) there are only 8, no more than in the second period. The noble gas, argon, closing the third period with only two states, has therefore 2, 8 and 8 electrons in its three shells, making in all 18, its atomic number. Potassium (19), next in the periodic scale, has four shells, with only one electron in the outer shell, giving the distribution 2, 8, 8, 1, (instead of 2, 8, 9 in three shells) and calcium (20) follows on with 2, 8, 8, 2 (instead of 2, 8, 10). In fact *k* never goes above 4 at any point in the periodic table, although at the top end there are any number of radioactive elements with 7 shells. Beyond argon (18), therefore, opening a new outer shell before all the inner shells are complete, is simply part of the system: Bohr and his colleagues in the early 1920s took some time to see what this fact implied.

The application of Pauli makes possible a simple code to designate the electron orbits of any atom. Now for esoteric reasons familiar to chemists, the 4 *k*-numbers actually occur define four blocks, *s*, *p*, *d*, *f*, corresponding to 1, 2, 3, 4, so that, for example, the three 3-states, 3_1, 3_2 and 3_3 in the preceding paragraph are known as $3s$, $3p$ and $3d$ ($3f$ is excluded by the rule $n > k$, but $4f$ and $5f$ can and do occur). This system allows the electron orbits of any atoms to be coded in a way which is extremely useful to the physical chemist, revealing both the physical properties of the atom (such as its power to conduct electricity), and its chemical properties (such as its reactivity with other elements). The codes can be very complex, such as $3d^{10}4s^2 4p^6$ for krypton (36), but then the last two terms, $4s^2 4p^6$, designate the noble gas that closes period 4, just as $5s^2 5p^6$ would designate that which closes period 5 (which happens to be xenon (54)). They also indicate that the outer shell is full, so that adding just one more electron must start a new shell, with its single orbiting electron in the *s*-state: this is reflected in the codes 5*s* for rubidium (37) and 6*s* for caesium (55), which, although very simple, tell all that needs to be known. In fact the codes 2*s* (lithium), 3*s* (sodium), and so on, are those of the highly reactive alkali metals (which used to cause so many explosions in early nineteenth-century laboratories – remember what happened when Sir Humphry Davy discovered sodium).

The position reached with calcium (20) is that of an atom with two electrons in its fourth and outer shell, but with no *d* electrons in the third shell. This deficiency is now made good, so that the next element, scandium (21), has one 3*d* electron, titanium (22), 2, and so up to zinc[45] (30), which has 10, as shown by its code, $3d^{10}4s^2$ (where $4s^2$ indicates the two *s*-electrons in the outer fourth shell). After zinc, in period 4, the number of 3*d* electrons is unchanged, with new 4*p* electrons being added to the outer shell, to end with the noble gas krypton (36) which has 8 – once again as shown by its code $3d^{10}4s^2 4p^6$ (noting that $2 + 6 = 8$).

This process is repeated in periods 5 and 6: a new outer shell is created with 2 *s*-electrons, and then the next inner shell begins to fill up with *d*-electrons, again up to a maximum of 10. When this process is complete *p*-electrons are added, one by one up to 6 (the maximum number), to the outer shell, which then closes with a noble gas – respectively xenon (54) (code $4d^{10}5s^2 5p^6$) in period 5 and radon (86) (code $4f^{14}5d^{10}6s^2 6p^6$) in period 6.

Looking back to the periodic table, one can see that all the elements in columns 1 and 2 have in common a new outer shell created by adding, respectively, 1 and 2 *s*-electrons. Then columns 3 to 12 are defined by having 1, 2, . . . , 10 *d*-electrons added, not to the outer shell but to the one next inside. Finally columns 13 to 18 are defined by having 1, 2, . . . , 6 *p*-electrons added to the outer shell. In this way, the table has clearly defined *s*-, *d*- and *p*-blocks. Furthermore, the *d*-block plainly interrupts the process of adding electrons to the outer shell, which stops with group 2 and is resumed with group 13. For this reason the *d*-block contains the so-called 'transition elements': the fact that these are all metals follows from the electron configuration, a result of fundamental importance.

The end of the last paragraph but one discloses a significant anomaly, which greatly concerned Bohr and the men around him. The atomic number of radon (86) is not found by adding 18 to that of xenon (54) (immediately above in group 18), but by adding 32: the question is, what are the extra 14 elements? The answer lies in that part of period 6 in block *f* of the periodic table. Of the 14 elements in this block, only two, cerium (58) and erbium (68), were known in 1869, when Mendeleyev established the periodic table. These are both rare earths, elements so difficult to discover and identify in nature that they are known as 'lanthanides' from the Greek for 'to escape detection'. Although beginning with lanthanum (57), the first of the series, all members have six shells, the two outer shells, numbered 5 and 6, together remaining substantially unchanged throughout. Since it is mainly the outer shells that determine the chemical character of an element, it is no wonder, then, that the rare earths remain hidden: they are, chemically speaking, a collection of look-alikes.

Bohr then showed how the building up of the series, after lanthanum, could be explained by starting the fourth state *f* in the fourth shell, leaving the two outer shells unchanged. Thus, cerium (58), which opens the series, would have 1 *f*-electron, praseodymium (59) 2, and so on up to lutetium (71) with 14. (Once more there is a small glitch, so that cerium in fact opens with 2 *f*-electrons, while ytterbium (70) also has 14. This corresponds to the loss of a 5*d* electron with cerium, which is regained by lutetium.)

The same principle extends to a second *f*-series, that of the so-called 'actinides', all with 7 shells, but with the 5*f* level being filled up across the series, from protactinium (91), with 2 5*f*-electrons, to lawrencium (103) with 14. The fact that the 5*f* and 6*d* levels are close in energy means that the process is far from smooth, and in any case all the actinides are radioactive, with the nine elements after plutonium (94) not occurring in nature.

The voyage of discovery narrated above was far from plain-sailing. There were any number of shoals to navigate. In 1922, when Bohr was ready to present his whole scheme in a lecture at the university of Göttingen, a French chemist named Georges Urbain claimed, after years of intensive chemical analysis, to have identified element 72 as a rare earth, which he named 'celtium'. Rutherford was persuaded, but Bohr saw that if Urbain was right then the whole build-up to 14 *f*-electrons must fail. If Bohr was right, then this element was a *d*-block metal, analogous to zirconium (40), so that both would be in group 4 of the periodic table.

This was demonstrated by two colleagues in Copenhagen subjecting zirconium-rich mineral samples, borrowed from a museum, to analysis by X-ray crystallography – a process (described in Chapter 6) analogous to spectroscopy. This disclosed a hitherto unknown metal fitting into slot 71 in the periodic table. After some confusion this acquired the name, 'hafnium', after *Hafniae*, the Roman name for Copenhagen. Moreover, hafnium – a thousand times more common than gold – proved to be not all that rare.

Bohr also won the Nobel prize for physics in 1922, the same year as Albert Einstein won the deferred 1921 prize. Bohr ended the obligatory Nobel lecture with his views on the periodic table, citing the recent discovery of hafnium: 'The theory is at a very preliminary stage and many fundamental questions will await solution.'[46]

The Nobel prize was a watershed in Bohr's life. Thereafter honours and appointments offered by different universities followed thick and fast. Bohr, although always willing to go abroad for short periods, was only ever to abandon his institute in Copenhagen when the fortunes of war, in 1943, made this inevitable. Not surprisingly, so long as peace allowed, many eminent scientists visited Copenhagen.

Niels Bohr in the Second World War

The story of Bohr in the Second World War is remarkable and is linked to one particular person, Werner Heisenberg (1907–76). Heisenberg (whose uncertainty principle is discussed on page 278) heard Bohr lecture in Göttingen in 1922, and as a result of making an objection was invited to a three-hour walk on the Hainberg, a local mountain. Following this walk, Bohr told his friends, 'he understands everything'. Heisenberg, then only twenty-one, was invited to Copenhagen, where he was to work on and off from September 1924 to June 1927, when he was appointed full professor at Leipzig University. Even before Heisenberg had arrived, Bohr had said, 'Now everything is in Heisenberg's hands – to find a way out of the difficulties.'[47]

Bohr's words were prophetic, but in 1924 neither he nor Heisenberg could foresee the insupportable strain in their relationship – almost that of father and son – that would follow as a result of the rise of Nazi Germany. When Adolf Hitler came to power in Germany in January 1933, the position of Jews in public life was threatened immediately, although no-one then foresaw the horrors of the Holocaust. As far back as 1922, an embittered professor, Philipp Lenard (1862–1947), had propounded 'Deutsche Physik'; at the same time Einstein, denounced for his 'Jüdische Physik', was unable to give a lecture at Leipzig.[48] Many German scientists, at every level, were Jewish, and from the beginning of the Nazi era colleagues outside Germany, including such eminent men as Rutherford, organised to help them. The result was an immense loss of scientific talent to Germany, matched by an equal gain in countries such as England, France, Sweden and the United States – where Einstein found a new home in 1933. Jewish refugees from Germany, Austria and, after 1940, occupied Europe are among the greatest names in mid-century science. Heisenberg, who was not Jewish, remained in Germany, and when the Second World War started in 1939, he was still at Leipzig. Then in 1941 he became professor in Berlin, and Director of the Kaiser Wilhelm Research Institute. By this time he was already conducting small-scale experiments focused on building a nuclear reactor.[49] Like any physicist of his day, he was conscious of the destructive potential of the atomic nucleus, and saw that the best means of releasing this power would be a bomb based on uranium. Given his status among German scientists, Heisenberg, in the course of 1941, if not earlier, had begun to realise that he himself would be involved in any construction programme. And as Heisenberg noted after the end of the war, 'From September 1941, we saw in front of us an open road to the atomic bomb.'[50]

Germany invaded Denmark on 9 April 1940, and within one day had occupied the whole country. Denmark, with a Nordic culture acceptable to Hitler, was then allowed a degree of autonomy not granted to any other country in occupied Europe. This did not, as Germany had hoped, make the occupation any more popular with the Danes. Bohr, in particular, spurned all contact with German officials, and Heisenberg, knowing that Bohr was half Jewish and head of an institute employing many Jews, recognised immediately that he was in some danger and did what he could to help. A contact with Cecil von Renthe-Fink, the German plenipotentiary in Denmark, was useful, and as long as he remained in office, that is, until November 1942, Bohr's institute was left alone as were the 8000-odd Danish Jews. Against this background, Heisenberg decided to visit Bohr in Copenhagen in September 1941, although there had been not a single contact between the two since April 1940.[51] The meeting proved to be one of the most enigmatic events of the twentieth century. No records were kept of what was said, so the dramatic reconstruction of the meeting between Bohr (and his wife) and Heisenberg in Michael Frayn's *Copenhagen* could be remarkably close to the truth. In any case, the timing was significant. The German invasion of Russia in June 1941 had radically changed the direction of the war, and although the United States was still neutral in September 1941, its nuclear weapons programme was already beginning to take shape. In 1942 the decision was taken to base the scientific research in Los Alamos, New Mexico, where, by the end of 1943, the programme would be fully operational.

Niels Bohr left Copenhagen in the final months of 1943, with Los Alamos as his ultimate destination. The story begins in January 1943, when James Chadwick, an old friend and colleague in England, helped by the Danish underground, got a letter through to Bohr, inviting him to come to England. The wording of the letter was not transparent; its key was in two sentences:[52]

> Indeed I have in mind a particular problem in which your assistance would be of the greatest help . . . You will . . . appreciate that I cannot be specific in my reference to this work, but I am sure it will interest you.

Bohr's reply left little doubt about his own feelings:

> I have to the best of my judgment convinced myself that in spite of all future prospects any immediate use of the latest marvellous discoveries of atomic physics is impracticable.

This proved to be a gigantic miscalculation, but in any case Bohr remained in Denmark – at least for the time being. Times, however, would change for the worse before the year was out. In mid-September Bohr learnt that the Germans were about to round up all refugees, and it soon became clear that he was also in danger. On 29 September the underground organised his escape by boat to Sweden across the Ålesund (where a bridge has now been built), and the next day he was on his way to Stockholm. The British, hearing of his arrival, immediately arranged a secret flight to Britain. He arrived in Scotland early on 6 October, followed a few days later by his son Aage; his wife and other members of the family remained in Sweden. During his one-week stay in Sweden, Bohr also arranged for asylum to be offered to Denmark's Jewish population, almost all of whom escaped.

Bohr did not stay long in England: he and Aage arrived in New York by sea on 6 December. By this time he had been fully briefed about the work being done at Los Alamos, and he was amazed. He himself arrived there before the end of the year, to become part of a programme in which many of the world's most distinguished scientists collaborated on the design and manufacture of weapons of unprecedented destructive power. Even so, as he was to say after the war, 'They did not need my help in making the atomic bomb.'[53] (This is not entirely true: he made an important contribution to the design of the initiator of the plutonium bomb, the neutron trigger essential for starting the chain reaction.)

Once in America, Bohr was immediately concerned about the political consequences of atomic weapons. He was well connected, not only in scientific circles, and believed that if he could meet Churchill and Roosevelt he would persuade them of the need for international cooperation in the field of atomic weapons. His first meeting with Churchill, on 16 May 1944, was a disaster. Lord Cherwell, Churchill's scientific adviser, was also present, and for much of the time the two were arguing with each other. Bohr, whose English was always hard to follow, got nowhere. As he told a friend the same day, 'It was terrible. He scolded us like two schoolboys.'[54] The moment was inopportune. Churchill was preoccupied with planning the invasion of France, and at that stage in the development of the atomic bomb there was nothing concrete to demonstrate its incredible destructive power. The first (and only) test came more than a year later.

A meeting with Roosevelt on 26 August went much better: Roosevelt later told a friend that he found Bohr one of the most interesting men he had ever met. This did not help his cause. On 18 September, Churchill and Roosevelt met to discuss Anglo-American policy relating to atomic weapons. They made two key decisions: first the whole programme should continue to be kept secret; second, the collaboration between the two countries would continue after the end of the war. (The British, to this day, still have privileged access to American nuclear installations, including the Nevada Test Site for nuclear weapons.)

The two world leaders also agreed that 'enquiries should be made regarding the activities of Professor Bohr and steps should be taken that he is responsible for no leakage of information, particularly to the Russians'. Behind this anxiety (which was mainly Churchill's) was a letter from Peter Kapitsa in the Soviet Union, which Bohr had collected from the Soviet Embassy in London on 20 April 1944. The letter invited Bohr to visit the Soviet Union: Bohr's reply was non-committal and was vetted by the British Secret Service before being delivered to the Embassy. It did, however, contain the words, 'I am hoping that I shall soon be able to accept your kind invitation.'

The Soviet Union, at that time, was an ally fighting hard to defeat Nazi Germany. Bohr himself still had every reason to mistrust its government. A friend and colleague, Lev Landau, had been imprisoned in 1938, and Bohr's appeal to Stalin did not help him. He knew well that the openness he looked for in the field of atomic weapons was not to be found within the Soviet Union. Indeed the Soviet programme after the end of the war was kept just as secret as the Anglo-American programme. On the other hand Bohr foresaw that the development of atomic weapons could lead to a terrifying new arms race, on a scale unprecedented in history. On this point he proved to be right, as Churchill himself lived to see.

Kapitsa, until his definitive return to the Soviet Union in 1934, had been a colleague of Rutherford at the Cavendish Laboratory, where his work was extremely productive. He was part of the international world of science where Bohr was a prominent figure. To Bohr he was a trusted colleague. It is not surprising then that Bohr saw some promise in a meeting between the two. His world was not Churchill's, and in the circumstances of war in 1944 scientists were subordinate to politicians. The day after the September meeting with Roosevelt, Churchill, worried about Bohr, asked, 'How did he come into this business?'[55]

This is still a good question. When Bohr left Denmark in September 1943, the Manhattan Project was already under way, and the team of scientists working at Los Alamos could hardly have been stronger. What then did Bohr have to offer? The answer – already related in his own words – is not all that much. None the less, Bohr's arrival in Sweden was seen as the escape of one of the world's most distinguished nuclear scientists out of the hands of the Germans. This explains the immediate interest of both the United States and the Soviet Union. (It could also explain Heisenberg's visit to Copenhagen in 1941.) The importance of Bohr is reflected in a report in the *New York Times* of 8 October 1943:[56]

> Dr. Niels H. D. Bohr, refugee Danish scientist and a Nobel prize-winner for atomic research, reached London from Sweden today bearing what a Dane in Stockholm said were plans for a new invention involving atomic explosions . . . The plans were described as of the greatest importance to the Allied war effort.

After that a news embargo kept Bohr's name out of the press. He remained in America until June 1945. In July the atomic bomb was successfully tested at Alamogordo, New Mexico, and in early August two Japanese cities were destroyed, in the first and only operational use of this weapon. Bohr returned to Denmark on 25 August, and, on 7 October, his sixtieth birthday, Copenhagen students marched in a torchlight parade to serenade the Bohr family at Carlsberg, their home. Bohr lived another seventeen years, devoting his time to lectures in almost every corner of the world on the causes closest to his heart – particularly that of free access to the work done by scientists. The tributes that followed his death, on 18 November 1962, reveal a man with an astonishing capacity to inspire both love and admiration. Appropriately, given Bohr's family connections with Elsinore, there was always something of Hamlet about him: he was a man who never quite knew how to come to terms with the world or with the men who counted in it. In the world of science he knew Rutherford, Bragg, Einstein, Pauli, Heisenberg, Kapitsa and Oppenheimer, and outside it, Churchill, Roosevelt, Nehru and Weitzmann, and yet there was always part of his message that failed to get across.

The neutron: a missing link

The discovery of the neutron by James Chadwick (1891–1974) was a major landmark in twentieth-century science. Like so many physicists of his time Chadwick had learnt his trade from Rutherford, first at Manchester and then in Cambridge. His early life was bleak. Bollington, the town where he was born, is in England's Peak District. As in many places in the region, the local economy was based on cotton-spinning. When Chadwick was four, his father, who worked in the industry, threatened by local unemployment, moved to nearby Manchester. The boy was for many years left with his grandparents, but he finished school in Manchester. There, one of the masters, recognising his ability in mathematics and physics, entered him for two university scholarships, both of which he won. The die was cast for a man from a town whose people were later described as 'independent . . . close-knit, dour, inbred, different'.[57]

Chadwick, who had enjoyed few loving relationships in his childhood, came to Manchester with many of these qualities, although later he was to outgrow them. He was, however, intelligent, loyal, hard working and very well mannered. He was also an atheist during his whole life. In his second year at Manchester, he heard Rutherford lecture on electricity and magnetism, and, in his third year, aged only nineteen, he began to research on his own. One thing led to another, and in 1913, with a first-class degree and an M.Sc. behind him, to say nothing of five research publications on radioactivity, the 22-year-old Chadwick was awarded an 1851 Science Research Exhibition.

There was, however, one condition: the research must not be carried out in Manchester. The choice fell on Hans Geiger's laboratory in Berlin: until 1912 Geiger had worked with Rutherford in Manchester, and was more than ready to welcome Chadwick. In Berlin, he chose to research the behaviour of β-particles, first discovered by Rutherford but much less well understood than α-particles. The research went well, but within a year war began to threaten. Geiger, who was called up as a reservist, not only advised Chadwick to return to England but also lent him 200 marks for the journey. Chadwick, as unworldly as ever, failed to leave on time. For three months he was able to continue working in the Berlin laboratory, but on Friday 6 November 1914 he was arrested as an enemy alien and interned for the rest of the war.

Prison consisted of converted racing stables, where in Barrack 10 Chadwick found himself part of an elite intellectual group, enjoying also the outside company of an earl, musicians and painters, jockeys and racehorse owners – a broadening of his social horizons that he could well do with. And although life, to begin with, was particularly hard, Chadwick, by the end of his time, had been able to organise a small laboratory and even correspond occasionally with Geiger and Rutherford. What is more, unlike many of his generation he survived the war, but when he returned home in 1918 there was no hero's welcome awaiting him. Dutifully, before the end of the year, he submitted to the Commissioners of the 1851 Exhibition a report of the research done in Germany, mostly in captivity.

Once back in Manchester Chadwick resumed his work with Rutherford, and when in 1919 Rutherford succeeded J. J. Thomson as Cavendish Professor at Cambridge, Chadwick, helped by a studentship at Caius College, was able to continue working with him. He went on to be appointed the first Assistant Director of Research at the Cavendish in 1924, a tenured position, which he held until appointed Professor at Liverpool University in 1935, the same year as he was to win the Nobel prize.

In 1925 Chadwick also had the good fortune to marry Aileen Stewart-Brown, the daughter of a prosperous Liverpool family, whom he had met while she was staying with friends in Cambridge. Peter Kapitsa, then a Cavendish colleague, would be best man.[58] The marriage, which was extremely happy, lasted until Chadwick's death, nearly fifty years later. Chadwick used the occasion to leave his early life behind him, so that none of his relatives were invited. This seems particularly hard on his mother, who had done her best, however modest, to support him while he was a prisoner in Germany. Later, when her son was appointed to the Liverpool chair, she presented a piece of embroidered lace to Rutherford, accompanied by a memorable letter:[59]

> Dear Lord,
>
> I am sending a small present in appreciation for the kindness you have given to our son during the time he has been under your supervision both at Manchester and Cambridge universities.
>
> We are proud that our son has had such a distingwished a gentleman to help him.
>
> I hope you will accept this small token as it is a piece of my work
>
> Yours gratefully,
>
> A. M. Chadwick

In the early post-war years, the Cavendish Laboratory produced many new advances in atomic physics. F. W. Aston's new mass spectrograph made it possible to show that many elements contained different isotopes, so that, for example, some three-quarters of all natural chlorine consists of chlorine-35 and one-quarter of chlorine-37.[60] But what was it, in the atomic nucleus, that accounted for the difference between the atomic number, 17, and the numbers of the isotopes, 35 and 37? Aston's results did not provide an answer.

The key to Chadwick's work was the structure of the α-particle, and his results, according to Rutherford, strongly supported 'the identity of the atomic number with the nuclear charge'.[61] This principle, now long taken for granted even in school science, was then still uncertain, simply because of its implications for the structure of the atomic nucleus. In June 1920, Rutherford, in his Bakerian lecture at the Royal Society, had suggested a 'neutral doublet', within the nucleus, formed by a proton[62] combined with an electron. He then added, prophetically, that 'its external field would be practically zero, except very close to the nucleus, and in consequence it should be able to move freely through matter. Its presence would probably be difficult to detect by the spectroscope, and it may be impossible to contain it in a sealed vessel.'

At the same time, Rutherford had suggested that, at close contact between nuclei (within a distance of 3×10^{-13} centimetres), enormous forces, much greater than those of gravity, operated. This foreshadowed the *strong interaction*, now fundamental in particle physics and accepted as the reason for the great stability of the atomic nucleus. Then, also, the neutron (another term coined by Rutherford) would 'greatly simplify our ideas as to how the nuclei of heavy elements are built up'.[63] Chadwick, in the course of long conversations with Rutherford – often in the darkness of the Radium Room of the Cavendish, while waiting to start counting scintillations caused by α-particles hitting their target – became convinced that neutrons must exist.

Neutrons became something of an obsession with Chadwick, which gripped him even when the interest of others, including Rutherford, waned. In 1928 Geiger offered Rutherford the much improved Geiger counter, which although designed for cosmic ray research could be adapted to α- and β-particles. By this stage the Cavendish experimenters were looking for something better than sitting in a dark room counting the sparks caused when such a particle hit its target.[64] The work could last for hours, but the whole significance of an experiment could turn on the observed frequency of such scintillations. Within a year the new Geiger counter, as adapted by the Cavendish, could register such events by a black mark on a moving paper strip – such as in an electrocardiograph. With the help of radio amplification, it could register the ionisation caused by a single α-particle in an ionisation chamber, as well as recording up to 500 such events per minute. Another model applied the same methods to counting protons.

The ionisation chamber itself, the core element of the Geiger counter, is a gas-filled chamber with positive and negative electrodes with an electric potential between them. Ionising radiation, such as that caused by particles, causes the gas atoms to split into positive ions and electrons, the former going to the cathode and the latter to the anode, so that a current, proportional to the intensity of the radiation, then flows. A single ionising event (e.g. from one α- or β-particle), however transient, can still be recorded. In Geiger counters for everyday use, the record is an audible bleep, whose frequency registers the intensity of any radiation present. The dosimeter, worn on nuclear sites, and about the size of a mobile telephone, is a modern black-box form of this device.

From as early as 1928 Chadwick had begun to base his strategy on the disintegration of beryllium atoms by α-particles. Behind this was a simple arithmetical rationale. Beryllium, in nature, has an atomic number of 4 and an atomic mass of 9. The corresponding numbers for α-particles, as helium nuclei, are 2 and 4. If, therefore, beryllium could be caused to disintegrate, then the result could be two α-particles and one uncharged particle.[65]

The way to disintegrate beryllium was to bombard it with α-particles. To begin with, this was a problem at the Cavendish since radium, its only useful radioactive source, produces both α- and β-particles: the alternative was polonium, which produces only α-particles, but the Cavendish's stock was minute, until, via the scientific grapevine, a steady supply became available from an American hospital.

This factor provided two German scientists, Walther Bothe (1891 1957) and Herbert Becker, who had an adequate stock of polonium, with the chance to get in ahead of Chadwick. Bothe and Becker were mainly interested in the emission of γ-rays as the result of bombardment by α-particles: at least when replicated at the Cavendish, the results obtained from a number of elements showed that two of them, lithium and beryllium, did not yield protons. The radiation energy of beryllium was particularly high. Chadwick, helped by the American polonium, had the German experiment repeated and found that, by a knock-on effect, particles were emitted from the beryllium target in the same direction as the α-particles bombarding it. What is more, these new particles were able to penetrate lead to a remarkable degree.

Chadwick was plainly on the right track. Having first received help from Germany, he was then to get it from France. Marie Curie's daughter, Irène, and her husband, Frédéric Joliot, experts in working with polonium, having confirmed the German results, then passed the radiation emitted from the beryllium through paraffin wax, to find its intensity on being tested in the ionisation chamber significantly increased. This they attributed to protons being released as the result of γ-radiation, but if this was correct then this took place at a level of energy – 4.5 million electronvolts – never previously recorded.[66] The Joliot-Curies published their results; Chadwick, although satisfied that their observations were correct, still could not accept the conclusions drawn from them. As with the German experiments, the results were best explained on the basis that the protons recorded in the ionisation chamber were the result of the paraffin wax being bombarded by neutrons.

This at least was Chadwick's belief; his problem was to prove it

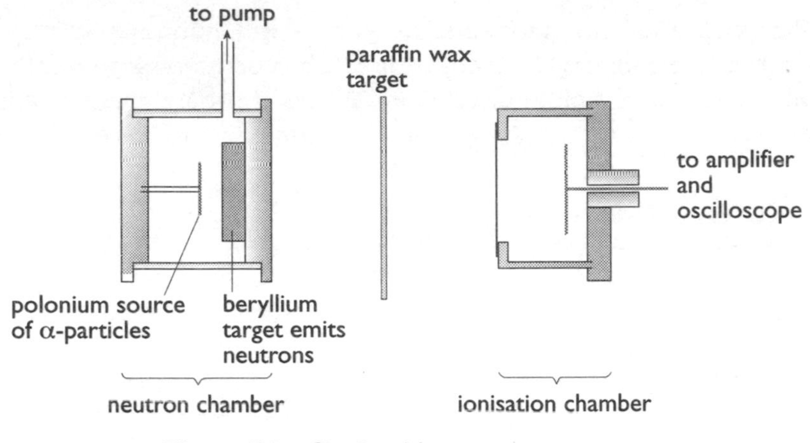

Figure 7.3 Chadwick's neutron apparatus.

experimentally. For this he designed a simple piece of apparatus – now to be found in the museum of the Cavendish Laboratory (see Figure 7.3 and P) – which looks like something the plumber left behind. It consisted of two chambers, separated into two chambers by a membrane. One of these, the neutron chamber, had an outlet for connection to a vacuum pump. This contained at the outside end the polonium source for α-particles, and at the other the beryllium target: beyond this was the paraffin wax target, for whatever particles were emitted by the beryllium. The paraffin wax defined the inside boundary of the ionisation chamber, which in turn was connected to an external amplifier. (This was essentially the same as the apparatus used by the Joliot-Curies.)

The apparatus operates in three stages: (1) the beryllium target is bombarded by α-particles; (2) the particles emitted by the beryllium bombard the paraffin wax; (3) the particles emitted by the paraffin wax activate the ionisation chamber in a way that can be measured, observed and recorded by an oscilloscope.[67] The question at issue was the identity of the particles in the second stage. (Those in the third stage would certainly be protons, emitted by the hydrogen component of the paraffin wax;[68] otherwise the ionisation chamber would not function.)

Chadwick, over a period of three weeks, worked night and day with this apparatus: the night hours were crucial, for then there would be no disturbances to the sensitive amplifier. On one occasion a short conversation was recorded:

> 'Tired, Chadwick ?'
> 'Not too tired to work.'[69]

Chadwick used his apparatus not only with hydrogen but with other light elements, helium, lithium, beryllium, carbon and argon. On 17 February 1932, he reported his results in a letter to *Nature*:

> They are very difficult to explain on the assumption that the radiation from beryllium is a quantum radiation, if energy and momentum are to be conserved in the collisions. The difficulties disappear, however, if it be assumed that the radiation consists of particles of mass 1 and charge, 0, or neutrons.

The renown due to Chadwick has sometimes been questioned: after all, others before him, like the Joliot-Curies, had produced results that should have led to the earlier discovery of the neutron. On the other hand only Chadwick was single-minded in his search for it. In a letter to a colleague in 1968, he wrote, 'The reason that I found the neutron was that I had looked, on and off, since about 1923 or 4. I was convinced that it must be a constituent of the nucleus.'

As for the Joliot-Curies, they 'had never heard of the idea of the neutron . . . As they were among the foremost workers in the field I think it is clear that the idea of the neutron was not so common as is now supposed to be the case.'[70] And in the words of Andrew Brown, Chadwick's biographer, 'The discovery of the neutron was not only a revolutionary event in physics; it would in time change the course of history.'

The cyclotron

Ernest Lawrence (1901–58) was an all-American star in the world of physics, now commemorated by lawrencium, element no. 103 in the periodic table, and the Lawrence Livermore National Laboratory in California – the home base of Edward Teller, father of the hydrogen bomb. He was fortunate in that his exceptional talents were almost always recognised, even though Canton, South Dakota, where he grew up, would not seem a likely part of the world for producing world-ranking scientists. Even so, while still at school, he acquired a local reputation for his skill in building and operating radios – then at a very early stage of development.

Academically equally gifted, he gained a college place while still sixteen, but St. Olaf College, in Minnesota, had little to offer him, and he only became a serious student when he entered the University of South Dakota in 1919, with his sights set upon becoming a doctor. There Lawrence had the good fortune to have his talents recognised by the Dean of Engineering, Lewis Akeley, so much so that on one occasion the Dean praised Lawrence in front of his fellow-students: 'Class, this is Ernest Lawrence. Take a good look at him, for there will come a day when you will all be proud to have been in the same class with Ernest Lawrence.'[71] By this time Akeley had converted Lawrence from medicine to physics.

Four years at South Dakota were followed by graduate study at the University of Minnesota, which offered him much wider horizons. There his enthusiasm for physics was unbounded, and his Master's degree paper was published. The next step was Ph.D. research in Chicago, where Lawrence made his own apparatus after the manner of Rutherford's string and sealing-wax at the Cavendish Laboratory. He was a particularly talented glass-blower. His subject, the photoelectric effect in potassium vapour, was esoteric, but followed on the photoelectric formula that had won Albert Einstein a Nobel prize in 1921. When Lawrence's apparatus was finally ready after months of painstaking work, it exploded as he made a final adjustment.

It was the same old story, and once again the scientist overcame his despair, helped by the encouragement of a young professor in the next-door laboratory. This was Arthur Compton, who by discovering the effect now named after him, added significantly to the particle theory of light: for all this he was awarded the Nobel prize in 1927. Lawrence had witnessed the key experiment that confirmed Compton's theory. His own work was also going well, but he would need another year to complete his thesis. But then the grant of a fellowship at Yale, for research in the Sloane Physical Laboratory, meant another move, and another chance for Lawrence to add to his renown. This also allowed him to witness totality with the solar eclipse of 24 January 1925, and life became even better later in the year with the award of a National Academy of Sciences fellowship, soon followed by his Ph.D.

At this stage Lawrence moved on to the area where he would make his name on the world stage. His object was to study the results of bombarding mercury vapour with an electron beam. What interested him was the probability that an electron would ionise an atom. His experiments produced precisely this result, and enabled him to measure the ionisation potential of the mercury atom with unprecedented accuracy. (This also produced the most accurate value, to date, of Planck's constant h.)

In 1927 Lawrence visited Europe for the first time, starting with England, where inevitably he made for Cambridge and the

Cavendish Laboratory. There he met another Ernest, Ernest Rutherford, whose career had been remarkably similar, if a generation earlier. The talents of both men, from unlikely backgrounds – South Dakota and New Zealand – had been recognised from an early stage, and they shared the same genius for improvisation. That two new elements, atomic numbers 103 and 104, are named after them (tentatively in the case of Rutherford) reflects the importance of being Ernest in subatomic physics. At Cambridge, Lawrence was also to meet J. J. Thomson, Kapitsa, Chadwick and Cockcroft – the company could hardly have been more distinguished. The visit to England ended, however, with anticlimax: Lawrence was in Yorkshire for the solar eclipse of 31 May 1927, but poor weather prevented him seeing the totally eclipsed sun. Lawrence's European travels continued in the same style, and included meeting Marie Curie in Paris and Schrödinger in Vienna. Needless to say, Lawrence returned home a confirmed American.

Once back at Yale, Lawrence soon learnt that the University of California, Berkeley, were anxious to appoint him associate professor, and after considerable agonising he accepted – mainly because of the facilities promised for research and experiment. Lawrence felt immediately at home, and Berkeley was to be home for the rest of his life. There he focused his research on the atom, and this at a time when it had become clear that further progress required bombardment at energy levels far higher than any available to Rutherford, working with α- and β-particles emitted by radium. European physicists were already in the race to achieve such levels. (In Germany three had tried to harness the energy of lightning by stretching a 700 metre chain between two mountains: one paid with his life, and the experiment had still failed because the gas-discharge tube had been unable to withstand the electrostatic potential.)

Lawrence's inspiration was to apply a principle that he called the 'multiple acceleration of ions': this would require a device that would continually boost the energy of charged particles in a circular trajectory. Two horizontal hollow copper electrodes are such as would be formed by cutting a hollow copper disc across a diameter. The ion source provides a stream of charged particles, ideally protons, i.e. hydrogen ions. The electrodes and the ion source are contained in a vacuum chamber. The electrodes, by being connected to an oscillator, provide an intermittent boost to the particles introduced by the ion source. The magnetic field created across the chamber, at right-angles to the plane of the electrodes, ensures that the particles maintain a constant angular velocity.

This is the basic design of Lawrence's cyclotron.[72] Each successive boost increases the energy of the particles. Because the angular momentum under the influence of the magnetic field is constant, every complete circuit has the same duration, regardless of the velocity of the particles. Since this increases with their energy, the result is that the circuits become successively longer, so that the path followed by the particles is a spiral. The process is analogous to that of a pendulum given a boost at the end of every swing: the swings would become progressively longer, but their period would remain constant, as proved by Huygens. The cyclotron, as opposed to the pendulum, operates with electromagnetic forces rather than gravity. As the particles approach the internal boundary of the electrodes, the spiral path allows them to be deflected by a secondary electrode on to a tangential path leading to the target.

The first successful use of this apparatus, which led to the resonant acceleration of hydrogen ions, occurred when Lawrence was away from Berkeley in April 1930. When he returned he had a linear accelerator constructed on the same principle, but in this case, if the oscillating frequency were to remain constant, every successive chamber had to be longer to accommodate the increased velocity produced by each successive energy pulse. This apparatus foreshadowed that developed a year later by Cockcroft and Walton in Cambridge, with which they succeeded in disintegrating the lithium atom. In 1932 Lawrence was to achieve a similar result with his cyclotron.[73]

Both designs had immense potential for future development, largely by massively increasing the strength of the electric and magnetic fields. The earliest models, in which the diameter of the combined electrodes was about 10 centimetres, could rest on a tabletop. Although the potential across the electrodes was only 160 volts, the final energy achieved was equivalent to 1300 volts, which in turn created 80,000 volt-protons.[74] This, however, was only the beginning. Lawrence immediately set his sights on much larger models of his 'proton merry-go-rounds', going up to an almost unimaginable 25,000,000 volt-protons. Perhaps without realising what he was achieving, he was setting the course for big science, with instruments built on a massive scale at unprecedented cost.

The versatility of the cyclotron, the first effective particle accelerator, was apparent at a very early stage. In March 1933, deuterons (nuclei of the heavy isotope of hydrogen, ^2H) were directed to a lithium target, to produce α-particles of unprecedented range and energy. This was just the beginning: a new large cyclotron allowed for twelve different targets within the chamber. There were also missed chances: when, in February 1934, Lawrence learnt how the Joliot-Curies in France had succeeded in inducing artificial radioactivity by bombarding boron with α-particles (for which they were awarded the Nobel prize for chemistry in 1935), he was able, within a few minutes, to produce similar results with the cyclotron. In the following twenty-four hours, twelve more elements were made radioactive.

In October 1933, Lawrence, the only American physicist invited to the annual Solvay conference[75] in Brussels, found himself as defendant in the case of the cyclotron, in a court presided over by the world's leading physicists: Rutherford, in particular, was extremely sceptical about an instrument that by-passed the Wilson cloud chamber[76] – although later at Berkeley a new improved model was linked to the cyclotron.

In spite of Rutherford, the cyclotron caught on almost immediately: Cornell, Columbia and Harvard all wanted one, as did Copenhagen, Paris and Cambridge, in a list which would eventually include research establishments throughout the world. In 1936, for the first time, radioactive isotopes were supplied for medical use. It was becoming clear that Berkeley was acquiring a reputation worldwide, which led to the Radiation Laboratory being upgraded to become a separate institution within the university, with Lawrence as its director.[77] Even so, an elderly Regent of the University of California, asking a colleague of Lawrence's about when to expect 'the cyclotron's practical effect on life', was answered, 'I'd estimate, optimistically, fifty years'. 'Five years' would prove to be a better answer.

The increase in the size and power of cyclotrons was phenomenal. In 1939 Berkeley's new 60 inch model produced α-particles with energy the equivalent of more than a ton of radium. When, in 1939, Lise Meitner and Otto Frisch reported how uranium bombarded with neutrons would split into two lighter elements (which process Frisch would call 'fission'), this result was replicated with the Berkeley cyclotron.[78]. Later in 1939, Lawrence was awarded the Nobel prize for physics.[79] By this time Berkeley was producing isotopes almost on demand, but it was only in 1940 that neutron bombardment produced ^{14}C – a long sought after carbon isotope with a relatively long half-life. Later in the year, the radioactive element astatine (atomic number, 85) was synthesised by nuclear bombardment, and even more important, plutonium (atomic number 94), although it was not definitively identified until February 1941.

By this time Lawrence was already working on the construction of a new 184-inch cyclotron. The costs were astronomical, but Lawrence, exemplifying the principle that nothing succeeds like success, raised the necessary funding. Since September 1939, the possibility of developing atomic weapons during the Second World War had added new urgency to such a project. The Rockefeller Foundation gave an unprecedented $1,150,000.

In the event, Lawrence, like so many other top-ranking physicists, became involved in the development of atomic weapons. Here, plutonium first identified at Berkeley would be critical as one of the two fission elements suitable for an atomic bomb. The light isotope of uranium, ^{235}U, would be equally important, and here the problem would be separating it from the much more abundant heavy isotope, ^{238}U. One way of achieving this was to adopt the already established principles of mass spectroscopy: the means for separating an element into its different isotopes depends on the difference in their atomic weights. If such particles are then accelerated by electromagnetic forces, such as are applied in a cyclotron, the heavier will separate from the lighter.

In March 1942, Lawrence agreed to the conversion of the 37 inch cyclotron at Berkeley to a giant mass spectrograph, in which ionised uranium chloride would separate according to the two isotopes of uranium, according to the scheme of Figure 7.4. This essentially would be one of the two processes involved in the production of ^{235}U in two massive plants at Oak Ridge, Tennessee, for the atomic bomb that in August 1945 would destroy Hiroshima. The production of plutonium (for the bomb that destroyed Nagasaki) depended on a quite different process, based on Enrico Fermi's atomic pile, the essential nuclear reactor, which is described later in this chapter.

There is nothing in the history of science comparable to the development of circular accelerators, of which Lawrence's 4 inch cyclotron of 1930 was the prototype. The end of the road may not yet be in sight, but by 1980 CERN's Large Electron Positron (LEP) Collider, just outside Geneva, was doing essentially the same work in a circuit whose length was 26 kilometres – an increase

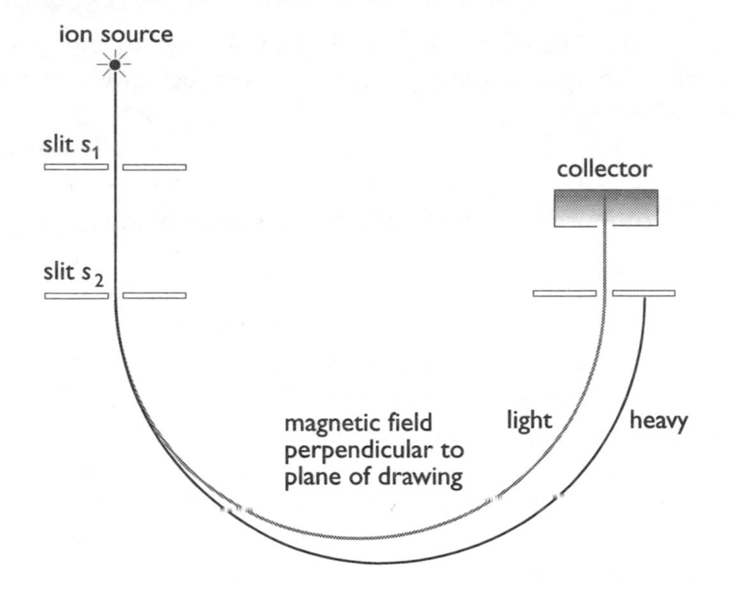

Figure 7.4 Cyclotron used as a mass spectrograph.

over fifty years by a factor of 400,000. When it comes to scale, the development of aircraft, starting with the Wright brothers in 1903, was, after fifty years, nowhere in comparison. With both inventions, however, the basic principle remained the same, but the increase in range transformed the worlds in which they were first realised.

Fermi's chain reaction

'It can now be assumed that a time will come when research workers, by producing or splitting atoms at will, can bring about transmutation of an explosive character, real chain reactions. One can then imagine the enormous release of available energy which would take place.'[80] These are the words of Enrico Fermi, born in Rome just over a hundred years ago, on 29 September 1901.

In physics, Fermi's chosen field, he was not only a supreme theorist but also responsible for setting up and carrying out one of the most remarkable experiments in the history of science. On 2 December 1942 Fermi's atomic pile, laboriously constructed in a doubles squash court on the main campus of the University of Chicago, went critical in the presence of more scientific talent of world rank than had ever assembled at any one place. In 1942, however, wartime censorship prevented this event being known to the public. None the less, it opened the way to the development of the atomic bombs that in August 1945 would destroy two Japanese cities, and the public then learnt that both science and warfare would never be the same again.

Fermi's father, Alberto, was a high-ranking railway official. In 1915, with the family mourning the death of their older son and brother, Giulio, the second son, Enrico, still only fourteen, turned all his energy to the study of science. The father was then persuaded by a friend and colleague to allow his son's sights to focus on entrance to the Scuola Normale Superiore in Pisa, which then offered the best science teaching in Italy. When applying for admission, the young Fermi stunned the entrance examiners with his solution to a difficult problem on sound waves. From this point he never looked back: in 1922, aged twenty-one, he graduated *magna cum laude*, and with a government grant awarded for his exceptional talents he went off to Göttingen in Germany to study with Max Born – one of the leading physicists of the day.

His time in Germany was not a success, except for the fact that it led to an introduction to two Dutch physicists, Paul Ehrenfest (1880–1933) and Samuel Goudsmit (1902–78). The result was that Fermi spent the last three months of 1924 at Leiden, where the physics department had become renowned for the work of two Nobel prizewinners, Heike Kamerlingh Onnes (see Chapter 9) and Hendrik Lorentz (1853–1928). This brought Fermi face to face with the quantum revolution.

The year 1924 proved to be a decisive one for Fermi: he met Laura Capon, his future wife, and at the end of the year he was appointed professor at the University of Florence. (There was considerable opposition from the old guard on account of his age.) The next year, 1925, his interest in quantum physics quickened after hearing of the Pauli exclusion principle. He then went on to develop the theory now known as the Fermi–Dirac statistic. Fermi's place at the top of Italian physics was assured, and in 1926 he was appointed to a new chair, in theoretical physics, which had been specially created for him at the University of Rome. It was there that he turned to experimental physics, helped by two very able contemporaries from the engineering school, one of whom, Emilio Segrè (1905–89), would later become his biographer.

Established in Rome, Fermi turned his thoughts to romance, and he and Laura Capon were married in 1928. She was herself a chemist, and to bring her up to date in physics, Fermi taught her about Maxwell's equations (see Chapter 4) during the honeymoon.

In his own work, he switched his focus to the atomic nucleus. Interest in this field was already considerable, particularly in Paris, where the Joliot-Curies were achieving the remarkable results described on page 249, while in Cambridge Chadwick isolated the neutron.

In 1934 the Joliot-Curies produced artificial radioactivity by bombarding aluminium with α-particles. The result, following the equation

$$^{27}\mathrm{Al} + {}^{4}\mathrm{He} \longrightarrow {}^{30}\mathrm{P} + \mathrm{n},$$

was a new unstable isotope of phosphorus plus a neutron (n). The phosphorus decayed immediately into silicon, according to the equation

$$^{30}\mathrm{P} \longrightarrow {}^{30}\mathrm{Si} + \mathrm{e}^{+},$$

significantly accompanied by the release of a positron, e^{+}, a possibility only recently discovered by Dirac.

Fermi's reaction was almost immediate, but it was also original: he began to research into β-radiation, caused not by α but by neutron bombardment,[81] which gave him the advantage of working with uncharged particles. Fluorine and aluminium were the first two elements selected for this process, and the desired β-radiation was the result. The use of radon,[82] the only naturally radioactive inert gas, to bombard beryllium (the set-up used by Chadwick to discover the neutron in 1932), provided Fermi with the required neutrons. Once the process proved effective, Fermi applied it to a host of elements: iron, silicon, phosphorus, chlorine, vanadium, copper, arsenic, silver, tellurium, iodine, chromium, barium, sulphur, cobalt, gallium, bromine and, finally, gold. The basic chemical technique of precipitation was used to test that the decay product was the element required by theory.

Fermi delegated much of the experimental work to the team he had recruited, and two of its members found that neutron bombardment produced radioactive silver with a half-life of 2 minutes 18 seconds. The apparatus consisted of a silver cylinder, just large enough for the tube containing the radon source and the beryllium target to be placed inside it. To maximise the bombardment of the silver by the neutrons emitted from the beryllium, the whole apparatus was housed in a protective lead casket.

Results, however, proved not to be constant, varying with unlikely factors, such as the position within the casket or the material surface of the table on which it rested. Because varying conditions showed that proximity of lead increased the effectiveness of neutrons, Fermi investigated the capacity of lead to absorb them. Fermi first had a lead screen placed between the neutron source and the silver, but then chanced to replace this with one made of paraffin wax. This greatly increased the radioactivity of the silver, an effect that Fermi saw as the result of the neutrons being slowed down by the hydrogen atoms in the paraffin.[83] Similar results were obtained when silver was replaced by aluminium.

Taking into account that the uncharged neutron would not be subject to electromagnetic forces, Fermi used ordinary momentum theory to show that the transfer of energy would be at its maximum when the neutron impacted on a particle of equivalent mass – as in billiards, which offers a useful model for this process. This equivalent particle could only be a proton. The ideal would be to bombard hydrogen, but as a gas this was unsuitable material for a screen: paraffin, with twice as many hydrogen as carbon atoms, was optimal, especially since carbon has a relatively low atomic mass.

On 10 October 1934 Fermi suddenly realised that reducing the speed of the neutrons had increased their capacity to penetrate silver. This could best be explained by the resonance of the particle: this is defined by its wave component, which in turn depends on its velocity. For optimal penetration, this must fit the resonance of the nucleus. Experiment then showed that water, with two hydrogen atoms to every one of oxygen, was the ideal decelerator, particularly if the heavy isotope, deuterium, replaced the common light isotope of hydrogen.[84] Fermi, without fully realising the implications – which would prove to be far-reaching – had discovered the 'moderator' as a key element in the 'nuclear reactor' (although it would be some years before these terms, and what they represented, became familiar in the world of physics).

Theory explained this process of retardation by making use of *effective cross-section*, a geometrical concept, based on how large a surface area a particle or nucleus should have for impact to be observable. With slow neutrons the effective cross-section proved to be much greater than the actual geometric cross-section; some atoms, such as those of barium and cadmium, had a particularly strong attraction for slow neutrons.

Fermi, with his newly acquired understanding of what was happening, decided systematically to bombard all elements. At the end of the line, uranium produced decay elements with β-radiations with different periods, 1 minute, 13 minutes, and even longer, with

13 minutes the most intense. New radioactive elements were clearly being formed, but their characteristics plainly excluded uranium itself and thorium. Further tests excluded protactinium, actinium, radium, francium and radon. The question was, had a new element, with atomic number 94, been created? If so, its place in the periodic table suggested that it would resemble rhenium or manganese.

Fermi, however, never having worked so far up in the periodic table, believed that he had created elements 93 and 94. In Germany, Liese Meitner (1878–1968) and Otto Hahn (1879–1968), having replicated his experiments, also concluded that they had created these elements.

In 1938 both physics and politics, as they affected Fermi, changed dramatically. Mussolini led Italy, whose war in Abyssinia had led to sanctions imposed by the League of Nations, into alliance with Nazi Germany. This led to new laws directed against the Jewish population, which decided Fermi, whose wife was Jewish, to emigrate to the United States (where Segrè had preceded him).

Fermi had already been offered a six-month visiting professorship at Columbia University in New York, which he planned to take up in the summer of 1939. In 1938, the American Embassy in Rome had already accepted him as an immigrant, after a procedure including a mental test in which he had been asked to add 15 and 27 and divide 29 by 2. The award of the Nobel prize for physics in 1938 then provided Fermi with the opportunity to get out ahead of time, and, after going to Stockholm with his wife and children to collect the prize, he went straight on to New York.

At just about the same time, in December 1938, Hahn and Strassmann, at the Kaiser Wilhelm Institut in Berlin, discovered that barium (56) and molybdenum (42) were decay products of uranium, following neutron bombardment, but being chemists they had done no more than identify these elements. By this time Lise Meitner, being Jewish, had been forced to leave the institute (in spite of Hahn's protests), to find an academic home in Sweden.

In Paris Frédéric Joliot-Curie, a physicist, realised immediately that, if Hahn and Strassmann were right, the energy then released must be gigantic.[85] He replicated the Berlin experiment by placing a neutron source inside a small metallic cylinder, coated on the outside with uranium oxide: this was then placed in a still larger bakelite cylinder. The bakelite then became radioactive, which could only be the result of decay elements of uranium being embedded in it. Precipitation then revealed radioactive barium. Joliot-Curie then confirmed this result by using barium oxide to coat the metal cylinder. The fact that the bakelite did not become radioactive could only mean that all the neutrons had been absorbed by the barium oxide – the result he had predicted. Once again, this experiment revealed enormous energy release.

In January 1939, Fermi, safe in New York with his Nobel prize money, heard from Bohr (who also happened to be there) what Hahn and Strassmann had achieved in Berlin, a month earlier. He also heard that their results had been communicated to Meitner in Stockholm, where she was being visited by her nephew, Otto Frisch (1904–79), who was married to Bohr's daughter.

Aunt and nephew discussed these results in the course of a long walk through the winter snow. Following all the experiments carried out by Fermi in Rome, the only explanation they could find for the Berlin and Paris results was that uranium, on absorbing neutrons, must split into two elements in the middle range of the periodic table. This process Frisch called 'fission', a term borrowed from biology. Meitner and Frisch then reported this conclusion in a letter to *Nature*.

This was only the beginning. Since the binding energy for two mid-range atoms is less that that of one large combined atom, such fission must be accompanied by a vast release of energy in the form of radiation. Even so, this would be only a small fraction of the force binding particles together in nucleus. This meant that Joliot-Curie's experiment, which projected atoms some 3 millimetres on to the walls of the bakelite container, was the first ever nuclear explosion.

Helped by strong support from Bohr, Fermi, who was building up a new team at Columbia, was granted permission to use the university's cyclotron to produce neutrons, as Lawrence had done at Berkeley. Fermi had become an experimenter *malgré soi*. Following on from Meitner and Frisch's letter in *Nature*, he showed that the fission of uranium by neutrons must lead to the release of other neutrons. This follows from the fact that going up the periodic scale the number of neutrons increases faster than that of protons. Since, however, uranium can decompose in a number of ways, it was uncertain just how many neutrons would be released: this is now known to average at 2.5.

If experiments with nucleus fission were to continue, the choice of the right moderator would be critical. This involved embedding the radioactive elements in solid material that would slow down the neutrons emitted as a result of the fission process: this was necessary for the so-called 'slow chain reaction', which attained its maximum effect of 'neutron capture' with a definite neutron velocity, which could be calculated in advance. In other words, the function of the moderator was to slow down the neutrons, but not too much. The quantitative relation between the moderator and the fission element had to be just right. Joliot-Curie, in France, chose heavy water, that is, water based on deuterium, the heavy isotope of hydrogen, as a moderator. Fermi, in New York, found it better to work with graphite – one of the three allotropic[86] forms of carbon – but at a level of purity not to be found in ordinary commercial supplies. But by this stage, Fermi and many others were coming together in another ball game, perhaps the best known and certainly the most devastating, in the entire history of science. The Manhattan Project, in spite of its name, found its scientific base in Los Alamos, New Mexico – far from New York.

The Manhattan Project

'It is still an unending source of surprise to me how a few scribbles on a blackboard or on a sheet of paper could change the course of human affairs.' With these words, the mathematician Stanley Ulam described the Manhattan Project.[87] In the United States its origins are to be found in a letter to President Roosevelt, written by Albert Einstein in September 1939, in which he related how recent discoveries by Joliot-Curie, Szilard and Fermi made possible the construction of a bomb of unprecedented destructive power.[88] The reason for his concern was that neutron bombardment appeared to be capable of setting up a chain reaction, with the energy released increasing exponentially at every stage. (This would be a fast chain reaction, governed by the factor of 2.5 mentioned above.) Given that the energy locked up in the atomic nucleus (such as was released by the experiments described in the previous section) is of an order greater than that of any chemical reaction by a factor measured in millions, Einstein was hardly exaggerating the destructive power of an atomic weapon – if it could ever be constructed. By the time he wrote his letter, this was beginning to look possible.

There were three key problems, which, although formidable, could apparently be solved – although the cost and effort would be simply tremendous. The first was to find an element that would sustain the reaction. Following the line of experiments initiated by

Fermi, the only possible element, occurring naturally, was uranium. This, however, was subject to a critical limitation: according to a paper published by Bohr in September 1939, only the odd-numbered isotope of uranium, ^{235}U, could produce a chain reaction.[89] The problem was that 99.275% of all natural uranium consisted of an even-numbered isotope, ^{238}U, with only 0.72% being ^{235}U.

There were two solutions to this problem, each of which would require entirely new apparatus. The first, obvious enough in principle, but appallingly difficult in practice, was to separate out the ^{235}U. This task was undertaken by Frisch in England, where he had arrived shortly before the beginning of the Second World War. There he worked with another Jewish refugee, the German physicist Rudolf Peierls (1907–2000), but the first measurable sample of ^{235}U was actually produced by a colleague of Fermi's at the University of Minnesota on 29 February 1940. This was immediately sent on to Columbia, where Fermi was able to prove experimentally that Bohr's contention relating to ^{235}U was correct. This still left Frisch and Peierls to continue their search for a viable method of large-scale separation.

The second solution was to apply to uranium the principle first developed in the early 1930s by the Joliot-Curies for creating new radioactive isotopes, so as to produce an odd-numbered isotope of a new element in sufficient quantity. This would have to be higher than uranium in the periodic table, since it was already known that no lower element could produce the required reaction. Here success finally came, in February 1941, to Glenn Seaborg (1912–) and Emilio Segrè, working with Lawrence's cyclotron at Berkeley. Used for neutron bombardment, this produced in minute quantities the odd-numbered isotope, ^{239}Pu, of plutonium – a radioactive element with no known natural source. At the beginning of 1941, therefore, both options were open: ^{235}U and ^{239}Pu could be produced, although for a nuclear bomb this would have to be on a scale millions of times as great as anything achieved in the laboratory.

This leads to the second problem – the one most familiar to the general public – that of critical mass. Once some initiating event starts a fast chain reaction at the centre of a mass of uranium, it will only continue for so long as a sufficient number of the neutrons, produced by reaction in each successive generation, find targets in the form of new uranium nuclei within the mass. Quite simply, if the mass is too small, too many neutrons will escape through its surface rather than find nuclei to fission. Anything less than the critical mass is too small: this is what defines it. Peierls, having calculated an effective cross-section for the uranium atom of 10^{-23} square centimetres, found that the time between successive generations would be about four millionths of a second. The energy released, mainly as heat, would increase exponentially to the point that, after some eighty generations – equal to one fifty-thousandth of a second – the mass of uranium would have expanded and vaporised, to blow itself apart in all directions. The chain reaction would then cease, leaving the greater part of the mass to be dispersed as radioactive dust, to come finally to rest as nuclear fallout. Uranium is some nineteen times heavier than water, and Frisch and Peierls calculated that critical mass would be of the order of 5 kilograms: this would mean a sphere about the size of a grapefruit. The actual chain reaction in the split second before disintegration would take place within a core smaller than a grain of rice. This is what destroyed Hiroshima on 6 August 1945.

The third problem was to find an initiator to set off the reaction. A lump of uranium the size of a grapefruit would produce massive α-radiation, but no internal reaction. Without neutrons to start the chain reaction, it would never take place. The initiator was a major part of the weapon technology: it was based on the same principles as Chadwick had used to discover the neutron in 1932. This was a problem only solved at a relatively late stage, at Los Alamos, after the Manhattan Project was well under way. So what then was this project?

The Manhattan Project, which began to take shape in the summer of 1941, was the culmination of the process of developing an atomic bomb which was taking place in both Britain and the United States. The British side was coordinated by the MAUD[90] Committee, established by the physicist G. P. Thomson (son of JJ) in June 1940. The Americans, on their side, reacting to Einstein's letter to Roosevelt, had set up an Advisory Committee on Uranium in October 1939, and by Presidential Order this was incorporated into the new National Defense Research Council, also in June 1940. In the summer of 1941, Mark Oliphant (1901–2000), representing the British side, flew to the United States with the final report of the MAUD Committee, which opened with the following words:

> We have now reached the conclusion that it will be possible to make an effective uranium bomb which containing some 25 lb of active

> material, would be equivalent as regards destructive effect to 1,800 tons of T.N.T. and would also release large quantities of radioactive substances.[91]

The report received the personal attention of Roosevelt on 10 October. By this time Oliphant had visited Lawrence at Berkeley, where he had learnt that converted cyclotrons could provide the means of separating ^{235}U electromagnetically. At a meeting in Washington on 6 December 1941, it was agreed that Arthur Compton would head a new Section, S-1, of the Office of Scientific Research and Development. Its remit was to apply the results of research to the development of actual weapons. The following day, 7 December 1941, the Japanese bombed the American Pacific Fleet at Pearl Harbor, and the United States entered the Second World War. The die was cast.

At this stage the timetable for development focused on the production of plutonium, by means of a slow chain reaction in uranium embedded in a graphite moderator. This was an extension of the work already being done by Fermi at Columbia. At the beginning of 1942, Compton summoned Szilard, Lawrence, and other leading physicists to a meeting at Chicago, at which it was decided that the S-1 project should be located there. Fermi, therefore, had to leave Columbia for Chicago, where he set up shop in a doubles squash court at Stagg Field, the university stadium. There he had the space to build an atomic pile, known as CP-1, consisting mainly of graphite, in which, by means of a slow chain reaction, uranium would be converted into plutonium by capturing neutrons. Fermi's previous experimental work had already shown that the chance of capture was optimal at a certain ratio between the volume of the graphite and that of the uranium to be embedded in it. This correct ratio, worked out in advance, would then determine the way the graphite pile would be constructed. The ideal was a sphere, built of graphite bricks, in which uranium slugs would be embedded in a regular three-dimensional lattice.

Once the reaction started, at the core of the pile, two categories of neutrons would be released at every stage. Those belonging to the first category would continue the chain reaction; those belonging to the second would be captured by uranium atoms (atomic number 92), to transmute, through a highly unstable isotope of neptunium (93), into plutonium (94). As the process continued, much of the uranium would become the required plutonium. This would have to be separated by a chemical process, which would take place

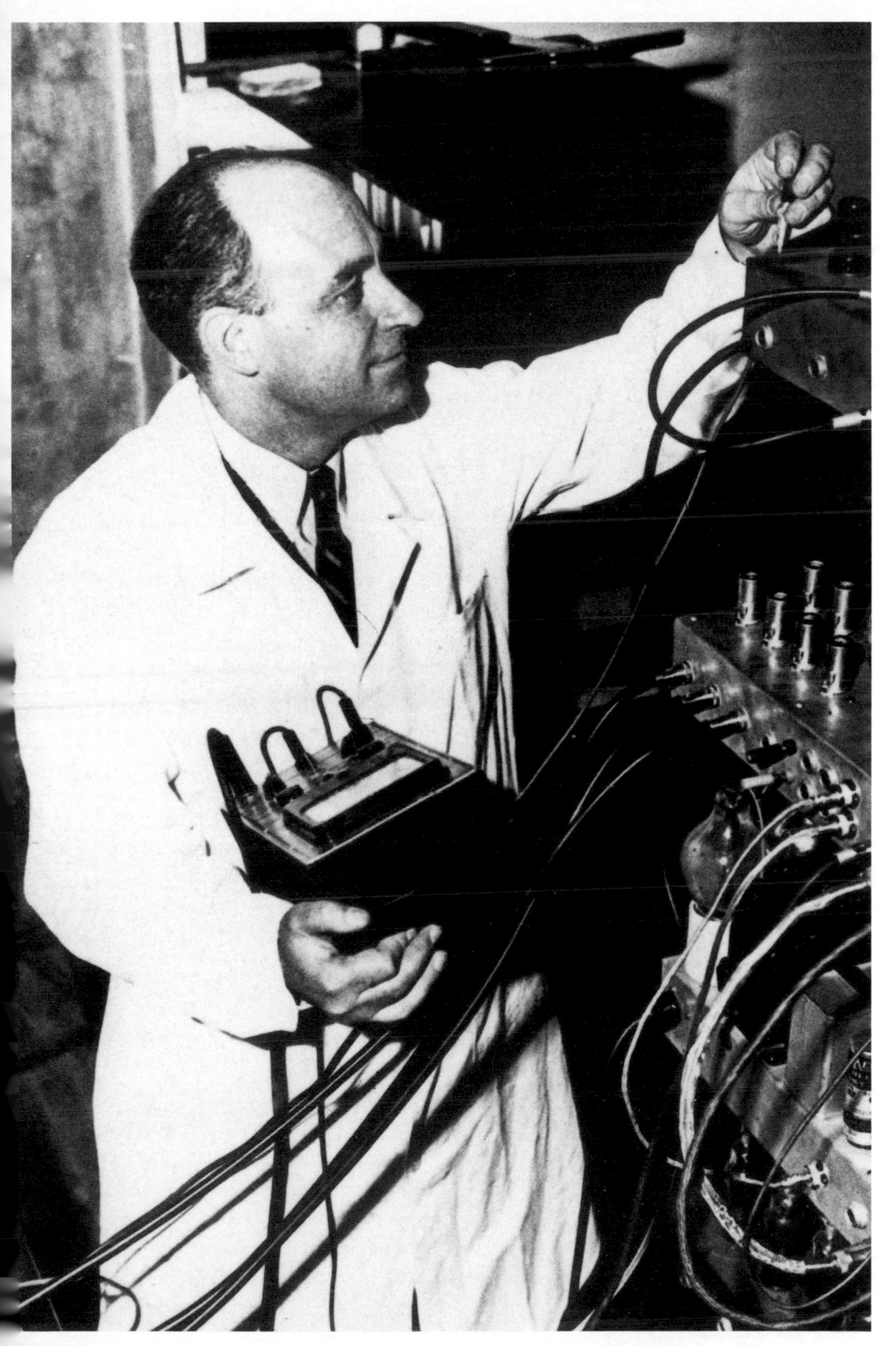

Enrico Fermi, 1901–1954.

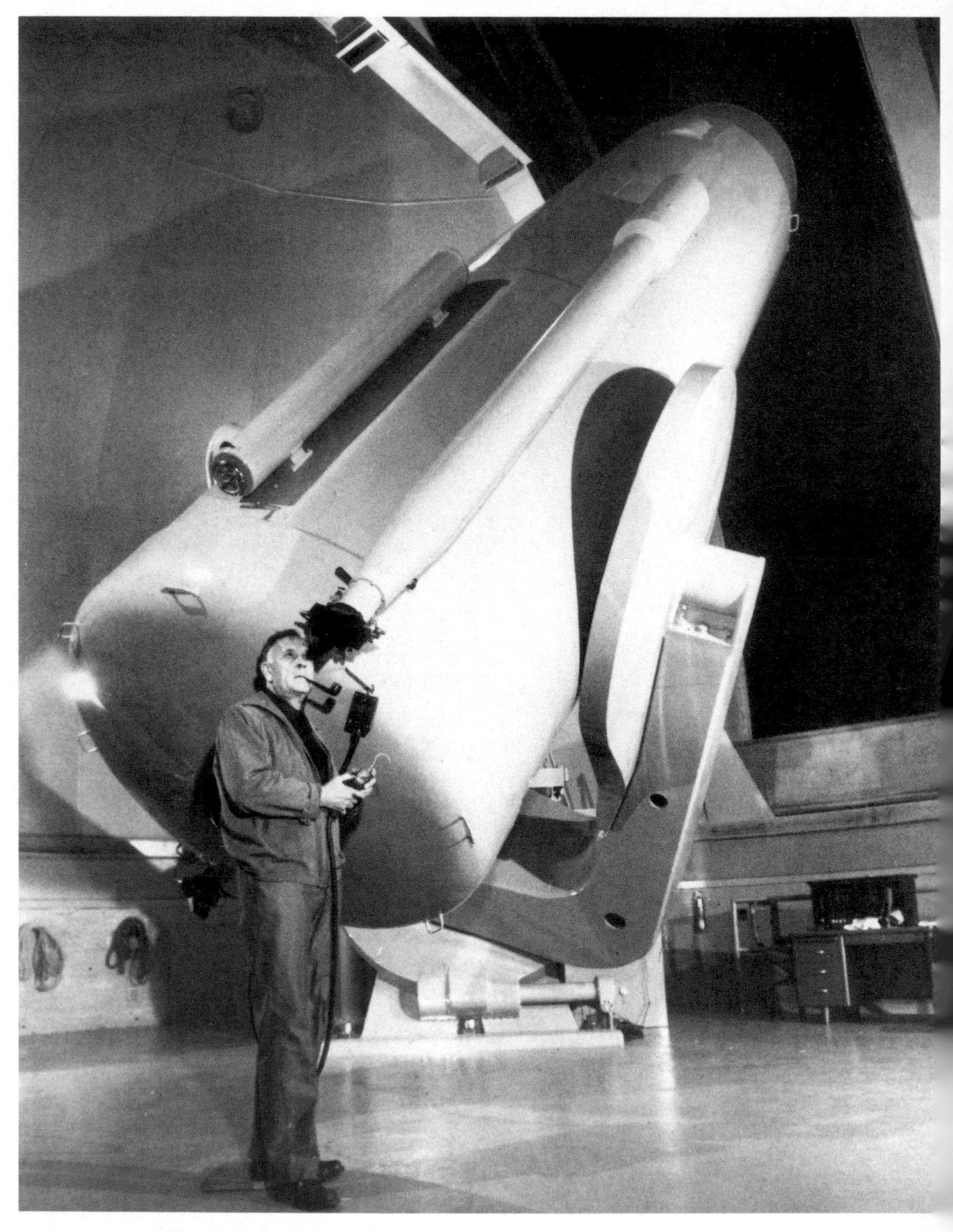

Edwin Hubble with the Schmidt telescope at the
Mount Palomar observatory, California.

Right: Max von Laue's first photograph illustrating the diffraction of X-rays by crystals.

Below: C. T. R. Wilson's original cloud chamber.

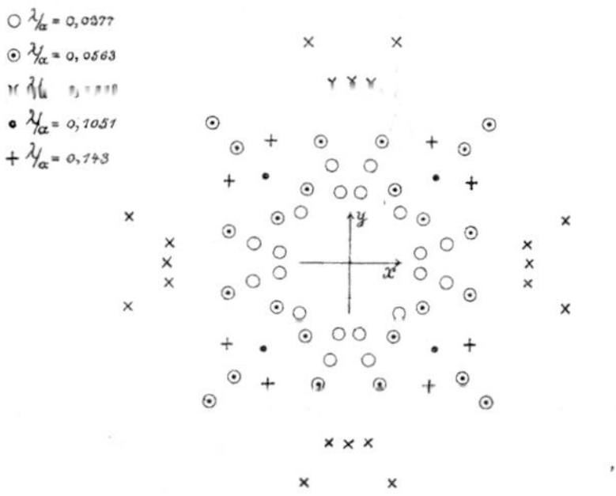

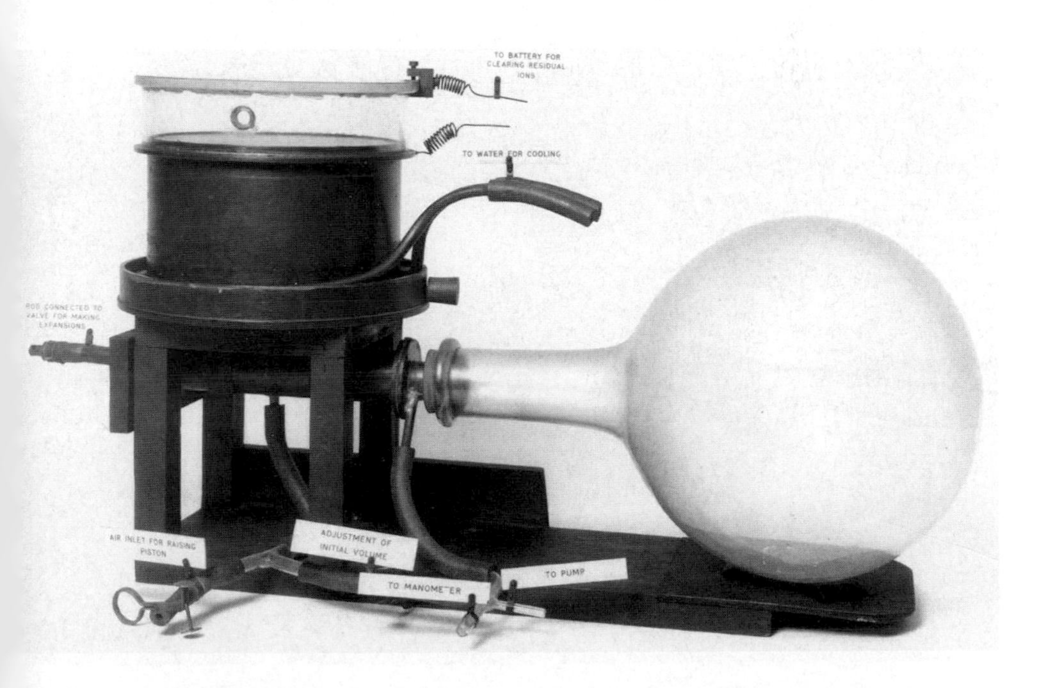

Francis Aston's first mass spectrograph, 1919.

James Chadwick's original neutron capture chamber, 1932.

An early model cyclotron with (from left to right) Donald Cooksey, D. Corson, Ernest Lawrence (the inventor), Robert Thornton, John Backus and W. Salisbury, with Louis Alvarez and Edwin McMillan on top.

Lord Rosse's Leviathan – the giant telescope built in the grounds of his Irish estate for observing nebulae.

The ANTARES neutrino detector.

The Hubble space telescope.

Left: The CERN tunnel.

Below: Aerial view of CERN – the white lin show the underground location of the tunnels

outside the pile, and quite independently of its operation. This was not really Fermi's problem.

What was his problem was to ensure that a sufficient number of neutrons of the first category would be available to continue the reaction. This was not at all simple, because any impurity in the graphite would lead to neutron capture, and thus reduce the factor k measuring the percentage of neutrons that continued the reaction at any stage. Plainly, for continuous production, k could not be less than 1: with sufficiently pure graphite, k would reach this critical threshold, provided the pile was large enough. The minimum size for criticality was again something that could be calculated in advance.

The fact that the pile, in its ideal form, should be a perfect sphere, did not make construction easy, particularly when graphite as a building material proved difficult and disagreeable to work with. The fact that a doubles squash court was necessary to contain the pile gives some idea of its size: given the sheer mass of the graphite, and the spherical shape of the pile, it had to be supported on a wooden framework.

In six weeks, working in sticky black dust from the graphite, an army of workmen, carpenters and students built the pile. In the event Fermi was able substantially to reduce the height of the pile, so that its final shape was that of the giant doorknob,[92] more than 6 metres high and 7 metres wide at the equator. It contained fifty-seven layers of graphite bars, weighing a total of 250 tons. Embedded in it were 6 tons of uranium, contained in small metal cylinders and placed in cavities drilled in one out of every four bars. (This took up relatively little space, since uranium is more than eight times denser than graphite.)

At every stage, k was measured: to begin with it was too low, but this could be corrected by replacing a few impure graphite bricks. As every layer was added, k increased, and as it came close to 1, control rods of neutron-absorbing cadmium were inserted in specially constructed tunnels in the pile. Finally, on the morning of 2 December 1942, with the final layer added, the scene was set for removing the control rods so that the pile would go critical. A balcony had been constructed to enable as many people as possible to observe what happened. The rate at which neutrons were produced was shown on a number of indicator dials and was also recorded, automatically, by a pen on a long roll of paper (such as is familiar in many measuring instruments).

In this first morning session, the operation was suddenly aborted, automatically, when a control bar set at too low a threshold had fallen back into the pile. Fermi could only say, 'I am hungry. Let's go and have lunch.' At lunch no-one talked about the experiment, and Fermi said next to nothing. Just after 2 p.m. the party returned to the pile, and by 2.20 p.m. the neutron acceleration rate had returned to the point reached in the morning. This time there was no automatic shut-down, and about an hour later Fermi, who had been all the time busy making calculations with his slide-rule, suddenly smiled and announced, 'The chain reaction has begun. The graph is exponential.' The first self-sustaining chain reaction was allowed to continue for 28 minutes, when Fermi ordered the control rods to be inserted. At 3.53 p.m. the experiment was over. The world would never be the same again.

Fermi's atomic pile was the first major achievement of the Manhattan Project. In the three months before that critical day in December 1942, the project had taken a new direction as a result of the appointment, first, of General Leslie Groves as military commander in September and then, a month later, of Robert Oppenheimer (1904–67) (Director of the Fast Neutron Laboratory at Berkeley and a colleague of Ernest Lawrence) as scientific director. Groves and Oppenheimer agreed that continued development of atomic weapons required a new dedicated laboratory. For this purpose they bought, for $440,000, the Los Alamos Ranch School, located at a height of 7200 feet on a mesa some 45 miles north-west of Santa Fé, New Mexico. The boys at the school could not come back after their Christmas holiday, and the premises were converted and extended to house a scientific research programme of which the sole purpose was to provide the technology to produce two atomic bombs, one based on ^{235}U and the other on ^{239}Pu.

On this remote site, more scientific talent than had ever been known in history to come together at any one place, set up shop. Of the names mentioned in this book Bethe, Chadwick, Cockcroft, Compton, Einstein, Fermi, Frisch, Goudsmit, Lawrence, Oliphant, Pauling, Peierls, Seaborg, Segré and Teller all worked under Oppenheimer at Los Alamos. (Dirac was one of the few to stay away, and even then his own research was focused on the needs of Los Alamos.) These were only the lead players; the supporting cast was just as formidable. Finally, as already related on page 242, Niels Bohr was spirited away from Nazi-occupied Denmark to join them.

General Groves, their military commander, described them as 'the largest collection of crack-pots ever seen'.[93]

Production of uranium and plutonium for the atomic bombs did not take place at Los Alamos. In both cases the industrial operation was on a colossal scale. ^{235}U was produced in a two-stage process, in two separate facilities, known as Y-12 and K-25, specially constructed at Oak Ridge, Tennessee. This was a remote site with abundant electricity available from the Tennessee Valley Authority, the best-known public works programme of Roosevelt's administration. At full operation, Oak Ridge consumed one-seventh of all the electric power produced in the United States. Y-12 is still operating, and still producing weapons-grade uranium; K-25 has become a historical monument. It is best seen from a viewing site about a mile away, where a plaque placed by the Tennessee Historical Commission contains the following words:

> K-25 PLANT
>
> As part of the Manhattan Project, the K-25 plant was designed to house work on separating U-235 from U-238 through the gaseous diffusion process. At the time of its construction, it was the largest industrial complex in history. Plant construction began in 1943 and was completed in 1945. Over 25,000 construction personnel worked on this plant. The main building exceeded 44 acres in size.

The plaque does not mention that K-25 produced absolutely nothing apart from the uranium for the bomb that destroyed Hiroshima on 6 August 1945. (In fact only half the process took place at K-25; the rest was at Y-12.)

Fermi's CP-1 was the prototype for the atomic piles that produced the plutonium for the bomb that destroyed Nagasaki on 9 August 1945. Production was located across a 560 square mile site, in a desert area in the state of Washington, with the Columbia River defining its eastern boundary. This was named after Hanford, the only centre of population, which is now a ghost town. The main plant consisted of three giant piles, spaced at 6 mile intervals along the river, whose icy water would be needed for cooling the reactor cores. The production cycle required uranium slugs, embedded in graphite, to be subjected to a slow chain reaction for a period of 60 days. For a further 100 days they would be placed, for cooling, into 16 feet deep pools of water, which would be suffused by an eerie glow, known as Cherenkov[94] radiation. Only then were the slugs fit to be sent to three special plants for the chemical separation of plutonium. All this was on a gigantic scale comparable to Oak Ridge. Finally, on 17 December 1944, the D-pile at Hanford went critical, followed by the B-pile on 28 December. General Groves reported to Washington that by the second half of 1945 he would have eighteen plutonium bombs.

All the time that Oak Ridge and Hanford were busy with manufacture, the galaxy of scientists at Los Alamos were designing the bombs. For both uranium and plutonium a key technical problem was to design an initiator for setting off the fast chain reaction providing the explosive force of the bomb. The neutrons that would initiate the whole process would be produced in essentially the same way as Chadwick had used in 1932 – only thirteen years earlier – when he had discovered the neutron by bombarding beryllium with α-particles from polonium.

On 17 July 1945 at 5.30 a.m., a test explosion of the plutonium bomb was carried out at Alamogordo, New Mexico, some 130 miles south of Los Alamos. This was a signal event in the history of science, whereby processes, otherwise known only to astrophysics, were orchestrated by man, and their consequences then recorded. It was appropriate that Hans Bethe, who in 1939 had discovered the carbon cycle that produces the energy of the stars, was among the witnesses. Oppenheimer recalled a phrase from the *Bhagavad-Gita*, 'Now I am become Death, the destroyer of worlds.'

Fermi's atomic pile was effectively the prototype for the reactor core of nuclear power stations. At Hanford water from the Columbia River was only needed for cooling; with nuclear power the heat generated by the reactor is used to produce steam to drive turbines. The nuclear reactor is also a research instrument, which can be used, selectively, for many different purposes: the Euratom High Flux Reactor, located on the North Sea coast of Holland and working with enriched uranium, is one example. There, the use of decay isotopes, produced by the reactor, for medical purposes, is a special research theme.

After the war Fermi returned to Chicago to continue nuclear research. An experiment carried out in 1952 was noted for the production of very short-lived fundamental particles. In the same year a new element, with atomic number 100, was identified in the fallout from the first hydrogen bomb explosion: this is now called 'fermium'. Later in 1974, the National Accelerator Laboratory (then recently constructed just outside Chicago for fundamental particle research) was renamed the 'Fermilab'. This belongs to the present era of big science, initiated by the Manhattan Project and first established at Los Alamos, and is still operating as a National Laboratory, funded on a scale that Fermi and the others working with him during the war could hardly have dreamt of before it all started in the early 1940s.

Quanta

Having followed the history of the atomic nucleus to its culmination in the Manhattan Project, it is time to follow another track. The idea of quanta was sprung on the scientific world in 1900 as a result of a new theory, propounded by the German physicist Max Planck[95] (1858–1947) to explain a phenomenon known as 'black-body radiation'. Quanta, in one form or another, have determined the character of physics even since. The underlying principle is simple enough. Change is essentially discontinuous and always occurs as a series of instantaneous events that can be measured – or perhaps, better, counted – in terms of single basic unit. This is the quantum. In every case in which it occurs it is extremely small, so that the cumulative effect of quantum change appears to be continuous and, as such, corresponds to the basic assumptions of classical physics. Some idea of the almost infinitesimal size of a quantum can be gained from the fact that in an electromagnetic field it corresponds to a single photon. This explains Abraham Pais's (1905–2000) definition of classical physics as 'that part of physics in which actions are large on the scale set by Planck's constant, velocities small on that set by the light velocity'.[96]

Planck's constant was proposed as a key component in a formula designed to measure the energy of an esoteric phenomenon, black-body radiation, which became known to physics in the closing years of the nineteenth century. In any absolute sense the phenomenon cannot even occur, since the black body itself is purely hypothetical. It is defined by the property of absorbing all radiation falling upon it (which is a characteristic property of the colour black). This in turn will be emitted (as can be observed from the heat radiated by a black object taken into a dark room after being exposed to sunlight). Although a perfect black body, in terms of the properties that define it, is purely hypothetical, it is possible to construct something very close to it in the laboratory. The way that electromagnetic radiation is emitted by it can then be studied, and this was the subject of a number of experiments carried out in the years immediately before 1900, notably by the Austrian physicist, Josef Stefan (1835–93).

The problem was that the law he proposed to explain the results proved in critical cases not to be compatible with them. It did, however, lead to the first satisfactory estimate of the sun's surface temperature.

Planck showed how the problem would be solved if the energy was radiated by a black body in small discrete bundles with a frequency-dependent intensity. This intensity is given by a formula that includes a constant h – now known as Planck's constant – which would come to dominate the new physics. At low values of radiation frequency, that is, at low energy levels, second-order terms could be neglected, so that h disappears from the formula, which then takes a much simpler form, called Rayleigh–Jeans[97] formula. This formula can be derived from classical physics and applies only at low energy levels, when the quantum nature of photons can be left out of account. The difference between Rayleigh–Jeans and Planck defines the watershed between nineteenth- and twentieth-century physics.

This only becomes critical at the microscopic level of fundamental particles (of which the proton is the largest). Recent research, published in the *Physical Review* for November 2000, suggests that the transition from classical to quantum physics is a question of phase, defined by an exact numerical boundary value (comparable to the freezing point of water at 273°K). This result, discovered by the Israeli physicist Dorit Aharonov, conflicted with the general view among quantum physicists that the transition is gradual.

It is significant that Planck defined his constant, and calculated its value, precisely in the year 1900. In its basic formulation, h simply states the ratio of the energy E of a quantum of energy to its frequency ν, so that $E = h\nu$. Since its value is an unbelievably small 6.262176×10^{-34} joules/second, it is perhaps not surprising that it only became known in 1900. None the less, twentieth-century physics would have got nowhere without Planck's constant.

Dimensions in the quantum domain, in which the laws of classical physics cease to apply, are expressed as unit multiples of Planck length, mass and time – a system in which the gravitational constant G, the speed of light c and the rationalised Planck constant $\hbar = h/2\pi$ are all equal to unity. (The significance of $\hbar$ introduced by P. A. M. Dirac (1902–1984) in 1925, and for this reason sometimes known as the 'Dirac constant', will become apparent later.) An idea of the domain of the Planck units is given by the fact that the Planck length is of the order of 10^{-35} metres, Planck mass 10^{-5} grams, and Planck time 10^{-43} seconds, which is also the time a photon, moving at the speed of light, takes to cover the Planck distance. Although not immediately apparent, whereas the Planck length and time are extremely small, the Planck mass is simply colossal. At 10^{-5} grams this may not be immediately obvious, but this is the mass of a single particle, where that of a proton is 10^{-27} grams. The equivalent energy of the Planck mass is 10^{19} gigaelectronvolts, where the highest energy attainable by the most powerful particle accelerators is of the order of 10^{3} gigaelectronvolts. In comparison CERN and similar facilities described in Chapter 10 are nowhere. But according to big bang theories of the creation of the universe, energies equivalent to that of the Planck mass did occur in the early stages, which may have lasted no longer than one unit of Planck time. Up to this point, only the quantum theory of gravity can describe what was happening.

This may be the cutting edge of today's astrophysics, but does all this relate to anything happening in the twenty-first century? The lesson of quantum physics, as it has developed since 1900, is that it does: this defines the leitmotiv of the rest of this section. In 1900 Planck unleashed a tiger, and it has not yet been put back in its cage. He himself did not know what sort of a beast it was.[98] It is time, therefore, to look at how the story unfolds.

Albert Einstein (1879–1955) was one of the very first people to 'realise that the advent of the quantum theory represented a crisis in science'. In his own words, 'It was as if the ground was pulled from under one.' According to Abraham Pais (1905–2000), 'the discovery of the quantum theory occupies a special position in that it signalled not only that new concepts had come, but also that old first principles had to go, or, better, had to be revised.'[99]

In 1905, Einstein, by applying *classical* physics (in the form of so-called Boltzmann statistics[100]) to black-body radiation, showed how, at a given frequency ν, it behaved like mutually independent energy quanta with energy h. It is now accepted that Einstein owed little to Planck in arriving at this result: on the contrary, in 1906 Einstein pointed out errors in Planck's original derivation of the black-body law, while at the same time the revised version implicitly adopted his light quantum hypothesis.[101] Even at this early stage, Einstein was questioning the assumption that there must be a strong conceptual distinction between light and matter, with light seen as waves travelling through a medium (whose existence Einstein rejected) and matter consisting of localised particles.[102]

Such was the origin of the photon, which put particles back into light (and all electromagnetic radiation). Maxwell's equations did not tell the whole story, and, as Sommerfeld noted in 1911, Newton's corpuscular theory of light was beginning to regain the ground it had lost to Huygens' wave theory. This, the first manifestation of particle–wave duality, could only cause massive disarray in the world of physics, so much so that 'Einstein's hypothesis of light quanta was not taken seriously by mathematically adept physicists for just over fifteen years.' In 1921 Einstein told his friend Paul Ehrenfest that the conflict with the world of physics was 'something fully capable of driving him to the madhouse'. It is now accepted that 'his paper of 1905 . . . marks the origin of the quantum theory as we understand it today'.[103] It is significant that it was for his work on light quanta, not relativity, that Einstein was finally awarded the Nobel prize in 1921.

In 1905 Einstein told only the half of it: in 1916 he added the key property that the photon was a true elementary particle, with its own elementary momentum $h\nu/c$ and energy. (The 1916 result was overshadowed by the observation on the occasion of the solar eclipse of 1919 of the gravitational effect of the sun on light emitted by stars, which confirmed the correctness of Einstein's general theory of relativity – and brought him world renown.[104]) It was, however, only in 1923 that the existence of the photon was confirmed by experiment, as a result of the Compton effect, described later in this section. By this time it was clear that Einstein's original classical approach was imperfect when it came to very high frequencies. In 1924, however, Satyendra Bose (1894–1974) in India succeeded in deriving the black-body radiation law without using classical theory. Einstein then generalised Bose's method to develop a new system of statistical quantum mechanics, now known as the Bose–Einstein[105] statistics, and was able to show how this coincided with the classical Boltzmann statistics at very high frequencies. The imperfection had been taken care of.

As Einstein, with his characteristic theoretical approach, was steadily adding to the list of quantum phenomena, a young American physicist, Arthur Compton (1892–1962), was following his own largely experimental agenda in radiation physics. The background to Compton's work was the new understanding of X-rays that followed from von Laue's discovery in 1912 (related in Chapter 6) that they act as ordinary light.[106] In the following ten years, not only was crystallography transformed, but there was endless discussion about the place of X-rays in the electromagnetic spectrum, particularly in relation to γ-rays. (The matter was resolved in 1922, when it became clear that γ-rays are X-rays coming from the atomic nucleus.[107])

Relatively little influenced by these developments, Compton calculated in 1922 that a quantum of radiation colliding with an electron undergoes a discrete change of wavelength. He then devised an experiment, based on X-rays, to confirm this result. The effect of X-ray scattering, known since then simply as 'the Compton effect', established the correctness of Einstein's theory of light quanta. The actual use of X-rays is critical, since they are the only practical source of electromagnetic radiation, with a wavelength shorter than what is now known as the 'Compton wavelength'.[108]

This result was critical for relating the physical constant *c*, the velocity of light in a vacuum, to its velocity in a transparent medium such as glass, which is much slower; otherwise there would be no refraction, no prisms, no telescopes. The quantum nature of the photon limits it to one velocity *c*. In a transparent medium such as water, the photon will be absorbed by a water molecule, which in the process will be raised to a higher energy level – a standard quantum effect. This energy will be dispelled, almost instantaneously, by the release of another photon, and so on until the light has passed through the transparent medium. The statistical effect of the constant delays built in to this process is to reduce the effective velocity of the light wave, confining it within limits defined by its spectrum.[109]

Einstein had already devised an experiment, carried out in Berlin by Geiger and Bothe, the results of which were incompatible with a pure wave theory of light, but he was, however, persuaded by Paul Ehrenfest that they were not decisive for his proposed light quanta. The Berlin experiment succeeded in so far as it failed to show a change in the wavelength of light – the so-called Doppler shift familiar to astrophysicists (see Chapter 8) – such as would be required by classical wave theory. Then, in 1922, Schrödinger showed the Doppler effect could be deduced from Einstein's quantum theory coupled with Bohr's frequency condition for atomic transitions described above. This is significant because Bohr himself had always held back from accepting Einstein's results. But if the wave theory was no longer necessary, was the alternative theory of light quanta sufficient?

This was the question that awaited an answer. The answer,

supplied by Compton, was reached by means of an apparatus consisting of an X-ray tube inside a large lead box. The spectroscopic character of the ray itself would be determined by the actual material of the target within the apparatus (which would be a metallic element, such as molybdenum) and also that of the crystal (which could be rock-salt or calcite). The X-rays, reflected off the target, are emitted from the tube at right angles to its main axis. They are then reflected off a calcite crystal on to a graphite block, which are the actual scatterers. Ionisation chambers then determine the consequences of the 'scattered' X-ray radiation impacting on the gas molecules within it: this was an essential technique for observing the 'Compton effect'. This was no simple matter, since the effect of such radiation, as detectable in a cloud chamber, is at first sight chaotic. It is a tribute to the quality and precision of Compton's apparatus (much of which he built himself) that the results he derived from it led conclusively to one theoretical interpretation.

Classical wave theory required that all electrons in the radiator are effective in scattering. This proved to be incompatible with certain observable effects, notably that some of the secondary radiation has a greater wavelength after impact than the primary beam.[110] Compton, after testing a number of possible explanations, was forced to conclude that only a small fraction of the electrons in the scattering material were responsible for the scattering, so that 'an electron, if it scatters at all, scatters a complete quantum of the incident radiation'.[111] On the other hand, 'according to the classical theory, each X-ray affects every electron in the matter traversed, and the scattering observed is that due to the combined effect of all the electrons.'[112] Compton's experimental results did not allow for this, but were fully compatible with the quantum explanation. The success of Compton's experiments, and his explanation of the results, did not come easily, and other, older physicists, such as Harvard Professor William Duane (1872–1935), produced contradictory results from their own experiments. With Duane this led to a series of debates, but at a meeting of the American Physical Society at the end of 1924, he conceded defeat, and the Compton effect (which was soon confirmed by any number of experiments carried out by others) became accepted as 'a fact of Nature'.[113] The German physicist Arnold Sommerfeld (1868–1951), who had extended Bohr's quantum model of the hydrogen atom to multi-electron atoms, wrote to Compton to tell him that the effect he had discovered 'sounded the death knell' of wave theory.[114] The theoretical consequences, which were not long in coming, turned the world of physics on its head.

In the present context, this upheaval (which opened a considerable can of worms) can only be presented as a series of historical events in the 1920s. The first lead figure is Louis-Victor Pierre Raymond, seventh Duc de Broglie (1892–1987), a French nobleman who affirmed his title even though the nobility had lost its estate some hundred years before he was born. Encouraged by his older brother, also a duke, de Broglie was interested in physics from a very early age, but it was war service as a radio operator on the Eiffel Tower that decided him to make it his career. Fascinated by Einstein's publications on the nature of light, de Broglie set himself the task of finding a physical interpretation combining its contradictory wave and particle properties. In November 1921, his older brother, Maurice, following experiments with X-rays, was able to show that his results could sometimes be 'described in terms of the wave theory, sometimes in terms of the emission theory of light'.[115]

In 1923, de Broglie confronted a paradox inherent in the interaction between Einstein's two main contributions to physics, the theory of relativity and the light quantum hypothesis. At first sight, the domains of the two theories could hardly be further apart. The phenomena following from relativity belong to the universe, as witness the observations made at the time of the 1919 solar eclipse. Quantum phenomena belong to the microcosmos of Planck's constant. In between is a world still governed to all intents and purposes by Newton's dynamics and Maxwell's wave theory. Even so relativity does not cease to be valid in the domain of quanta, and vice versa, so if the two theories are incompatible, one must be mistaken. And yet this seems to the case.

In relativity, the fact that the time dimension shortens as the velocity of an observer increases[116] must mean that the observed quantum frequency of a mass point must decrease. Because of the direct linear relationship between energy and frequency required by quantum theory, it must increase.

In 1923 de Broglie was able to resolve the paradox when he realised that a theory he was developing for synthesising the particle and wave characteristics of light quanta would apply in an equivalent form to quanta of matter.[117] In other words, there was a formal and physical symmetry between light and matter. This concept was extremely fruitful, for it provided a physical interpretation for the discrete electron orbits in the Bohr atom, while Bohr's own explanation was purely mathematical. In de Broglie's own words (spoken in 1924), 'This beautiful result . . . is the best justification we can give for our way of addressing the problem of quanta.'[118] He had shown that 'if there is particle-like aspect to light there must also be a wavelike aspect to matter.'[119]

In 1927, G. P. Thomson, by establishing electron diffraction[120] on principles similar to X-ray diffraction, opened the way to verifying de Broglie's wave–particle hypothesis experimentally. In 1929 de Broglie became the first and so far the only duke to be awarded a Nobel prize. (He died in 1987, aged ninety – more than sixty years after his most important work.)

De Broglie started a revolution in physics, which culminated with Bohr's statement of the principle of complementarity on 16 September 1927. In the meantime Schrödinger had discovered wave mechanics, and almost immediately his equation for the wave function of a particle was accepted as fundamental. Pauli had discovered that the electron possesses a fourth quantum number, and his 'exclusion principle' – the name given by Dirac – meant that there could never be more than two electrons sharing the same first three quantum numbers. In 1927 Heisenberg stated his 'uncertainty principle' according to which there was a fundamental limit to accuracy at quantum level, so that the product of the factors defining the position of a particle and its momentum at any one time must always exceed the rationalised Planck constant $\hbar$.

Finally, the mid-1920s witnessed the first appearance in the world of physics of one of its most remarkable scholars, Paul Adrien Maurice Dirac (1902–84). In spite of his name, Dirac was English, although his Swiss father, who taught French at a school in Bristol, would only allow him to speak in French. The result was that he became a taciturn loner,[121] devoted to the study of mathematics, which his father much encouraged. The results were remarkable. He graduated from the University of Bristol when he was nineteen, and two years later, in 1923, he was at Cambridge, where he immediately felt at home in the world of Rutherford, Bohr, Heisenberg, Schrödinger and Einstein, in which he was quite undaunted. Already, in 1927, Einstein referred to 'Dirac to whom in my opinion we owe the most logically perfect presentation of quantum mechanics'.[122] In 1932 he became Lucasian professor at Cambridge, succeeding, aged thirty, to the chair once held by

Newton.[123] The following year he shared with Schrödinger the Nobel prize for physics.

Although Dirac – undoubtedly one of the greatest pure intellects in the history of science – was recognised as such at a very early stage, it would need a very talented spin-doctor to sell him to the general public. He never conducted an experiment, and his science was so rarefied that during the Second World War he remained on the margins of the Manhattan Project. Yet, in 1928, his relativistic wave equation[124] (now engraved on his memorial in Westminster Abbey) corrected the failure of Schrödinger's equation to explain electron spin, which was discovered by Uhlenbeck and Goudsmit in 1925 and is essential to the fourth quantum number. This equation had negative solutions that could only be interpreted in terms of 'antimatter', while at the same time Dirac predicted the production of electron–positron pairs by a photon of sufficient energy, a possibility confirmed experimentally by Carl Anderson's (1905–) discovery of the positron in 1932. This opened up a whole new realm of particle physics, which in the second half of the twentieth century largely defined big science, as known from the world of giant accelerators. This is turn is governed by the Fermi–Dirac statistics,[125] in which only one particle, now known as the fermion, can occupy each quantum state. This is complementary to the Bose–Einstein statistics introduced on page 274.

Fermi, Dirac, Bose and Einstein, these are the landmark names that define the quantum world, in spite of all that has been done since their day. Einstein, the oldest of the four, was only twenty-three years older than the youngest, Dirac. Between them, they created a new world. The time was the 1920s, and this new world opened up vast new prospects in what is now known as particle physics, with considerable feedback from astronomy. Next to molecular biology (which lies largely outside the scope of this book), this is pre-eminently the field that has defined big science. This is the story of Chapter 10.

8

Astronomy: 1542–2001

Cosmologists are often in error but never in doubt.[1]
Lev Landau

Celestial dynamics: Newton's legacy

TODAY'S ASTRONOMY is the end-product of a scientific revolution which can be taken to start with the publication of Copernicus's *De Revolutionibus Orbium Coelestium* in 1543 and to end with that of Newton's *Philosophiae Naturalis Principia Mathematica* in 1687. Until well into the nineteenth century, astronomy was, in today's scientific jargon, positional. Its main concern was to work out the positions of the known heavenly bodies in relation to each other, given Kepler's three laws of planetary motion and Newton's inverse-square law of gravity, and applying to them the mathematical tables (of logarithms and trigonometric functions) developed during the sixteenth and seventeenth centuries.

The key instrument for observers was the telescope. A refracting telescope, with an optical system based on lenses, was first used in astronomy by Galileo. The alternative was an optical system based on mirrors, and it was a reflecting telescope invented by Newton that first brought his work to the attention of the Royal Society.[2] Both systems had their weak points: lenses suffered from the fact that the refractive index of light varies across the spectrum, to produce images distorted by chromatic aberration. Mirrors, made from metal, suffered from distortions produced by changes in temperature. On balance, these proved to be the lesser of two evils, so for some 150 years from Newton's time – a critical period in the history of astronomy – reflecting telescopes held the day.

The position was reversed in the course of the nineteenth century, when improved lenses in innovative optical systems overcame chromatic aberration, to restore to favour refractive telescopes. Then at the end of the century, the invention of silvered glass mirrors restored the balance in favour of reflecting telescopes: this is the present position, and for more than a hundred years optical telescopes (including the Hubble telescope now orbiting the earth) have been reflectors. The advantage, quite simply, is that mirrors can be constructed on a scale far beyond any possible lens.

The development of accurate and reliable telescopes, combined with the mathematical base necessary for processing observations (invariably recorded in numerical terms), led to the establishment of national observatories, whose main remit was to catalogue the position of stars – largely for the benefit of navigators. One further instrument was essential, Christiaan Huygens' pendulum clock, invented in 1657 to bring unprecedented accuracy to timekeeping. The first national observatory was founded in Paris in 1667: the Royal Observatory at Greenwich followed in 1675. Both became dominant in astronomical research, and for nearly 300 years the Astronomer Royal at Greenwich led the field in Britain. Then, in 1951, the Royal Greenwich Observatory moved to Herstmonceaux in Sussex, and, in 1989, after selling out to a property speculator, to Cambridge. There, in 1998, it disappeared, almost unnoticed, and although there is still an Astronomer Royal, the office is purely honorary. In France, where national institutions are cherished, the Paris Observatory is still in business.

Returning to the seventeenth century, what were the prospects for future research when the national observatories were founded? The main focus was on accurate observations to be used for calculating the dimensions, first of the solar system and the heavenly bodies comprised in it, and second of the whole cosmos that lay outside it. As to the solar system, considerable progress, going back to ancient times, had already been made: outside it, practically none. Although one means for calculating the distance of stars, known as parallax to astronomers, consists of applying the process of Snel's triangulation with the baseline defined by two opposite points in the earth's orbit, it was only in 1838 that the German astronomer Friedrich Bessel (1784–1846) became the first to estimate the distance of a star with any accuracy.

This story comes later (see page 297). Although the problem Bessel solved had certainly interested astronomers, before him progress was effectively confined to positional astronomy within the solar system, relying on observations made with reflecting telescopes. The state of the art, when this process started in the seventeenth century, was comparatively simple. Accepting Copernican astronomy, it was long known not only that the earth was round but roughly how large it was. This was the result of Eratosthenes (*c.* 276–194 BC) measuring shadow lengths at midday in two widely separated points in Egypt; he also measured the earth's obliquity, that is, the angle between the plane of the equator and that of the ecliptic, the plane of the earth's orbit round the sun.[3] Some two to three centuries earlier Anaxagoras (*c.* 500–428 BC) had correctly explained both solar and lunar eclipses, while Hipparchus (*c.* 180–125 BC) became the first to record the position of stars according to their latitude and longitude on a celestial globe. This led him to discover the precession of the equinoxes, the phenomenon according to which the position of the stars at any given time of year varies over a very long cycle – now known to be some 26,000 years. He also estimated the relative distances of the sun and moon, based on the fact – observable with solar eclipses – that to an observer on earth they both appear to have the same size.[4]

Throughout the seventeenth century, observations made possible by the invention of the telescope greatly extended knowledge of the solar system. For one thing Copernican astronomy established that Mercury and Venus, the two planets closer to the sun than the earth, must occasionally pass between them. Such an event, known as transit, is defined by one heavenly body passing across another much larger one, in the observer's line of sight. This phenomenon, in various forms, is extremely useful to astronomers, even at the furthest reaches of the universe.

Although the periods of Mercury and Venus are less than a year – that of Mercury is only 88 days – transit occurs much less frequently. The reason is that the orbits of the two planets do not lie in the ecliptic. Therefore, for transit to occur, the planet must cross the ecliptic – an event occurring twice in every complete orbit – at a time when it would also be in alignment with the sun and the earth. Although, in the case of Venus, more than a century can pass without this happening, in the 1630s, Pierre Gassendi (1592–1655) was able both to predict and observe the transit of Mercury in 1631 and that of Venus in 1639.

In the 1670s Giovanni Cassini (1625–1712), in France, and the first Astronomer Royal, John Flamsteed (1649–1719), in England, both used parallax to calculate the distance of Mars from the earth. Cassini used a baseline with one end in Cayenne, on the Atlantic coast of South America, and the other at various points in France. The problem here was to ensure that observations, on both sides of the Atlantic, took place at the same time, for no known timepiece would maintain its accuracy during an Atlantic voyage. Flamsteed avoided this problem by relying on diurnal parallax, which establishes the baseline according to the distance that a given location travels, overnight, as a result of the earth's rotation. The problem here is that a planet will also move, relative to the earth, in the same period.

Both Cassini and Flamsteed were able to overcome their problems, at least to the extent that the recorded observations, when used for calculating the distance between the sun and the earth, produced results with better than 90% accuracy. Cassini's result (confirmed by Flamsteed's) was that the distance was 87,000,000 miles. Considering that Kepler's calculation, at the beginning of the century, was 14,000,000 miles, this was a vast improvement. What is more, Kepler's own third law then made it possible to calculate the distance of all the known planets from the sun, subject to the same margin of error.

While Cassini and Flamsteed were busy with Mars, the Danish astronomer Ole Roemer (1644–1710) was observing eclipses of the moons of Jupiter. These, as most eclipses, must occur with almost perfect regularity, so that, in principle, successive occurrences should provide an accurate measure of time. Roemer, however, noted that his times varied according to the changing distance between earth and Jupiter, as the two planets continued in their respective orbits. Assuming, then, that this discrepancy could be explained by the time taken by light to cover the distance, he was able to calculate the velocity of light at 140,000 miles per second – a figure about 75% accurate (in part explained by the inaccuracy of Cassini and Flamsteed's measurements).

The advance of the astronomy of the solar system in the 1670s was remarkable, and in this same decade an extraordinarily gifted young man first came to notice. Among the great and the good in the history of astronomy, Edmond Halley (1656–1742) has few equals. A Londoner, he had the advantage of being born into a prosperous and enlightened family at a time when the world of science was being transformed. In his first ten years Halley lived through the restoration of the monarchy (1660), the Great Plague (1665) and the Great Fire of London (1666), which destroyed both his home and his school, St. Paul's. No matter, the city was rebuilt under the masterful eye of Christopher Wren (who would become one of Halley's many friends) and the institutions that were to count in Halley's life (such as the Royal Society) flourished.

With an extraordinary eye for significant scientific discovery, Halley was one of the first to realise that the work done by Isaac Newton, at Cambridge, held the key to the universe. Not only did Halley apply Newton's laws, but he ensured the publication of Newton's *Principia* in 1687, and paid for it out of his own pocket. For this alone, science must be eternally grateful. By this stage, Halley, applying Newton, had already achieved much in his own right. He went up to Oxford when he was seventeen, and while there wrote three original papers and a book about Kepler's laws. He left, however, in 1678, without a degree: one reason was that Flamsteed, impressed by Halley's book, encouraged him to go to the island of St. Helena, in the south Atlantic, to compose a catalogue of southern hemisphere stars. This task took two years: Halley, aged twenty-two, returned to England in 1678 and was immediately elected fellow of the Royal Society. His *Catalogus Stellarum Australium* was published a year later.

Although Halley was well known in his day, he left neither memoirs nor many letters revealing his own character. To judge from his achievements he had an incredible capacity for hard work, combined with ability to enlist others' collaboration. At the same time, like many a young man, he sowed wild oats, and suspected adultery was to lose him, for good, the friendship of Flamsteed, whose extreme piety had led him to holy orders. (Writing to Newton on 7 February 1695 he referred to Halley's 'history which is too foule and large for a letter'.[5])

To the young Halley, contemplating the state of the art in science, the new learning and the methods that supported it suggested a wealth of promising topics for research. A whole new world was opening up, and it was clear from a very early stage that Halley's ability to explore it was unrivalled. The criticism recorded of him may simply reflect the jealousy of less able men in the same field.

The discovery of a new comet in November 1680 directed Halley's attention to that field, which, above all others, is associated with his name. The new comet was the first ever to be discovered telescopically. It took everyone by surprise, and after becoming, every night, brighter and brighter (and clearly visible to the naked eye), it finally disappeared, one twilight, behind the sun. Then, some two weeks later, it reappeared, again at twilight; until Flamsteed put him right, Newton, who had been following it closely from Cambridge, failed to realise that it was the same comet.

When the comet reappeared, Halley was on his way to Paris, where he would meet Cassini: his interest in comets was aroused. It was reawakened by the appearance of another bright comet in August 1682. By this time, Halley had returned home from travelling in Europe and married Mary Tooke, a child of a wealthy family. Helped by a substantial dowry, he set up his own observatory and began a long-term programme observing the moon every night for more than eighteen years, the period of one complete cycle of the plane of the moon's orbit round the earth. In spite of Halley's dedication to a science which required endless night-time observations, his marriage, which lasted for fifty-four years, was extremely happy, and produced three children.

Halley's observations of the new, 1682, comet became known to Newton, who shared his interest. This led to the first meeting between the two, probably in London. What then followed is related in Chapter 2; here it is sufficient to note that the relationship that developed was critical, in astronomy, both for Halley's future and for Newton's reputation. When it came to comets, Newton failed to apply his celestial dynamics to any actual instance, although (as he would note in Book II of the *Principia*) the 'force of gravity propagated to an immense distance, will govern the motion of bodies far beyond the orbit of Saturn'.[6] His mistake was to focus on the comet of 1680; Halley, by persevering with that of 1682, was to make his name one of the best known in astronomy.

Following the publication of the *Principia* in 1687 Halley and Newton continued to write to each other about comets, and the longer the correspondence continued, the more certain Halley became that he must focus on 1682. The problem was that Newton's law of gravitation required an elliptical orbit, while observations required a mean distance from the sun far greater than that of any planet: applying Kepler's third law, this would mean an orbit of extremely long duration. From the observations made of the comet of 1680 Halley calculated a period of 575 years. (Newton had failed here, since a numerical mistake, corrected by Halley, had led him to a parabolic orbit, which would have meant that the comet would only once ever enter the solar system).

For 1682 Halley required the most accurate possible observations. These had been made by Flamsteed as Astronomer Royal, the last man likely to help him. Finally, in 1695, Halley asked Newton (whose request Flamsteed would hardly refuse) to act for him. His letter to Newton was prophetic: 'I must entreat you to procure for me of Mr Flamsteed what he has observed of the Comett of 1682 . . . for I am more confirmed that we have seen that Comett now three times, since the yeare 1531, he will not deny it you though I know he will me.'[7]

Halley's intuition proved correct: Flamsteed's observations suggested a period of about 75/76 years, and going backwards in time this fitted in with bright comets observed by Kepler in 1607 and Peter Apian in 1531. Working from observations recorded at a time when Newton's and Kepler's laws were unknown, the orbits of the two earlier comets proved to be similar to that of 1682. The extreme generality of these laws was decisively confirmed. Halley, however, only published *A Synopsis of the Astronomy of Comets* in 1705, a delay explained by the lengthy calculations required for a table of the parabolic elements of twenty-four comets going back to the year 1337. This shows, in particular, how close the recorded figures for the years 1531, 1607 and 1682 are to each other – just one more product of Halley's astonishing industry.

Halley's research in astronomy went far beyond comets. In 1704, a year before his *Synopsis* appeared, he became Savilian Professor of Geometry at Oxford, in spite of Flamsteed's bitter opposition. In the following years two of Halley's achievements stand out. The first related to the eclipse of the sun, accurately predicted for 3 May 1715. The path of totality crossed the greater part of England and included London – the last time the city had this experience.[8] The weather was fine, and, as Figure 8.1 shows, Halley arranged for observation from fifteen different locations, of which six were close to the edges of the path (which was 304 kilometres wide). From each of these locations, the sun would be totally hidden by the moon for a period of up to three minutes.

The relevant times were recorded with unprecedented accuracy, revealing an error of only four minutes in Halley's own predictions. The recorded results have stood the test of time, subject only to small corrections. They have provided evidence for a small change

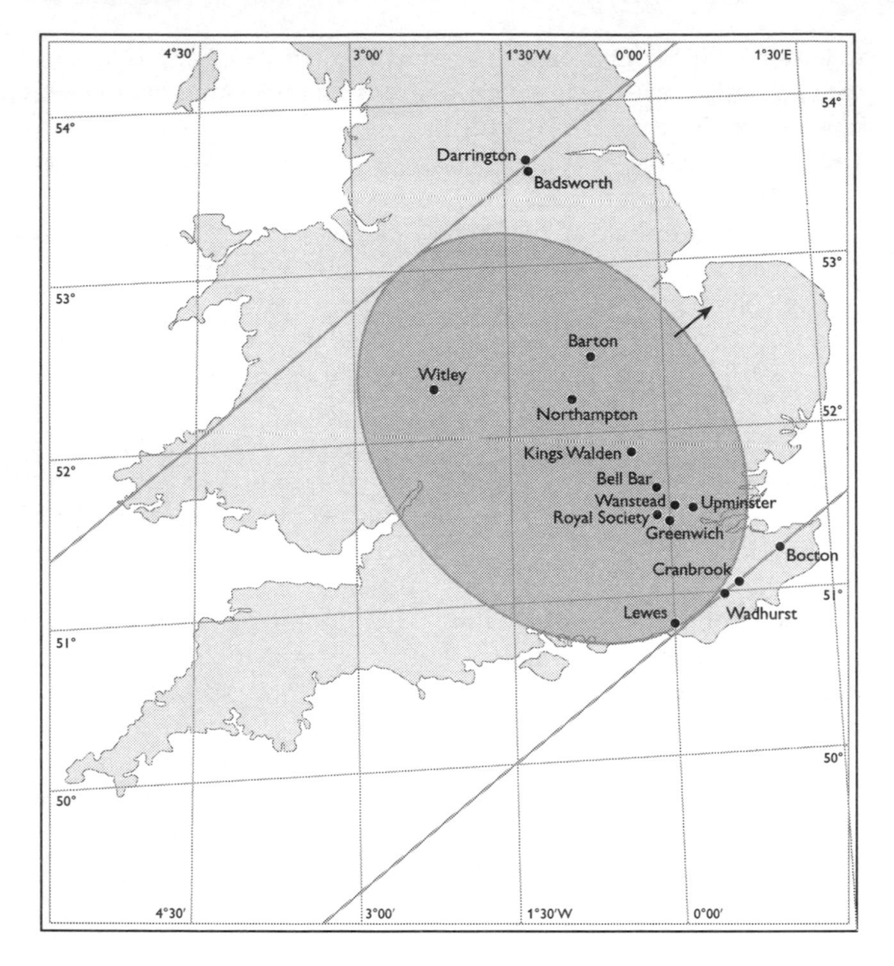

Figure 8.1 Solar eclipse of 3 May 1715: the path of totality.

in the diameter of the sun over the intervening three centuries, but this is but one instance of their usefulness in positional astronomy.[9]

In 1718 Halley realised that the stars in the sky, as observed from the earth, are not fixed in relation to each other. The positions of three bright stars, Sirius, Procyon and Arcturus, had changed by more than a full degree of arc since the time they were first recorded in antiquity. This phenomenon, known as 'proper motion' and first observed by Halley, would then play a key part in determining both the distance of stars and their velocity. It was also incompatible with Aristotle's principle that all stars are at the same distance from the earth – a significant point 300 years ago.

On the last day of 1719 Flamsteed died. Ironically, Halley succeeded him as Astronomer Royal – an office he would hold, together with his Oxford chair, until his own death in 1742. This came quietly while he was sitting in his chair, savouring a glass of wine; towards the end of his life he reportedly 'never eat Thing but Fish, for he had no Teeth'.[10]

Halley's immense legacy to astronomy was confirmed by the following appearance of his comet, first observed telescopically by a German farmer on Christmas Day 1758. This was the result of Alexis Clairaut (1713–65) applying Halley's methods, taking into account, as Halley had also tried to do, the disturbances to the comet's orbit when it was in the proximity of the two large outer planets, Jupiter and Saturn. Clairaut was already a convinced Newtonian, first because of a scientific visit to Lapland where his measurements confirmed that the earth (as required by Newtonian gravity) was flatter at the poles, and then as a result of his observations of the moon (which confirmed Newton's inverse square law). By predicting the date, 13 March 1759,[11] of the perihelion of the returning comet (when the comet would be closest to the sun), with an error of only 32 days, Clairaut finally established Newtonian dynamics in France (and then the world at large), at the same time discrediting Descartes' theory of vortices – a decisive watershed in the history of science.

This, however, was only part of Halley's legacy: he had also insisted that the transit of Venus (which he had correctly predicted for 6 June 1761 and 3 June 1769) would provide a unique opportunity for calculating the length of the astronomical unit (AU), which measures the mean distance between the earth and the sun. In alignment with the earth and sun Venus would be as close to the earth as any planet could ever be. If then the duration of transit could be measured from different points of the earth's surface, the results, after involved calculations, could then produce the necessary parallax measurements for determining its distance from the sun.

Halley's recommendations were made in a speech to the Royal Society in 1716.[12] After his death in 1742, the French astronomer Joseph Delisle (1688–1768) took great pains to organise astronomers from the whole Western world to make the required observations in 1761. Although this was in the middle of the Seven Years War, with France and England on opposite sides, the French contribution, with thirty-two observers, was the largest – a remarkable tribute to an astronomer from the enemy camp; the British, with only eighteen observers, were in fourth place. Even so the whole 1761 operation was a failure: the observations, such as they were, did not meet the necessary standard, although in many instances bad weather was responsible.

All was by no means lost, since there would be second bite at the cherry in 1769. The Royal Society presented the British government with an ambitious programme, including, for the first time, observations in the southern hemisphere and from the Pacific Ocean. Following the recommendations of Nevil Maskelyne, the Astronomer Royal, the choice finally fell on Tahiti, and at the end of the day, to the surprise of many, Captain James Cook was appointed to command the expedition. The Admiralty provided the ship, but King George III made a personal grant of £4000 to the Royal Society to cover the scientific costs.

Although the observations of the transit made from Tahiti are the best known, in 1769, just as in 1761, observations were made from many other parts of the world: they were essential to the programme's success. This proved to be less than had been hoped for, largely as the result of an optical effect, known as the 'black drop', which frustrated accurate observation of the beginning and end of transit. Even so, a more accurate calculation of the earth's distance from the sun did follow: 95,000,000 miles.

The transit of Venus always occurs as a double event, with eight years between the two occurrences. The longer period is well over a hundred years, so that since the 1760s there have only been two more transits, in 1874 and 1882. In the intervening 125 years positional astronomy had advanced far beyond the eighteenth-century programme, and the missing dimensions of the solar system had been discovered by other methods of much greater accuracy. The twenty-first century will open with two more transits, in 2004 and 2012, but the observations then to be made will have a very low profile among astronomers, although there could be considerable public interest.[13]

It is time to return to the eighteenth century, and the discovery by James Bradley (1693–1762), who would succeed Halley as Astronomer Royal, of a new astronomical method for measuring the speed of light. Its basis was stellar aberration, first observed by Bradley in 1729. Any star, if observed with sufficient accuracy over the course of a year, appears to orbit in a small ellipse: Bradley realised that this phenomenon was the result of the difference in time taken by the light from the star to reach the earth at different points in its orbit. His observations led him to calculate the speed of light as equivalent to 308,300 kilometres/second, a substantially better result than Roemer's in 1675.

In the second half of the eighteenth century, astronomy finally came to terms with the structure, composition and diversity of the universe beyond the orbit of the outermost known planet, Saturn. This was essentially the realm of stars, which with much improved telescopes would become infinitely better known, and prove to be much more complex. James Bradley's catalogue of 60,000 stars, completed just before his death in 1762, was only a beginning. The man who would really open up this new world was William Herschel (1738–1822), one of the most romantic figures in the history of astronomy.

Herschel was born in Hanover, the son of a musician in the band of the Royal Footguards, who was also an admirer of astronomy. Both music and astronomy were to shape the young Herschel's life; following his father, he became an oboist in the military band, but in 1857, aged nineteen, he left Hanover after it had been occupied by the French at the beginning of the Seven Years War. His destination was England (whose king was also King of Hanover), where he arrived penniless.

After some years as an itinerant musician, Herschel was appointed organist of the Octagon Chapel in Bath in 1766, one of the best jobs in English music. By this time his interest in the theory of harmony had led him to study first mathematics, and then optics and astronomy. Once established at Bath astronomy became his passion, to which he devoted all his free time. In particular, he set about constructing reflecting telescopes with unprecedented accuracy and power or resolution.

Herschel was hopelessly overcommitted. To help lighten the load, he asked his sister Caroline, twelve years younger, to join him and work as his assistant. He went over to Hanover (where Caroline was already setting out on a promising career as a singer) to fetch her, and on the days spent travelling back to England did nothing but talk of astronomy. This set the pattern for a very long-lasting relationship: William Herschel lived to be eighty-four, and Caroline, who became a noted astronomer in her own right, to ninety-eight.

A letter written in her first year in Bath gives some idea of what Caroline had to put up with: 'He used to retire to bed with a bason of milk or glass of water, and Smith's *Harmonics and Optics*, Ferguson's *Astronomy*, etc., and so went to sleep buried under his favourite authors; and his first thoughts on rising were how to obtain instruments for viewing those objects himself of which he had been reading.'[14] This described the summer routine; winters were mainly for music, in which Herschel was just as diligent, composing twenty-four symphonies, as well as seven violin and two organ concertos, for concerts in Bath.

It was with his telescopes that Herschel would make his name. Although he was never to succeed in the first task he set himself, finding the actual distance of a star, his discoveries transformed astronomy. In particular, he was 'one of those who did most to complete the transition in astronomy and make development in time – evolution – a familiar working concept'.[15] The great breakthrough came in 1781, when he discovered a new body moving across the sky. This he showed to be not a comet (whose discovery would not have been all that remarkable) but the first 'new' planet to be discovered in all recorded history. This was called Uranus, the Greek word for 'heaven'.[16]

The world took notice. King George III was so impressed that he appointed Herschel his personal astronomer, which meant moving house to Datchet, near Windsor. This was a mixed blessing: although Herschel was able to build a 40 foot reflecting telescope, with a 48 inch mirror, good viewing nights were often taken up with showing it off to the court. (On one occasion, the King said to the Archbishop of Canterbury, 'Come, let me show you the way to heaven.'[17]) At the same time the giant telescope was so unwieldy that Herschel often preferred to work with one of only 20 feet.

Herschel's great interest was in nebulae, vast luminous clouds in the night sky, observed from ancient times, but whose true nature was still unknown. Herschel started in 1781 by observing the nebula in the Orion constellation, one of only four known to him. His scope for observation was then considerably enlarged by studying a catalogue of sixty-eight nebulae and star clusters, recently compiled by the French astronomer Charles Messier (1730–1817).

The key question with each nebula was whether it consisted of myriads of individual stars or was simply a vast luminous cloud of gas. In the course of the 1780s Herschel confirmed the first possibility, but in 1790 he found a cloud of luminous gas surrounding a single star. Both were important discoveries. Herschel went on to extend Messier's catalogue to include more than a hundred nebulae, and explained the Milky Way as viewed from earth as part of our galaxy seen from the inside. After he died, Caroline returned to Hanover and, using his recorded observations, prepared a catalogue containing 2500 nebulae and star clusters. The Royal Astronomical Society rewarded her with a gold metal.

When it came to separate stars, Herschel started off on the wrong track, by assuming that all stars were of the same absolute magnitude. Any observer can see vast differences in apparent magnitude. These have long been the basis for systematic cataloguing, with the stars in every constellation being listed in order of brightness with the letters of the Greek alphabet. (This may be satisfactory for stars visible with the naked eye, but, with only eighty-odd constellations, this system is nowhere near sufficient for dealing with the tens of thousands of stars observable with a telescope.) Herschel's original assumption was that a bright star was necessarily closer to the earth than a faint one, with their respective luminosities being the measure of their distance.

The truth of the matter came to Herschel as a result of the discovery, made in 1782 by the eighteen-year-old John Goodricke (1764–86), that a well-known star, Algol (β Persci), was an eclipsing binary. This meant that it was not one but two stars, orbiting around their common centre of mass and in turn eclipsing each other when observed from the earth. The fact that the two stars had different luminosities was decisive in Herschel's change of mind.

In 1784 Goodricke, deaf and mute from an early age, went on to discover that the luminosity of two stars, δ Cephei and β Lyrae, varied in a fixed cycle, whose period could be measured. From the end of the nineteenth century this was to lead to the discovery of a whole class of 'Cepheid' stars, whose variable luminosity proved to be a good measure of their distance from the observer.

The way was open for study of the heavens in three dimensions, with no previous assumptions about the direction and speed of movement of its different components, to say nothing of their age, composition or state of evolution.

All these were matters that would redefine astronomy in the nineteenth century. Even so, there was still room for new discovery in the classic Newtonian universe. A key figure here was Henry Cavendish (1731–1810), one of the great originals in the history of science. He was born into a wealthy and aristocratic family: both his grandfathers were dukes, but it was from his father, Lord Charles Cavendish, a younger son, that he acquired his love of science.

Here Cavendish's interest was obsessive, so that his friends were few, his social life limited, everyday comforts spurned, religion banished from life: little would then distract him from experiment and discovery in science. It was once said of him that he had 'uttered fewer words in his life than any man who had ever lived to be eighty'.

Until he was over fifty, Cavendish lived in the London house of his father, who died in 1783. He then moved to a house near the British Museum, but he also had two others, one of which, in Clapham, was stocked with his scientific apparatus, but otherwise offered little in the way of comfort. There was also a large tree in the garden, which Cavendish used to climb for his meteorological and astronomical observations. Some idea of his approach to life at this stage comes from an incident relating to his banker, who simply wanted to inform him that he had a credit balance of some £80,000: Cavendish replied that he did not wish to be 'plagued' about it, and if it 'was any trouble to the banker he would remove it'.[18]

Ever since Newton's day, scientists had been concerned to find an exact measure of his gravitational constant: he had proved that the gravitational attraction between two bodies was proportional to the product of their masses, divided by the square of the distance between them. To find the constant factor in this equation was one of the great unsolved problems of the eighteenth century. This was much easier said than done, since in any experiment involving two separate masses, the gravitational force between them would be completely overshadowed by that of the earth acting on both of them. Taking the earth as one of the two masses would not help, since without knowing the value of the constant, there was no way of knowing the mass of the earth.

As earlier as 1735 a geodetic survey of South America, organised by the Académie Française, included tests with both pendulums and plumblines in the vicinity of a high mountain, Pinchincha. The results, which were inconclusive, appeared in a book, *La figure de la terre*, published in 1749 by Pierre Bouguer, a member of the survey team. Cavendish, having read this book, accepted that a plumbline or a pendulum offered the only '2 practical ways of finding the density of the earth'.[19]

The problem remained intractable. In 1772, following the recommendation of Maskelyne, the Astronomer Royal, the Royal Society set up the Committee of Attraction to look at it again. In 1774, Maskelyne himself led an expedition to Schiehallion, a mountain in Scotland; its methods were essentially the same as those of the French in South America, but better equipment made for much improved results. (Working in Scotland must also have been easier than in Ecuador.)

Cavendish, following the work of the committee and subsequent research with great interest, also made many useful suggestions. In 1798, he became convinced that he would do better to experiment with two separate masses, neither of them part of the earth or its surface. The idea came from one of his few friends, the Reverend John Mitchell, who had recently died. Mitchell had 'contrived a method of determining the density of the earth, by rendering sensible the attraction of small quantities of matter; but . . . he did not complete the apparatus until a short time before his death, and did not live to make any experiments with it.'[20] After Mitchell's death, the apparatus was given to Cavendish, who reconstructed it in his house in Clapham. Its new form is illustrated in Figure 8.2.

The scale was considerable, and the apparatus would have filled nearly a whole room in Cavendish's house. Some idea of its size is given by the fact that the large spheres were 12 inches, and the small, 2 inches in diameter. The large spheres were suspended from a heavy beam, which, by means of pulleys, could be rotated about its vertical axis: the small spheres were connected by a light rod, suspended by a silvered copper wire, which had sufficient elasticity

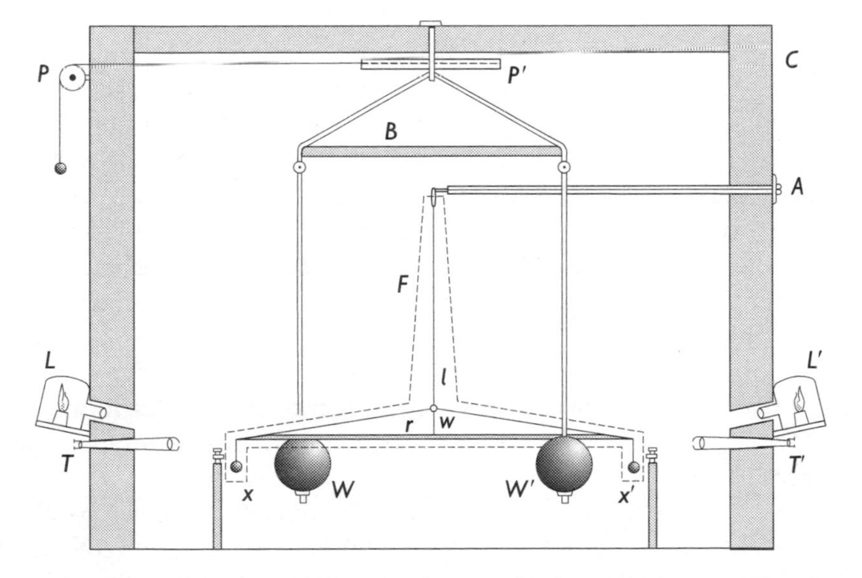

Figure 8.2 Henry Cavendish's torsion apparatus: *C*, outer casing; *PP′*, pulleys for rotating the beam *B*, from which the two large balls, *WW′*, are suspended; *F*, inner casing; *A*, torsion balance adjustment; *l*, torsion wire; *r*, torsion rod, steadied by wires, *w*, supporting two small balls, *xx′*; *LL′*, lamps; *TT′*, telescopes.

to allow some rotation. This wire had a torsion couple, essentially a measure of the torque required to twist it from its equilibrium position. This explains the name 'torsion balance'.

The mode of operation was to measure, according to the torsion couple whose quantity was known in advance, the gravitational attraction of the large upon the small spheres. This would be indicated by the amount the beam connecting them rotated as a result of this force: different measurements could be taken according to the position chosen for the two large spheres. The result of this force would be to create a period of oscillation, equivalent to that of a pendulum, so that the two light spheres would move back and forward like the balance wheel of a watch. Both the density of the earth and Newton's gravitational constant could then be calculated according to a formula incorporating the dimensions of the apparatus and the circumference of the earth (first measured by Eratosthenes – see page 283), together with two variables, the period and angle of rotation of each oscillation.

The results achieved by Cavendish with the torsion balance were his most important contribution to science. The main obstacle was simply that all the spheres were subject to the massive force of the earth's gravity, some 50,000,000 times that which they exerted on each other. To avoid every possible disturbance the apparatus was set up in a dark hermetically sealed room, with the different settings and measurements controlled from outside. The displacement of the two smaller spheres was read to an accuracy of better than a hundredth of an inch; Cavendish had to use a telescope to read the vernier scale which was illuminated by a narrow beam of light directed from outside.

Cavendish's figure for the density of the earth was 5.48 times that of water:[21] this is remarkably close to today's figure (obtained by methods unthinkable in Cavendish's day) of 5.52.

Although eighteenth-century research into the earth's gravity, culminating with Cavendish's experiments in 1798, was strictly earthbound and related only indirectly to astronomical observations, its result was critical for measuring gravitational attraction throughout the universe. The mass of the earth,[22] once known, is a baseline for determining that of any other body in the solar system and derivatively, by methods already explained, in the realms outside it.

In 1838, Friedrich Bessel (1784–1846) finally succeeded in measuring the distance to a star. He relied, as others before him, on parallax, but he had the insight that the stars with the largest proper motion were likely to be those closest to the earth. With hindsight this seems obvious enough, but in 1838 few had the experience of relative motion observable by any train traveller. In any case, Bessel, having chosen 61 Cygni, a star with a large proper motion (5.2 arc seconds per year), was able to measure an annual parallax of 0.3136 seconds – sufficient to calculate its distance at 11.2 light-years. This was another significant breakthrough.

The last great triumph of Newtonian astronomy was the discovery of the planet Neptune in 1846. This was the result of both competition and cooperation between astronomers in England, France and Germany. In England, John Couch Adams, confronted with unexplained irregularities in the orbit of Uranus, attributed them to an unknown body outside its orbit. In France, Urbain Le Verrier (1811–77), working with the same problem, derived both the mass and the precise orbit. This enabled Johann Galle (1812–1910) in Germany to search a specified area of sky on 23 September 1846, to discover the 'new' planet.

The classical era of positional astronomy ended with an impressive display of megalomania. Behind it was a wealthy Irish landowner, the Earl of Rosse, with a passion for astronomy. On his estate in Ireland (whose wet climate was ill suited for observing the skies) he built progressively larger telescopes, ending with a mirror 6 feet in diameter. With this giant he was first to observe a spiral nebula, known as M51 from its place in the Messier catalogue. M51, or the type of phenomenon it represented, was to provoke a still continuing astronomical debate about the constitution of galaxies. Lord Rosse also gave the name to the Crab nebula, after being the first to observe filaments vaguely resembling the pincers of a crab. He was unable to identify its true nature as a glowing cloud of gas and dust. This, like the explanation of much of what Lord Rosse observed – including M51 – had to await the methods of astrophysics, described in the next section.

Finally, on the side of pure theory, the *Mécanique céleste* of the Marquis de Laplace (1749–1827), published over a period of twenty-six years (1799–1825), was a work to rival Newton's *Principia*. Laplace, who after keeping his head during the French Revolution – helped perhaps by his humble origins – was ennobled in 1817 under the restored monarchy of Louis XVIII, was prophetic in his theories of planetary origins. Napoleon, always interested in astronomy, was particularly impressed by Laplace's achievements, and asked him what part God had played. Laplace is still remembered for his classic answer: 'I have no need of that hypothesis.'

Astrophysics

In 1835, a well-known French philosopher, Auguste Comte (1798–1857), discussing celestial objects, stated that 'never by any means, will we be able to study their chemical composition, their mineralogical structure, and not at all the nature of the organic beings living on their surface'.[23] Never was a prophecy so quickly proved mistaken. Since Comte's day progress has been made in all the fields mentioned – considerable when it comes to chemical composition, limited, in relation to extraterrestrial organic beings.

Unwittingly, Comte sketched out a programme for astronomical research which would be underway almost within his own lifetime. The breakthrough came in 1859 when Kirchhoff, by looking at the sun's spectrum through a yellow sodium flame, observed, as described in Chapter 6, the D-line of sodium, which he interpreted, correctly, as proving the presence of sodium vapour in the sun's atmosphere. The scope of this new method was vast, so that within thirty years some fifty elements had been identified in the sun's atmosphere. Of these hydrogen, first observed by A. J. Ångstrom[24] (1814–74) in Sweden, was to prove decisive for the new science of astrophysics.

It soon became clear that spectroscopy could also be used for stars, and, in the four years from 1863 to 1867, Angelo Secchi (1818–78) in Italy classified, according to four different types, the spectra of some 4000 stars. Type was primarily determined by the predominant colour in the spectrum, from blue/white to red. The transition from red to white heat, observable in the laboratory as the temperature of metals increased, indicated that that of stars, as classified by Secchi, decreased from type I to type IV. This, the first indication of the temperatures of different stars, laid the foundations for the theory of stellar evolution, on the basis that in the course of time a star must become cooler.

In the course of the nineteenth century, photography transformed astronomical observation:[25] the word itself was coined by John Herschel (1792–1871) (son of William, and himself a distinguished astronomer), who in September 1839 photographed his father's 40 foot telescope, then in an advanced state of decay. The obvious advantage of photography was its power to create a permanent record. There were early landmarks, such as a photograph taken of the eclipse of the sun in 1851, but it was only in the 1870s that photography became really useful.

This was the result of the development of dry gelatine plates, allowing for much shorter exposure times. These were first used for astronomy by William Huggins (1824–1910) in England in 1876. New results came thick and fast. Henry Draper (1837–82), in New York, had already in 1872 photographed the spectrum of the bright star, Vega, and, in 1879, Huggins did the same for other stars, using the new dry plates that also recorded part of the ultraviolet spectrum. He had already in 1864 analysed the light from a nebula in the constellation Dracula, to find spectra similar to those of luminous gases but not to those of stars. Using the new photography, he was also the first to record the 'red shift', although the phenomenon, together with the corresponding 'blue shift', had already been observed in France, in 1848, by Armand Fizeau (1819–96).[26]

The 'shift' phenomena are extremely important. They are the result of the Doppler effect, discovered for sound waves in 1845 (see page 88). In the case of light from stars, the wavelength observed becomes longer for a receding star, so that the spectral lines 'shift' in the direction of red: this is the normal case, in which the receding star is evidence of an expanding universe. A blue shift indicates a much less common approaching star. Even before using photography, Huggins had used the red shift in 1868 to determine the actual radial velocity of a star (which, if combined with that of its proper motion as described on page 288, will give its true velocity). The use of photography enabled very small shifts in spectral lines to be recorded for later study, making the phenomenon one of the most useful diagnostics in astronomy.

One of the most significant photographs was that of the Andromeda nebula taken by Isaac Roberts (1829–1904) in 1888. This, the most distant object visible to the naked eye, appears as a faint patch of light in the Andromeda constellation. In the eighteenth century Immanuel Kant (1724–1804) suggested that this might be just one of many complete star systems beyond the Milky Way, but Roberts's photograph lacked the resolution to show its true nature. (Already in 1885, a star flaring up in Andromeda and increasing in brightness until its light equalled that of one-tenth of the entire nebula had captivated astronomers.) In 1899, a photograph of the spectrum of the nebula indicated a 'cluster of sun-like stars', contrary to the light and dark bands earlier observed by Huggins. Individual stars were not observed until the 1920s.

At the turn of the twentieth century, the plethora of new recorded observations of nebulae had led to a great deal of theorising about their true nature. The fundamental question was whether the Milky Way, or simply the Galaxy, had any rivals in the form of other galaxies. The received wisdom, in 1900, was that the Galaxy stood alone, and that nebulae, whatever they were, were subordinate to it.

Once again human wisdom was waiting to be dethroned. The process was protracted, and the true nature of nebulae was only discovered by Edwin Hubble (1889–1953) in the 1920s. The full story comes later, for in the meantime there were other key developments both in astronomy and in the instruments it used. Observatories, and their chosen locations, followed an entirely new line. To enjoy the full benefit of spectroscopic observation, recorded photographically, an observatory should have an unprecedentedly powerful optical system, and a location without atmospheric disturbance. The Lick Observatory, completed in 1888, went far to satisfy both conditions – at least by the standards of the day. Its giant 91 centimetre lens was the largest ever installed, and only the 101 centimetre lens of the Yerkes Observatory near Chicago (founded some ten years later) would ever exceed it. Lick, however, had a far better location, 1283 metres above sea level at the top of Mount Hamilton in California. This means that a substantial layer of the atmosphere lies beneath it, a factor that has long dominated the choice of locations for major observatories. Yerkes marked the end of the big refractors, and there is now a powerful reflector, with a 61 centimetre mirror, on the same site. California, however, remained the location of choice. The Mount Wilson Observatory, with a 1.5 metre reflector, was completed in 1908; the Hooker telescope, with a 2.5 metre reflector, was added in 1917. These were the telescopes used by Hubble. In the end, even the Hooker was not powerful enough, and its proximity to Los Angeles also caused trouble from light pollution. California's answer was to build the Hale Telescope, with a 5.08 metre mirror, located at an altitude of 1710 metres on Mount Palomar. The Hale, completed in 1946, was for nearly fifty years the most powerful optical telescope ever built. Then, in 1992, at an altitude of 4200 metres, the Keck telescope, part of the Mauna Kea Observatory in Hawaii, was completed, with a giant 9.8 metre reflector comprising thirty-six hexagonal mirrors, each 1.8 metres across. Its site, at the top of the mountain – probably the best in the world for astronomical observation – is shared with a variety of other specialised telescopes.

Returning to the beginning of the twentieth century, the problem, stated simply, was still largely one of finding the distances of stars and other distant objects – such as nebulae, whatever they were – and then relating the result to magnitude. The starting point is the apparent magnitude to an observer on earth, but without knowing the distance, this is little help in calculating absolute magnitude, that is, the actual size. The stars whose distance can be measured by parallax methods, as first applied by Bessel in 1838, suggest a paradigm, but, given that much the greater part of what can be observed in the heavens (particularly with the most powerful telescopes) is much too distant for any parallax measurement, something more than this is required.

A hundred years ago, 'knowledge of distances was limited to a tiny local pool of stars within the as yet unfathomed ocean of the Galaxy'.[27] The challenge was immense, but, almost from the beginning of the twentieth century, astronomers were ready to take it up, and they did so with results that were to transform their science. At the turn of the century, observed magnitudes of stars could be measured with some accuracy by using such methods as comparison with a standard artificial star. Since antiquity brightness had been measured on a scale with six magnitudes (starting at 1.0 for Sirius,[28] the brightest star), and in 1856 the English astronomer Norman Pogson (1829–91) had transformed this into a precise measure by equating 5.0 (the difference across the scale from 1.0 to 6.0) to a ratio of 100 to 1.

In the following years the scale would extend far beyond 6.0 as powerful telescopes revealed ever fainter stars. To start with, good quality photographs meant that the assignment of stars to different spectral types could become much more sophisticated, so that Secchi's four types were extended to ten, known as O, B, A, F, G, K, M, R, N, S; beyond this there were subtypes, so that the sun is a star of spectral type G2. The classification is important, since the spectral type relates directly to the colour and the surface temperature of the star. In 1913, the American astronomer H. N. Russell (1877–1957) used a graph to plot all the stars for which he had reasonably reliable distances against their spectral type. The result is shown in Figure 8.3 and is known as the H–R diagram.[29] In a band from top left to bottom right, a remarkable concentration of these stars defines the so-called 'main sequence' (a category that includes the sun). With ever increasing accuracy in spectral classification, any star – assuming that it belongs to the sequence – can

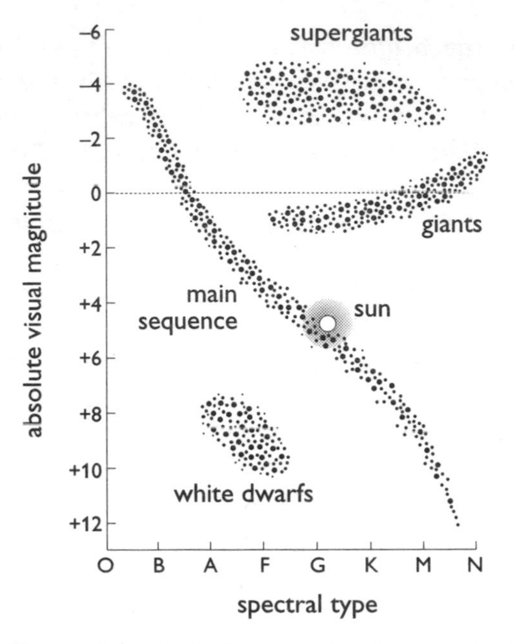

Figure 8.3 H–R diagram of stellar magnitude.

be placed in the band according to its spectral type, so that, for example, an F-star would measure +3 on the magnitude scale. Comparing this with its apparent magnitude gives a measure of its distance. This, the method of 'spectroscopic parallax' works for stars far beyond the range of stellar parallax. When it comes to dimensions, distance is only half of the problem, for even within the main sequence it gives little indication of the mass of a star. Once again, the problem would be solved if, first, for given stars the mass could be calculated, and, second, for a sufficient number of such stars, the mass could be consistently related to magnitude. The first condition was satisfied, because the mass of certain stars, the eclipsing binaries, could be calculated. In binary systems, which are extremely common,[30] two stars in close proximity orbit around their common centre of gravity. With eclipsing binaries the earth lies within their orbital plane: in this relatively uncommon case the two stars, as observed from earth, eclipse each other with complete regularity, so that the period of their orbits can be timed. This establishes, using Kepler's third law (see page 44), the distance separating them; then, by Newton's law of gravitation, their respective masses can be calculated. By the 1920s such measurements had accumulated to such an extent that A. S. Eddington (1882–1944) was able to plot a graph relating mass to absolute magnitude or luminosity. The number of points recorded was so great that their distribution along a single curve could hardly be coincidence: Eddington, therefore, had discovered how luminosity related to mass and in such a way that the most massive stars were the most luminous. His results extended from stars with a fifth of the sun's mass to others with twenty-five times as much.

Eddington worked not only with binaries (including so-called spectroscopic binaries), but also with another intriguing type of star, the 'Cepheid', already noted on page 293. In 1902 Henrietta Leavitt (1868–1921), head of photographic photometry at the Harvard College Observatory, began to study the Small Magellanic Cloud from photographs taken from the observatory's southern station in Arequipa, Peru. The two Magellanic Clouds, Large and Small, are vast aggregations of stars only known to the world following the sixteenth-century voyages of discovery in the southern hemisphere: this explains their name. There is nothing like them in the northern hemisphere. Henrietta Leavitt's research focused on stars with variable luminosity. The pattern was always the same: such a star would increase rapidly to maximum brightness, and then gradually become fainter to a point when the process restarted. Leavitt then noticed that the length of the period, which could be anything from a day to a month, increased with the brightness of the star. Since this was what Goodricke had long before observed with δ Cephei, such stars were named 'Cepheids'. In 1912, Leavitt showed that a graph relating the logarithm of the length of the period to apparent magnitude was, quite simply, a straight line. This was a result of enormous potential for measuring the distance to stars, for Cepheids (which include the polestar) occur throughout the cosmos.

The only problem was that no Cepheid's distance had ever been measured. Then, in 1913, Ejnar Hertzsprung (1873–1967), using a method known as statistical parallax, estimated the distance to the Small Magellanic Cloud at 30,000 light-years. (He was wildly out: the actual distance is 169,000 light-years, but even an incorrect result is useful as a starting point.) Cepheids were clearly in, and in 1914 a young astronomer, Harlow Shapley (1885–1972), who had hypothesised that they were binaries, was invited to join the staff of the Mount Wilson Observatory. There, with the help of the world's most powerful telescope, he tested his hypothesis by looking at globular clusters, which proved to be a rich source of Cepheids.

A globular cluster – a distinctive feature of the night sky – is a densely packed ball of stars, which can be numbered in millions. Halley had observed the brightest, ω Centauri, when he visited St. Helena in 1677; the Herschels discovered many more in their search for nebulae, with John Herschel noting their concentration in the direction of the constellation Sagittarius. Shapley then had the idea that globular clusters he had observed could define a vast sphere, encompassing the Galaxy. Hertzsprung's statistical parallax applied to Cepheids occurring in such globular clusters would then measure their distance from the solar system. Shapley, on the reasonable assumption that the brightest stars always had the same magnitude, noted that globular clusters belonging to a clearly defined group were balanced on either side of the Galaxy. From this he concluded that they were actually part of it, but so far away that if their centre defined that of the Galaxy, the sun would be way off centre. This would also explain the concentration in Sagittarius, noted by John Herschel. Shapley's reasoning led to a figure of 300,000 light-years as the diameter of the Galaxy, whereas the correct figure is about a third of this. In fact Shapley's figure led inevitably to certain conclusions, which if true would establish the Galaxy as a sort of island universe (including the Andromeda nebula), without any rival in the cosmos. The solar system might have been dethroned, but not the Galaxy.

This all threatened to be more than the astronomical community could take. The first round in the battle, fought out in 1920 in a debate in Washington organised by the National Academy of Science (and attended by Einstein), was between Shapley and Heber Curtis (1872–1942) of the Lick Observatory. Curtis, in rejecting Cepheids as a standard of measurement and insisting that the solar system was at the centre of the Galaxy, was mistaken. On the other hand he proved right when insisting that the size of the Galaxy, claimed by Shapley, was much too large – by a factor of about three – accepting, at the same time, the necessary consequence that the Andromeda nebula was far outside it. To Shapley, this could only mean that the 1885 nova, given its astonishing brilliance, was simply gigantic. So be it: Curtis could accept this, and he was right. What had been observed in 1885 was a supernova, not so much a star but a violent event signifying the death of a star. Supernovae, however, were not understood in 1920. Curtis, in another context, had also conceived of interstellar dust, which later was to prove a major factor explaining the error in Shapley's calculations of distance.

The time was ripe for a major breakthrough, in which the leading protagonist would be the American Edwin Hubble (1889–1953). The key to his research came in his interpretation of nebulae, originally a generic name for the many fuzzy patches of light observable in the night sky. In 1845, Lord Rosse, with his giant telescope, observed one, near the tail of the Plough, with a spiral structure. This is now known as M51 from its place in the Messier catalogue. In the course of time, the term spiral nebula came to apply to any regularly shaped nebula, circular or elliptic as well as spiral, with a characteristic bright centre; the well-known Andromeda nebula was one example.

The question soon arose as to whether spiral nebulae were galaxies, comparable to the Galaxy, defined for most observers by the familiar Milky Way (the actual shape of the Milky Way in the galactic plane is closer to a disc, bulging at the centre), which, following Shapley's observations, was confined largely to a single plane cutting the sphere containing the whole Galaxy. To begin with, the distribution of spiral nebulae made it difficult to accept them as galaxies. There were many more on one side of the Galaxy than the other, and practically none close to the galactic plane. On the other hand, painstaking spectrographic analysis of spiral nebulae, made by the American V. M. Slither in the period 1912–17, revealed the characteristics of starlight, which suggested that they must be galaxies. Then, at much the same time, Curtis suggested – correctly as it proved – that there were galaxies close to the galactic plane, but that the light from them (that would otherwise open them to observation) was scattered by interstellar dust. This would also explain the non-spiral nebulae observable within the Galaxy.

In October 1923 Hubble started a systematic search for novae in the Andromeda nebula: this led almost immediately to a star, which from photographic records going back to 1909 proved to be variable. Intense observation of the same star in February 1924 showed that it was a Cepheid, with an exceptionally long period indicating high luminosity. Since the star was so faint, this meant that it was very far away. Its distance, estimated at some million light-years, showed conclusively that the Andromeda nebula was far outside the Galaxy.

This meant also that the 1885 nova was in fact a supernova. Hubble first called Andromeda an 'extragalactic nebula'; he went on to discover a large number of similar spiral nebulae, to the point that he was confident enough to refer to them as galaxies. In 1929 he showed that all of them, save some close to the Galaxy such as Andromeda, were receding with velocities proportionate to their distance from the observer. The announcement of this result, which meant a continually expanding universe, was a bombshell.[31]

There were still problems. All these galaxies had much the same characteristics as the Galaxy, except for one, size. The Galaxy, according to Hubble's Cepheid measurements, which he refused to abandon, was always much greater. If it was just one among myriads of galaxies, what explained its disproportionate size? At the same time Hubble's estimated distances, related to the velocity with which other galaxies receded from the Galaxy, suggested that the universe had been in existence for a much shorter period than would be acceptable to geologists. In 1948, when Hubble was just short of sixty, Walter Baade (1893–1960), working with photographic records of the Andromeda nebula compiled at Mount Wilson during the Second World War, observed that the stars in the spiral arms were blue and white, while the rest were red and yellow. Baade designated stars in the first category 'Population I' and those in the second, 'Population II' – and in so doing established a principle applicable to all stars.

In 1948 Baade, working with the new Hale telescope on Mount Palomar, found that red–yellow Cepheids in the Andromeda halo were four times fainter than the blue Cepheids in the spiral arms used for Hubble's measurements. In effect there were two Cepheid scales, one for Population I stars and the other for Population II stars. Applying this result meant doubling the distance to Andromeda (now estimated at 2,250,000 light-years) and reducing the diameter of the Galaxy to 100,000 light-years. The result was to cut the Galaxy down to size and establish it as but one of many galaxies. At the same time, the estimated size of the universe was doubled, which meant that it became much older (as the geophysicists had long insisted).

The process of discovering an ever greater universe continued with Allan Sandage, a former student of Baade, who went on to be Hubble's assistant at Mount Palomar during the last five years of his life. Sandage, staying on after Hubble's death, discovered in 1958 that bright Cepheids in distant galaxies, used by Hubble to measure distances, were in fact nebulae lit by many stars. This discovery tripled the size of the universe, and increased its age to 13,000,000,000 years. At this point it must be clear that the universe, or the galaxies that compose it, contain many different components of varying sizes and ages and all in a process of continuous, and sometimes dramatic, change, starting at a single point, whose actual location in the space-time continuum is uncertain.

In the thirteenth century, King Alfonso X of Castile, a patron of astronomy, stated: 'If the Lord Almighty had consulted me before embarking on the Creation, I would have recommended something simpler.' These words, quoted in the introduction to Herbert Friedman's *The Astronomer's Universe*,[32] would resonate with any contemporary astronomer. The medieval king did not know the half of it; indeed most of what is known today in astronomy was discovered in the last hundred years. According to the *Cambridge Illustrated History of Astronomy*,[33] 'The majority of astronomers who have ever lived are alive today.' One of them is Hans Bethe – at the time of writing, ninety-five years old – who more than sixty years ago 'proposed the first detailed theory for the generation of energy by stars'. This is the very heart of the matter, and Bethe's work is a good starting point. The man himself is remarkable. Born in Strasbourg (then part of Germany, now part of France) in 1906, he made a name in astrophysics before he was thirty, only to lose his professorship at Tübingen because his mother was Jewish. Germany's loss was America's gain: Bethe went to Cornell University in 1935, where he was to remain professor of physics until 1975, making a major contribution to the Manhattan Project during the war years. His creative powers never left him: his solution of the problem of solar energy came when he was just over thirty, while he proposed a solution to the derivative solar neutrino problem (see page 321) when he was eighty – attracting considerable media attention. This was no problem for a man so articulate and warm-hearted as Hans Bethe, but what then did he discover?

Accepting an expanding universe, with the corollary that it all started with a 'big bang' some billions of years ago, the question is, what processes are at work to generate or consume the vast energy observed in stars and other components of the cosmos? According to Stephen Hawking (1942–) the big bang was a 'gravitational singularity', quite different from the cosmic processes that would follow it – the point is fundamental. These processes, such as they can be observed in main sequence stars (including the sun), consist of fundamental nuclear reactions. The most fundamental of all is the fusion of hydrogen atoms (atomic number 1) to make helium atoms (atomic number 2): the hydrogen, or thermonuclear, bomb is a man-made device of unprecedented power that demonstrates this process. The difference, when it comes to astrophysics, is that the process – or others like it – continues indefinitely.

The key is the proton–proton reaction proposed by Bethe in 1938 as the source of the sun's energy. The protons are simply hydrogen nuclei. In a mass of hydrogen, any two such protons will repel each other, since they both have a positive charge – as is common to all atomic nuclei. If, however, the protons come sufficiently close, they can interact as a result of a quantum effect known as 'tunnelling'. In about 95% of all cases, after three intervening stages, four protons produce one atom of the common isotope of helium, or helium-4 (^{4}He). This requires only two protons: the other two are ejected, with the potential to continue the process. (The 5% deviant cases produce little energy, but do relate to the solar neutrino problem.)

The proton–proton reaction, which is instantaneous, occurs only once with every 10 billion trillion (10^{22}) collisions, but, even so, in the sun it is sufficient to convert 5 million tons of mass into energy every second. This can only happen at a temperature of some 15,000,000°K, which the process itself maintains. (This is true of almost any form of combustion.) The energy generated in the process is continually sending the single electron of the hydrogen atom into outer orbits, and as it jumps back to an inner orbit, a photon, the quantum packet of radiant energy, is released: this electromagnetic radiation will be experienced on earth as heat (largely infrared) or light.

The sun has consumed about 4% of its original hydrogen stock over a period of some 4.5 billion years. In terms of mass it consists of 70% hydrogen, 28% helium and 2% heavy elements – which means everything else in the periodic system. (The proportions are quite different if the number of atoms is counted.) Heavy elements are important in astrophysics, since they came into existence after the big bang, in the interior of the stars created as a result of it. The big bang produced only hydrogen and helium, but this was enough to found the universe, with the process of nuclear synthesis accounting, in many different and often dramatic scenarios, for the remaining elements, right up to plutonium at the far end of the periodic table.

The proton–proton reaction is only the first step: it accounts for the energy of the sun and main sequence stars whose mass is equal to or less than the sun's. The energy of heavier main sequence stars comes from the carbon cycle, also discovered by Bethe.[34] The principle is the same, but the tunnel effect takes place between a proton and the nucleus of the common carbon isotope, or carbon-12 (^{12}C). Once the process is initiated, it continues through several stages, involving isotopes of carbon, nitrogen and oxygen (atomic numbers 6, 7 and 8), and producing, at its end, not only a replacement carbon-12 isotope, but an α-particle. In this case, also, there is a loss of mass, converted into energy, but the combustion process takes place at a much higher temperature, 20,000,000°K, which is the reason why it is characteristic of the heavier main sequence stars.

Although the carbon cycle illustrates how one element transforms into another as part of the process with which a star produces energy, these transition elements are all unstable: at the end of the day no new elements are added to the star. How then did the elements beyond helium emerge as permanent long-term components of heavenly bodies?

In principle this question had a simple answer. The energy of main sequence stars is more than sufficient for achieving the fusion of the abundant helium-4, but the result, beryllium-8 (^{8}Be) is an unstable isotope with a half-life of 10^{-19} seconds.[35] This effectively blocks any further reactions in the direction of the heavy elements. The way to surmount this obstacle was published in 1957 by a team led by Fred Hoyle, the *enfant terrible* of British astronomy.

Hoyle, who died in August 2001, was born in 1915, at a time when his father was serving as a private soldier in the Machine Gun Regiment, notorious in the First World War for its poor chances of survival. Machine-gunner Hoyle did survive, on his own account by being able to calculate where German shells were likely to fall, and then be somewhere else. He returned to his Yorkshire village to bring up his turbulent but brilliant son, Fred, who despite a talent for antagonising potential benefactors, passed through school and, in 1933, on to Cambridge at a time when financial support for working-class boys was very problematic. Starting with a college fellowship, awarded in 1939, Hoyle (except for the war years devoted to radar research) remained at Cambridge until 1973. Plumian Professor of Astronomy since 1958, Hoyle became in 1967 first director of the Institute of Theoretical Astronomy, which he had himself founded, to resign all his posts six years later after a dispute about the future of astronomical research.

Hoyle was immensely creative and often wrong, at every possible level. A popular study, aimed to demonstrate that Stonehenge was built to predict solar eclipses, is way wide of the mark,[36] and in serious astronomy Hoyle, in spite of all the evidence, long opposed the big bang theory. He himself coined this name as a term of derision, but then, in the early 1960s, two American radio engineers, A. A. Penzias (1933–) and R. W. Wilson (1936–), monitoring short-wave radio signals received by a balloon satellite, discovered low-level radiation, which was uniform in every part of the sky. The only explanation for this cosmic microwave background radiation was that it was a relic of the big bang, going back to a time when the universe was only 300,000 years old. This discovery, described by Bernard Lovell as one of the two most important astronomical events of the twentieth century[37] – the other was Hubble's expanding universe – had actually been predicted twenty years earlier by the Russian-American physicist, George Gamow (1904–68).[38] So what is there now going for Hoyle, apart from some science fiction and a musical comedy?

The answer is his 1957 paper published in collaboration with three other astrophysicists (one of whom, less renowned, but almost certainly more diplomatic, was awarded the Nobel prize in 1983). The team was known as B^2FH, after the initials of its members.[39] Hoyle deserves the main credit, because the paper would never have appeared without his solution to the beryllium-8 problem. This is the triple alpha process.

The idea seems too good to be true: instead of two helium-4 nuclei, i.e. α-particles, combining to form the short-lived beryllium-8, why not have three combine to form completely stable carbon-12, the most abundant carbon isotope? The idea first occurred to Edwin Salpeter (1924–) in 1952, but he could not make it stick. Plainly, the triple collision could not take place at one and the same instant of time; what had to happen was for two α-particles to combine to form beryllium-8, and then within the 10^{-19}-odd second available, have a third α-particle – hence the triple alpha – collide, so as to create carbon-12.

For Salpeter this collision would simply blow the beryllium-8 nucleus apart. Hoyle suggested that carbon-12 had a resonance that would allow the third α-particle to be absorbed, and this was soon confirmed experimentally. Once more the process released energy, and there was no reason why it should not continue, with further α-particle collisions, with carbon-12 as a baseline, creating successively oxygen-16 (^{16}O), neon-20 (^{20}Ne), magnesium-24 (^{24}Mg), and so on down the line, with the atomic mass increasing by 4 at every step. Significantly all these elements are relatively common, but, even so, at every stage radioactive decay can lead to the creation of other elements or isotopes. Nor need an α-particle always be part of the process, so that, for instance, two silicon-28 (^{28}Si) nuclei can combine to form one of iron-56 (^{56}Fe, the most common isotope), or of the unstable radioactive nickel-56 (^{56}Ni) and cobalt-56 (^{56}Co).

These 'iron peak' elements are the end of the road. Quite simply, the process that creates them (as described above) releases less and less energy at every successive stage, to the point that, beyond the iron peak, energy must be added in ever increasing quantities to continue it. The image is of a valley, with gentle slopes at the bottom and steep slopes at the top. The peak then refers to the relative abundance of iron as a result of the whole process: one can compare it with the accumulation of debris on the floor of a valley as the result of rockfalls. On the far side of the valley, energy is released not by fusion but by fission, with the highest levels at the extreme end of the periodic scale characterised in nature by the eleven radioactive elements from polonium-209 (^{209}Po) to plutonium-244 (^{244}Pu).[40] (It is not for nothing that hydrogen and plutonium, at the extreme ends of the periodic system, are the elements of choice for nuclear weapons.)

Starting with hydrogen and helium, and ending up, say, with iron, the question is, how can the process continue so as to create heavier elements? Two quite different processes, slow (s) and rapid (r), can achieve this result. The s-process turns on high-energy neutrons (abundant at stellar temperatures) which are captured by heavy elements: if, following capture, a β-particle is emitted from the nucleus, the atomic number increases by 1, and a new element replaces the original one. The s-process, starting with iron, can create all elements up to bismuth (83), but if bismuth captures a neutron, it immediately emits an α-particle, which, as a helium nucleus containing 2 protons and 2 neutrons, reduces the atomic number by 2 and the atomic mass by 4. The result is that bismuth-209 (^{209}Bi), the only natural isotope, converts to thallium-205 (^{205}Tl), the most common isotope of thallium. This is not so surprising, since all elements beyond thallium in the periodic system are radioactive. The s-process can in fact only produce twenty-eight isotopes, but these tend to belong to common elements, such as copper (29) and lead (82). The elements below bismuth that it misses out are correspondingly rare: platinum (78) and gold (79) are examples. The s-process, steady and slow, is characteristic of stars known as 'red giants' – about which more later.

The r-process, a main research focus of the B^2FH team, accounts for all isotopes and elements outside the scope of the s-process: it depends on one of the most dramatic of all cosmic events, the explosion of a supernova.

Although such an event is readily observed by telescope, it has been visible to the naked eye only four times in the past millennium. Fortunately all are well recorded, the first of the four, in 1054, by Chinese chroniclers, who had noted a brilliant new star in Taurus. Its explosion is now known to account for the origins of the Crab nebula, first observed in detail by Lord Rosse in 1844. Although some phases in the explosion lasted only a fraction of a second (in contrast to the millennia of the s-process), for three weeks, the star, much brighter than Venus, could even be seen in daylight.

This 1054 supernova has never been equalled, but in 1572 and 1604 two more were observed by the leading astronomers of the day, Tycho Brahe and Johannes Kepler. A generation or two later, the telescope would have revealed much more. In 1987, a fourth supernova, visible to the naked eye, appeared, and this time – as described on page 314 – the full battery of state-of-the-art astronomical instruments was turned upon it. All four of these supernovae were in the Galaxy, which because of dust in the disc, means that they were closer than about 20,000 light-years.

Given the vast size of the universe, this leaves open the possibility of any number of supernovae outside the Galaxy, only observable by telescope: the first of these, in Andromeda, was found in 1884. Fifty years went by before any systematic search. Then, in 1934, Fritz Zwicky (1898–1974), a Swiss astronomer working in America, started one, based upon the panoramic view of the night sky made possible by the new Schmidt telescope, which he had installed at Mount Palomar. Within a few years he had discovered twelve supernovae in a large cluster of galaxies in Virgo; by now the total number observed, in all the millions of galaxies, is more than 400. If one knew where to look, such an event could be observed every 10 to 20 seconds, so there are a lot of fireworks in outer space.[41] The question remains, how does a star become a supernova? And what then causes it to explode – the defining event of the whole class?

Before answering this question, it is as well to consider how it came to be asked in the first place. This followed the development of radio astronomy in the years following 1945. The principle is simple enough. Optical telescopes, working with the light that comes from stars, nebulae, and so on, relate to only one narrow band of the electromagnetic waves produced by the reactions described above. These arrive at the point of observation not only as a wave but as a stream of photons, the basic quanta of all such waves. Strangely enough, until 1945, few astronomers considered that a radio receiver, picking up such waves far outside the light frequencies, could provide significant information about the character of their source. In the early 1930s, Karl Jansky (1905–50), having noticed such waves as background noise to radio transmissions, was able to show that they originated from a part of the Galaxy outside the solar system, but few astronomers were interested, and before the Second World War only one instrument, built by an amateur astronomer, Grote Reber (1911–), was set up to investigate them. He was years ahead of his time, and the astronomical establishment still took no notice.

One problem is that at most frequencies electromagnetic waves cannot penetrate the earth's atmosphere: X-rays are an example, which explains why they are recorded from satellites (so that the SAS-1, launched in 1970, provided material for a catalogue of 161 X-ray sources). Satellite observation now extends across the entire electromagnetic spectrum, but serious radio astronomy began much earlier. As in much of big science, the impulse came from the military technology (in this case, radar) of the Second World War. Radio astronomy's main limitation is that it is only useful for wavelengths between 1 centimetre and 30 metres – a comparatively small part of the electromagnetic spectrum.

In Britain, always strong in this field, surplus radar equipment was available after the war for building the telescopes, which hardly resemble optical instruments. Of these, Jodrell Bank, developed by Bernard Lovell – one of the radar pioneers – would become the best known. The most spectacular work, however, was based on the Ryle radiotelescope, just outside Cambridge. This, built by Sir Martin Ryle (1918–84) in the 1950s, consists of a line of four fixed and four movable dishes, each 13 metres in diameter, along some 4.6 kilometres of a straight level stretch of railway (which was originally part of the line to Oxford). By changing the location of the movable dishes and recording observations over a period of several days, this instrument is effectively equivalent to a telescope of enormous size.

In 1962 Ryle's interest was aroused by a radio source which was three times subject to occultation by the moon. Observations from the Parkes radio telescope in Australia then revealed a double radio source, of which one component was a thirteenth magnitude blue star: this faint star, observed from Mount Palomar, showed a spectrum with a very pronounced red shift. This meant that it was receding from the Galaxy at a speed which, given the apparent magnitude, could only mean a star with enormous luminosity at a vast cosmological distance. This was the first quasar, or 'quasi-stellar radio source'.[42]

At much the same time, another Cambridge astronomer, Antony Hewish (1924–), constructed a radio telescope consisting of a network of wires, carried on poles, in a lattice covering a site of 2 hectares. This led to the discovery by a young research student, Jocelyn Bell (1943–), of a weak source, emitting regularly spaced pulses, separated by a period of just over a second. This so-called 'pulsar' proved to be a rotating neutron star, the first actual observation of a type of star whose existence had been predicted in 1934 by Baade and Zwicky, shortly after James Chadwick's discovery of the neutron in 1932.

A supernova was then conceived of as representing the transition of an ordinary star into a neutron star, but then in 1934 the neutron star was hardly more than a theoretical construct. Actual discovery by Hewish and Bell in 1964 therefore provided an anchor for new research into supernovae.

The question still remains as to what processes precede, indeed cause, the supernova apocalypse. One difficulty is there are at least two possible scenarios, Type I and Type II: the focus here will be on Type II, because the best ever recorded supernova, SN 1987A, first observed on the night of 23–24 February 1987, belongs to it. The explosion observed occurred some 160,000 light-years away in the Large Magellanic Cloud, and old photographs indicated that this was the end of an already known star, Sanduleak-69°202. Brightness, during the actual explosion, was roughly 10,000 times that of the old star, and the fact that every possible astronomical instrument in the southern hemisphere recorded the process meant not only that existing theories were tested, but also that new facts emerged to be interpreted. SN 1987A is, therefore, in a sense, canonical.

It came into existence about 11,000,000 years ago, with a mass about 18 times that of the sun. It started off by burning hydrogen, the primordial element in the universe. The outer layers expanded, to create a supergiant, shining 40,000 times brighter than the sun. In the core, fusion (the process explained by Bethe: see page 308) finally converted all the hydrogen into helium, which in turn converted into carbon, and then into neon, magnesium and oxygen, until a point was reached that silicon was the key element in the reaction process. The time taken at every stage was but a fraction of that of the preceding one, so that the final silicon stage lasted just about a week (compared with 1,000,000 years for the helium stage).

The silicon stage, in principle, should be the end of the line, since it could lead only to the creation of iron peak elements by nuclear synthesis. However, at a certain point in the process the core of the supernova collapses, from being about as large as the sun to become a lump measuring a few kilometres across. The gravitational energy converts into heat, represented by a flood of high-energy photons – which, being the basic quanta of light, explain the brilliant spectacle observed from earth.

This 'photo-disintegration' was a bright idea of Hoyle (and one or two others) in the 1960s. Its first result was to break down all the heavy elements produced, at rapidly increasing pace, before the event: the process continued through a succession of ever smaller nuclei, to the point that electrons were forced into protons – reversing, under the force of gravity, the familiar process of β-decay. The final result was a 'neutron ball', with about 1.5 times the mass of the sun but with a diameter something over 100 kilometres.

This cannot last: another explosion follows, sending out neutrons in all directions. By this time the collapse of the core has provoked that of the outer layers, so that, instead of expanding, they rush inwards – at a speed about a quarter that of light. The exploding core meets the imploding outer layers and wins the battle of Armageddon, with neutrons, or almost unbounded energy, ready to create heavy elements – far beyond the iron peak – in the culmination of the r-process.

For a few seconds part of the core continues to shrink, to reach the final destination, the neutron star. In the process it sends out a cascade of neutrinos, elementary particles – first identified definitively in 1956 – carrying no electric charge and moving with the speed of light. The neutrinos encounter the r-process, but while most of them are unaffected, some are taken up in it, to give the final push to the process of disintegration.

This is remarkable, since neutrinos have almost no mass. This tends to equate them with photons, which have a rest-mass of zero – which means that they would weigh nothing if they ever stood still. Otherwise they could not move with the speed of light in a vacuum, which is more or less the property that defines them. Whereas photons become apparent in electromagnetic waves (including light) there is nothing equivalent with neutrinos, so it is not surprising that they took so long to discover.

At all events, the neutrinos, through sheer weight of numbers – of quite literally astronomical size – turned the scales on the side of the outgoing shock wave, which then blew its way through the outer layers of the supernova, with the great majority of the neutrinos passing right through it into the outer universe. With SN 1987A a few were eventually detected on earth. In this process, fragments of the outer layer are scattered in all directions: such a collision between the exploding core and the imploding outer layer of a supernova is, in the present state of the art, the only known source of heavy elements in the universe.

This is the realm of star wars, or so it seems. What evidence supports this dramatic scenario? In the end it is a question of balancing one hypothesis against another. Each one was the product of involved mathematics, with predicted results following from computer processing. The problem is then to find results that can be checked by observation.

There are a fair number of recorded observations, going back to Tycho Brahe in 1572 and Johannes Kepler in 1604, but with SN 1987A, the astronomers knew exactly what they looking for – like the detected neutrinos. Theory required that nuclear reactions in the high-pressure shock wave should produce many heavy elements up to the iron peak. Among the first such products would be nickel-56, which (with a 6-day half-life) should decay into cobalt-56 (with a 77-day half-life), and this in turn becomes iron-56, which is stable.

With SN 1987A, in its first 100 days, 93% of the energy was produced by cobalt-56 decay, a very unlikely result if the theory was false. Most of this was the result of looking at spectra, the method pioneered by Kirchhoff in 1859. One astronomer referred to 'the most important and exciting observations concerned with the origin of the elements, confirming the theoretical model of nucleosynthesis is broadly correct'.[43] At the same time, the stream of neutrinos was also essential to the theory, which their arrival on earth then confirmed.

To the observer, the most spectacular result of the explosion of a supernova is the brilliant light, but in terms of energy this is but one-tenth of that carried by the material driven into space, and one- or two-hundredths of that of the neutrinos, produced within the core at a temperature of 48,000,000,000°K.

The supernova defines only one of the many stellar phenomena known to today's astronomers: they are equally at home with white dwarfs, red giants and black holes,[44] but a history encapsulated in a single chapter must draw the line somewhere.

Astrochemistry and solar physics

Astrochemistry and solar physics are both outside the mainstream of astronomy. Although they can be quite simply defined, their characteristic phenomena are problematic and extremely significant. Astrochemistry, as a research field, follows from the discovery of molecules in clouds of gas and dust between the stars. Inside a star heat and pressure combine to strip atoms of electrons to create a level of positive ionisation, in a state known as plasma, in which normal chemical bonding is impossible. This effectively rules out all chemistry inside stars, but not in interstellar space. The problem there is to find a means of detecting the molecules.

By the end of the 1930s, one or two simple carbon compounds had been discovered by optical spectroscopy, but this well-established procedure produced no further results. The way forward was found in radio microwave spectroscopy: the principle is that molecules have distinctive spectra at microwave frequencies that can then be detected by radio telescopes (which means that the whole field has only developed in the last half century). In 1968 Charles Townes (1915–), a pioneer in laser technology at Berkeley, discovered water (H_2O) and ammonia (NH_3); hydroxyl (OH) and formaldehyde (H_2CO) soon followed. Things became really interesting in the 1970s when organic compounds, such as ethyl alcohol (C_2H_5OH), were discovered in molecular clouds. Then, in the early 1990s, the NASA Ames Research Station detected features in a spectrum consistent with pyrene ($C_{12}H_{10}$), a protein body associated with the storage of starch, and in 1994 the amino acid glycine (NH_2CH_2COOH) was detected in a remote region of star formation. Since amino acids are the basic building blocks of proteins, some (including Hoyle) interpret this discovery as evidence of life elsewhere in the universe.

Carbon is once again the key element. Nucleosynthesis within stars (in the processes described on page 309) produces it in large quantities, with the so-called carbon stars being an important source. In interstellar space the carbon commonly occurs as grains of graphite, which absorb light from distant stars (which is one reason why telescopic observation in certain directions is almost impossible). In such dark clouds young stars and any associated planets may well absorb complex molecules at an early stage: such may be the beginning of chemistry as we know it on earth.

Solar physics brings astrophysics to a main series star, the sun, whose proximity allows measurements, and observations, impossible beyond the bounds of its planetary system – to which earth (the planet from which the measurements and observation are made) belongs. The sun, long a distinct research area in astronomy, is a yellow G2 dwarf star, at an average distance of 149,597,870 kilometres from the earth. Light takes 8 minutes and 19 seconds to travel this distance, which is equivalent to one astronomical unit (AU). The closest known star, Proxima Centauri, 4.224 light-years from the sun, is therefore some 267,132 times further away, so it is hardly surprising that so much more is known about the sun than any other star, nor that all other stars appear to be no more than specks in the sky.[45]

The sun, with an apparent magnitude of −26.7, is much the brightest object in the sky. The source of the sun's energy, as already explained on page 308, is mainly the proton–proton reaction, which steadily converts hydrogen into helium in the core, which because of its high density, compresses half the sun's mass into 1.55% of its volume. There the temperature is some 13,000,000°K. The core is separated from the photosphere, the visible bright surface of the sun, by a broad convection zone, in which hot material from the core rises to the surface, cooling in the process, while cooler material moves in the other direction. The result of this process is that the photosphere, with an effective temperature of 5800°K, is much cooler by a factor measured in thousands. This is the source of actual sunlight, but because of the intense particle activity within the sun, the photons comprising it take millions of years to reach the surface, in contrast to the solar neutrinos which cover the same distance in a matter of seconds. The photosphere, in spite of its intense brightness, is that part of the sun normally observed by astronomers, helped by powerful filters that reduce the glare. This is the home of sunspots, first observed by Galileo, and now studied intensively for their possible effect on weather here on earth, including global warming.[46]

Paradoxically, for all the highly visible profile of the photosphere, it is surrounded by an atmosphere, consisting first of the chromosphere, and then, beyond it, the corona consisting of turbulent clouds of gas. Observation of both is difficult, because of the intense luminosity of the photosphere, but since ancient times it has been appreciated that this is no problem when the surface of the sun is hidden from view during a total solar eclipse. Then the chromosphere and the corona are clearly visible to the naked eye. From Halley's day, both the path of the sun's shadow during an eclipse (which is never more than 200 kilometres wide), and the timing of total darkness along it, have been predicted in advance, with steadily increasing accuracy. Since solar eclipses occur about seven times in every decade, this provides reasonable opportunity for observation – given the uncertainty of weather. By the end of the nineteenth century, astronomers have been able to organise expeditions to whatever parts of the world offer the best conditions for viewing. The equipment used is modest compared with that of the vast mountain-top observatories, since it must be capable to being transported to and set up in almost any location. On the other hand, the close proximity of the sun more than makes up for the loss of power.

Until 1860, astronomers were uncertain as to whether the corona belonged to the sun or the moon, but in that year observations taken from points 400 kilometres apart confirmed that it belonged to the sun. From then on, intensive spectroscopic analysis revealed ever more information about the elements contained in it, and it was observation of the 1868 eclipse from India that first suggested the presence of an unknown element, which later proved to be helium.

A much greater problem came with the 1869 eclipse, observed best from North America. The spectrum contained Fraunhofer lines unrelated to any known element, and by the eclipse of 1927 there were sixteen such lines – and that at a time when the periodic table was almost complete. Following photographs taken in West Africa during the 1893 eclipse, a new element *coronium* was conjured up as the explanation, and its existence (entirely based on spectrographic evidence from eclipse observations) was accepted for nearly 50 years.

Finally, in the 1940s, the Fraunhofer lines were related to the presence of iron ions, Fe X, which means that the iron atoms had lost nine[47] shell electrons – something only possible, given the intense energy required to reach this state, at the temperature of the interior of stars. The corona was not so much a gas as a plasma, with a temperature of a million-odd degrees kelvin. This result confirmed also Niels Bohr's quantum theory of the atom described in Chapter 7.

Finally the eclipse of 29 May 1919 was remarkable for photographs, taken at the time of totality, confirming Albert Einstein's general theory of relativity according to which gravity affects light in a way that can be accurately predicted mathematically. The method was simple. During totality the sun would be among the bright stars of the Hyades cluster in Taurus, and if Einstein was right, their position, as observed and recorded during totality, would be affected by the deflection of the light coming from them by the sun's gravitational field. For this purpose, expeditions equipped with portable telescopes and all related paraphernalia went out to Principe, an island off West Africa, and Sobral in north-east Brazil. The photographs taken would then allow the position of the stars to be compared with that recorded when the sun was far from Taurus. The reference photographs were taken at Greenwich in January 1919, and, once the comparison was made, the deflection of light then revealed was of 'the amount demanded by Einstein's general theory of relativity'.[48] The results were described by the President of the Royal Society, J. J. Thomson, as 'one of the greatest achievements in the history of human thought . . . the greatest discovery in connection with gravitation since Newton enunciated his principles'.

The results did not entirely justify the hype, and it can be convincingly argued that they did not 'come down on Einstein's side in an unambiguous way'.[49] There was nothing special about the equipment, and as often with eclipses there were problems with the weather during totality. Similar observations were made with subsequent eclipses, right up to that of 30 June 1973, observed in Mauretania. The results confirmed those of 1919, but Einstein (although he was not present at the special meeting of the Royal Society) then won the day because he had the unequivocal support of Sir Arthur Eddington, the Astronomer Royal, backed up by J. J. Thomson. In the 1980s, variations observed in binary pulsars[50] (which can keep time to an accuracy of one part in 10^{10}) have confirmed Einstein's general theory to a level of accuracy far beyond that possible with any eclipse observations. This was something Einstein and Eddington never even dreamt of.

The solar neutrino problem is at the interface between astronomy and particle physics: the mass of neutrinos[51] – only recently proved to be greater than zero – is of the order of a hundred-thousandth of that of an electron, so they are very small indeed. Yet, in their trillions, they cover distances measured in millions of light-years with very little to block their path. The journey from the sun to earth should therefore be no problem for neutrinos, but they do pose a considerable problem to solar physicists, since they are far fewer in number than required by the standard model, originating with Bethe, of the way energy is generated inside the sun. At the same time, their origin means that they provide evidence of the reactions taking place in the sun's core.

The practical problem is simply to detect neutrinos in the first place. The detector must be extremely sensitive, which means that it must be outside the range of anything else likely to activate it, such as cosmic rays. This problem was solved in 1964 by the American Ray Davis, whose detector consisted of a tank containing 400,000 litres of perchlorethylene (C_2Cl_4), a chemical otherwise used in dry-cleaning. After the removal of 7000 tons of rock, this was located 1500 metres underground in a gold mine in South Dakota.

Chlorine-36 was chosen, because neutrinos very occasionally interact with the nucleus of the isotope chlorine-37, which occurs in one case in four. When this happens a neutron in the nucleus converts into a proton by losing an electron in β-decay. The result is then to move one step up in the periodic system to produce argon-37. Because this is a radioactive isotope – with a half-life of 34 days – the amount present at any one time can be measured, to produce a figure for the number of neutrino interactions.

Most of the neutrinos produced by Bethe's standard proton–proton reaction lack the energy necessary for the chlorine reaction, but one possible track involves the conversion of boron-8 to beryllium in an instantaneous process that involves the emission of a high-energy neutrino. These boron neutrinos constitute about 5% of the number reaching the earth, but, even so, every second some 3,000,000 penetrate each square centimetre of the earth's surface. According to Bethe's standard model, this should lead to about twenty reactions per month in the perchlorethylene tank. In fact, records kept for twenty years only recorded about nine, a figure which had physicists tearing their hair out.

Better results were hoped for from two new neutrino detectors of a different type based on gallium and capable of detecting some standard-model neutrinos.[52] Both were built underground, one under a mountain in northern Italy and the other in the Caucasus. Different methodology still produced the same results.

The problem could have been solved by allowing the temperature at the centre of the sun, 15,000,000°K to be reduced by some 10%, but what the solar physicists already know – the accepted model – does not allow this. Another, more ingenious, solution proved to be correct. This is based on the MSW effect (so-called because of the initials of those who conjured it up). Some two-thirds of the electron neutrinos on their journey from the sun must then convert into either muon- or tau-neutrinos (denoted ν_μ and ν_τ respectively), which the existing apparatus could not detect. Early in 1998, such conversion, which is known as 'oscillation', was revealed by the Japanese Super-Kamiokande detector, working with neutrinos produced by the action of cosmic rays high in the earth's atmosphere.

The phenomenon observed also required that the masses of the neutrinos involved could not be equal to one other, so that logically, in any one event, at least one neutrino must have non-zero mass. There are three known classes (as shown in table in Appendix B), e, μ and τ,[53] and it is almost inconceivable that one of these is an exceptional case, with zero mass. The results from Japan are therefore accepted as proof that neutrinos do have mass, thus resolving a dispute that had engaged particle physicists for many years.[54] In Canada, the Sudbury Neutrino Observatory has now confirmed the Super-Kamiokande results for solar neutrinos. The target at Sudbury consists of a tank containing 1000 tons of heavy water, located at a depth of 2 kilometres in a disused nickel mine. Because the electrons released as a result of the interactions caused by solar neutrinos move at a velocity greater than that of light in water, they produce a visible effect known as Cherenkov radiation, whose intensity can be measured by light detectors on the sides of the tank. The results achieved at Sudbury then confirmed the theory, stated above, that oscillation caused the apparent loss of solar neutrinos.

Although the problem of neutrino mass has been solved for particle physics, astronomers still have any number of unsolved problems in neutrino detection. The problem is by no means confined to neutrinos originating in the sun: the vast new ANTARES[55] detector for stellar neutrinos is now being constructed on the sea-bottom, at a depth of 2.5 kilometres, some 50 kilometres off the Mediterranean coast of France, near Toulon. At this depth, so-called 'muon-neutrinos', occasionally colliding with neutrons or protons, will release a particular fundamental particle, the muon. Once again, its presence will be revealed by Cherenkov radiation, measured by a complex of some thousand spherical light sensors, fixed at 500-metre intervals on cables anchored on the seabed and suspended from buoys. The recorded data will then be transmitted to shore by glass-fibre cable, where its significance will be revealed by computer processing.

9

Physics: ground zero

The arctic regions in physics incite the experimenter as the extreme north and south incite the discoverer.[1]

Heike Kamerlingh Onnes

The dimensions of heat

Sir James Dewar
Is cleverer than you are.
None of you asses
Can condense gases.[2]

Heat, or its absence, is fundamental not only to humankind's experience of the environment but also to its attempts to change it. In everyday life, heat operates within very narrow limits, and our tolerance for heat, determined by body temperature, is even more constrained. Just think of all the investment in heating and air-conditioning; even so, except in the Sahara or Siberia, humankind has little direct need to cope with temperatures above 40°C or below −10°C. On a scale starting with absolute zero (explained on page 143) and extending to the temperature inside the hottest stars, this range is minuscule, but, even so, it defines the temperature boundaries of the life sciences (although certain specialized organisms survive outside this range[3]).

As Chapter 1 shows, temperatures above the upper limit of 40°C have always been part of human experience. The use of fire as a means for transcending nature is fundamental, so that cooking used temperatures up to about 300°C, while pottery-making and metal-working went well above 1000°C. Even so, only twentieth-century technology was up to working with temperatures above the highest melting points of metals, such as tungsten at 3410°C. But then, by the end of the century, instruments such as the tokamak produced temperatures above 10,000,000°C.[4] This, the level of plasma physics, defines an entirely new realm, at least for earthbound experimental scientists. For astronomers, such temperatures are routine: their problem is to measure them and work out, at distances measured in light-years, what they tell about the physics of stars.

The problem of measurement is critical. Science, in any temperature range, must work with some property or state of matter that, in a way that can be calibrated, changes when heat is applied. The familiar mercury thermometer is based on expansion and contraction of metal according to temperature. Apart from problems of accuracy, the range is limited by both the freezing and boiling points of mercury: conveniently, these (−38.9°C and 356.6°C respectively) are safely beyond everyday requirements.

Even within the range of the mercury thermometer, there are any number of alternatives; outside the range (for instance, in pottery kilns) one or other of these is essential. By convention, instruments measuring very high temperatures are known as pyrometers (from the Greek *pyr*, fire), while cryometers (from *kryos*, frost) measure very low temperatures.

The thermocouple, consisting of two wires made of different metals and joined at both ends, is often the key component in both instruments. In 1822 Thomas Seebeck (1770–1831) discovered that maintaining the two ends at different temperatures created a measurable electric potential, varying according to the difference in temperature. One junction of the thermocouple is then maintained at a standard reference temperature, while the other is exposed to the temperature to be measured. For some purposes a single thermocouple is sufficient, but thermocouples connected in series, to form a thermopile, make a more sensitive instrument.

Pyrometers operate on the principle that high-temperature sources emit radiation, which can be either infrared or visible light. One instrument – the total radiation pyrometer – uses a concave mirror to focus such radiation on to blackened foil connected to a thermopile, allowing the temperature of its source to be measured electrically.

Cryometry, to make any sense, first requires some principle governing the ultimate limits of low temperature: the simple question as to how cold can it get, proves to have a precise answer, although this was not appreciated before the eighteenth century. The situation changed in 1703, when Guillaume Amontons (1663–1705) discovered experimentally that for a constant volume of air any increase in pressure was directly proportional to the rise in temperature. He then calculated the 'extreme cold' of his apparatus at the point where there would no pressure: the result was purely theoretical, since he knew no way of ever attaining this state, which we now know is ultimately unattainable. Even so, the mathematics pointed to a constant, if unattainable, level below which no temperature could go.

In the nineteenth century the principle of an absolute zero was generally accepted, but estimates of its correct value varied widely. Scientists were taken by the idea of a remote and inaccessible point, known as *ultima Thule*,[5] an apotheosis of everything that made the little-known polar regions so terrible and forbidding. Only extreme scientific curiosity could lead anyone to explore this domain (which was first named, in 310 BC, by the Greek navigator Pytheas).

For one thing, the problem of attaining ever lower temperatures was quite different from that of high temperatures. The reason is simple: there are any number of sources of heat, many, such as fire, known since earliest times. There are no comparable sources of cold, although gradually over time processes were discovered where heat was lost so as to make a substance colder. One such process is evaporation: this occurs at the surface of a liquid when molecules with the highest kinetic energy escape into the atmosphere and enter the gas phase. The result is that the average energy of the molecules remaining in the liquid is lowered: this corresponds to a loss of heat.

The reasons for this, as shown below, only became clear in the nineteenth century: the principle was applied in antiquity to cool water jars. If the walls were porous, they would become saturated and the water on the outside surface would evaporate. The more heat that reached the surface, the more rapid the process of evaporation, and the more efficient that of cooling. This makes the system well suited to hot dry climates, such as North Africa and the Arabian peninsula – where earthenware jars are still used in this way. The price is paid in loss of the liquid contents: a quart of warm water becomes a pint of cold water. This is the earliest historical instance of practical refrigeration.[6]

Paradoxically, although a substance on becoming colder loses energy, the process requires energy to be supplied (as by the sun in Africa), so that the end result is a net expenditure of energy (which is why refrigerators add to the electricity bill). Although, once more, an explanation had to wait until the nineteenth century, artificial refrigeration in the laboratory began half a century earlier. A significant development was Martinus van Marum's discovery, in 1787, that ammonia (recently discovered by Priestley) became liquid when subjected to a pressure of five atmospheres. This is a general phenomenon: every gas will become liquid if subjected to sufficient pressure, provided that it is below a critical temperature.[7] This varies from one gas to another and is always substantially higher than the temperature at which the gas becomes liquid at standard pressure.

For ammonia, the boiling point in the liquid phase is only −33°C, which makes it useful in refrigeration. The same is true of carbon dioxide (−78°C), a gas which converts directly from the solid to gas,[8] a process known as 'sublimation'. (The lowest boiling point, that of helium, is −268.9°C, but this was not known until the twentieth century.) Faraday, after liquefying both ammonia and the even more recently discovered chlorine in 1823, was the first to see that the change from the gas to the liquid phase radically lowers the temperature – establishing a property extremely useful for practical everyday refrigeration.

The thermodynamic breakthrough

In 1824, Sadi Carnot (1796–1832) published his *Reflections on the Motive Power of Fire.*[9] Carnot, who had trained as an army officer, had concluded that British economic power was the result of technological supremacy, particularly in the use of steam power. This provided the motivation for research that would prove to be a key resource in the new science of thermodynamics. Carnot died young, and never tasted success.

Ironically, it was a Scotsman, William Thomson (1824–1907), later Lord Kelvin, who made the most use of Carnot's research. This equated heat with work in such a way that every unit increase in temperature represented a corresponding unit of work. The relation between heat and work had already been established, notably by Count Rumford, but Thomson's insight showed that the fixed points on the temperature scales in use since the early seventeenth century were defined by special cases: the melting point of ice and the boiling point of water had no general significance, and depended in any case on external factors such as pressure. On the other hand the Celsius scale was well established, so Thomson used it to define the units of temperature measured from absolute zero. The Frenchman Victor Regnault had already calculated this to be −272.75°C, a figure which Kelvin accepted. Then, on his scale (now the standard Kelvin scale), ice melted at 272.75°K and water boiled at 372.75°K[10] at standard atmospheric pressure.

This very important advance established the benchmark for all low-temperature physics. (For high-temperature plasma physics, with temperatures measured in millions, 273° does not make much difference one way or the other.) In 1851 Thomson formulated two laws governing thermodynamics, the science that studies the conversion of energy from one form to another, the two forms being heat and work. The laws prescribe the direction in which heat will flow and the availability of energy to do the work.

Although the laws were conceived of in terms of formal mathematics, they can be stated in ordinary language. According to the first law, the total amount of energy in an isolated system is always conserved, so that appearances to the contrary are always the result of conversion from one form to another. The second law states that the general tendency in nature is for heat to flow from hot to cold.

The first law is that of the conservation of energy, and the second, that of entropy. Rudolf Clausius, in Germany, had stated his own version of both laws in 1850 and proved the second by showing that if it did not hold then a perpetual motion machine could work to pour energy continually back and forth between hotter and colder bodies. In the present context, the combination of the two laws requires an input of energy from outside to maintain any refrigeration system. At the same time, the laws had important consequences for the age of the solar system: the energy of the sun was continually being lost by being radiated outside the system, and, indeed, if the rate of loss could be measured, it would provide the means for estimating the age of the solar system and also for predicting a period of time after which the earth would become too cold to sustain life. The theological implications were as drastic as those of Darwin's theory of evolution and just as difficult to accept for a man such as Thomson, with a strong religious commitment to the truth of the Bible. It was also the end to the 'balanced, symmetrical, self-perpetuating universe',[11] as conceived of in the seventeenth century by Boyle, Newton and their contemporaries.

Thomson's next step was to introduce electricity into thermodynamics. In 1834, experiments carried out in France by Jean Peltier (1785–1845) showed how, at the junction of two conductors, one made of antimony and the other of bismuth, an electric current could produce a rise or fall in temperature according to the direction in which it flowed. In 1851 Thomson showed that this was just one instance of reversal in so-called thermoelectric circuits. By the end of the century, this phenomenon was already being applied in refrigerators.

The liquefaction of gases

In spite of early progress in the first half of the nineteenth century, which had witnessed the liquefaction of chlorine, ammonia and carbon dioxide, in 1850 the six permanent gases, among them oxygen, nitrogen and hydrogen, still remained a problem. Attempts were made to solve it by increasing the pressure to new levels – up to 200 atmospheres.

The breakthrough came towards the end of the nineteenth century with a rush of activity and conflict, with the main players in England, France, Poland and, above all, Holland. Plainly the key to the problem was what happened to a gas as it hovered around the temperature threshold at which it became liquid. This proved to be not at all simple.

Johannes van der Waals (1837–1923), the son of an Amsterdam carpenter, started off in life as a primary school teacher, but, as he pursued his career, to the point of becoming a headmaster in the Hague, he took up the study of physics in Leiden – some 15 kilometres away. There he became interested in the liquefaction of gases by pressure, particularly in relation to the recently discovered critical temperature. He focused on an equilibrium area in which gas and liquid coexisted, and introduced a new factor, molecular density, into the established pressure–volume–temperature equation, the classic equation of state going back to the time of Robert Boyle 200 years earlier. Primarily a theorist, he developed new mathematical equations to fit the results of recent experiments in this area.

Essentially, van der Waals' equations take into account the fact that gas molecules have a finite volume, while at the same time the attractive force between them reduces the pressure on the walls of the container. For Maxwell in England their importance justified the study of 'the low Dutch language'. The equations are still fundamental (although Dutch scientists now publish in English[12]). They describe a laboratory threshold in which a gas–liquid mixture contains both low-density high-temperature molecules (gas) and high-density low-temperature molecules (liquid). This suggests the physical separation of the two states, with the immediate result that the liquid, by a process similar to evaporation (long used for refrigeration), would become much colder. Practical application came in the form of a cascade process, whereby, at each successive stage, liquefaction became possible at increasingly lower temperatures. Starting with a series of gases G_1, G_2, $G_3 \ldots$, with critical temperatures $T_1 > T_2 > T_3 > \cdots$ descending towards 0°K, G_1 in its liquid state is the starting point for the liquefaction of G_2, G_2 for G_3, and so on, getting colder and colder down the line.

The first stage can be taken to illustrate the process. The apparatus consists of a glass tube containing G_2, surrounded by another containing liquid G_1. The result is to reduce the temperature of G_2 to the level of G_1: if this temperature is less than T_2, then G_2, if subjected to sufficient pressure, will become liquid. This is more or less what the French physicist Louis Cailletet (1832–1913) did in December 1877. Choosing for G_1 sulphur dioxide (which is liquid at −29°C) and for G_2 oxygen (which, according to his calculations, should become liquid at around −200°C) Cailletet subjected the oxygen, in the inner container, to enormous pressure, at the same time cooling the sulphur dioxide to its liquid state. Then, by suddenly releasing the pressure on the oxygen, its temperature – conforming to the equation of state – dropped to −200°C. The result was an oxygen mist that condensed in droplets running down the walls of the container. Although this was no more than a momentary result, oxygen had been liquefied for the first time – the actual boiling point is −183°C.

In the new year, 1878, Cailletet went on to liquefy nitrogen (boiling point −195.8°C), but even so he had failed to find any means of creating a permanent reservoir of either gas in its liquid state – an essential step for the process to be really useful.

Success was first attained, in 1883, by two Polish physicists at the University of Kraków, Szygmunt von Wroblewski (1845–1888), a brilliant theorist, almost blind after six years' hard labour in Siberia (the price paid for joining a student protest against the Russians in 1863[13]), and Karol Olszewski, a practical chemist. Adapting

Cailletet's apparatus, they not only liquefied oxygen, but succeeded in keeping a measurable quantity 'boiling quietly in a test-tube'.[14]

This, unfortunately, was the end to the collaboration between two very strong personalities: von Wroblewski, left to conduct his own experiments, went on to hydrogen – a much greater challenge, given a boiling point more than 50° below that of nitrogen. Late one night, he was severely burnt after knocking over a kerosene lamp in his laboratory: he died three weeks later, aged only forty-three. Olszewski, after suffering several laboratory explosions of glass tubing, designed new apparatus using metal containers, which, on the continent was recognised as 'the greatest progress in the field of the liquefaction of gases'.[15]

This claim brought an immediate hostile reaction from the other side of the English Channel. Although he had many admirers, James Dewar (1842–1923) must be counted as one of the most quarrelsome men in the history of science. The son of a Scottish innkeeper, he was crippled for life after falling through the ice on a frozen pond when he was ten years old. Even so, he entered Edinburgh University at the age of seventeen, where he won many prizes and became acquainted with many eminent scientists.

At Edinburgh Dewar concentrated on chemistry, and worked with such distinction that in 1875, aged thirty-three, he was appointed Jacksonian Professor at Cambridge (where, according to the terms of the endowment, he should find a cure for gout). In spite of the fact that the Cavendish Professor, one James Clerk Maxwell, was another Scotsman, Dewar and Cambridge were not compatible. Not surprisingly then, Dewar, after being appointed Fullerian Professor at the Royal Institution in 1877, moved to its premises in London, remaining there for the rest of his life – but never giving up his chair at Cambridge (which he held for forty-eight years).

Once at the Royal Institution, Dewar, largely inspired by the success of Cailletet's experiments in Paris, decided that he was cut out for research in the same field. At the Friday evening lectures, Dewar followed the style of Faraday – helped by talking with his ghost, late at night, in the darkened building. Unfortunately, for all his undoubted flair as a lecturer and demonstrator, Dewar never came close to Faraday's sweet nature. On the contrary, hearing of Olszewski's success as a follower of Cailletet, Dewar immediately condemned him in the most bitter terms.

Dewar is particularly remembered for two remarkable achievements, the invention of the vacuum flask and the liquefaction of hydrogen. As to the former, Dewar realised that low-temperature research would profit enormously from any means of storing liquid gases at extremely low temperatures. For one thing, the cascade could start with a liquid gas at a far lower temperature, G_3 instead of G_1, or, in practical terms, nitrogen instead of sulphur dioxide. Dewar knew from bitter experience that volatile liquids, such as liquid oxygen, exploded ordinary glass vessels and caused metals to shatter. He found the solution in 1892, when he conceived of a glass vessel, with inner and outer walls enclosing a vacuum. After many setbacks, he produced a flask, coated on the inside with a thin layer of silver or mercury to reduce radiation losses, which perfectly met his requirements. The first model was presented to the Prince of Wales at a public meeting at the Royal Institution. The Dutch physicist, Heike Kamerlingh Onnes (1853–1926) described it as 'the most important appliance for operating at extremely low temperatures'.[16] The familiar Thermos flask, for household use – for some years actually known as a 'Dewar' – was produced almost immediately by a German manufacturer.

Following the success of Olszewski and von Wroblewski with oxygen in 1883, the liquefaction of hydrogen was the goal that everyone was aiming for. Within a year both of the Poles, working independently with different methods, claimed to have produced colourless droplets running down the side of a container, the same result as Cailletet had achieved with oxygen in 1877. Dewar, in an article in the *Philosophical Magazine*, commended Olszewski's work, noting that it opened the way to 'an accurate determination of the critical temperature and pressure of hydrogen'.[17] But then Dewar's own interest was fired, and in the race for success he never had another good word for Olszewski.

The race was longer than Dewar had expected. Already, in 1894, he seemed to be on the brink of success, but he reached the finishing line only in May 1898. Dewar's apparatus first produced liquid oxygen by a cascade process, which was then used to cool hydrogen down to a temperature of –205°C. This, under a pressure of 180 atmospheres, was released in a continuous stream through a jet into a vacuum vessel maintained at the same temperature. The process continued into a second vessel, contained within a third. Within five minutes this system produced 20 cubic centimetres of clear colourless liquid hydrogen. The jet then froze up as a result of air contaminating the hydrogen. Liquid oxygen, introduced into the liquid hydrogen in a glass tube, froze immediately, so confirming that a far lower temperature had been reached.

Characteristically for the low-temperature game, one thing led to another, and in 1899 Dewar, working with liquid hydrogen, succeeded in liquefying fluorine. Here he collaborated with Henri Moissan, the man who had discovered it. The problem was not the low boiling point (which is only −188.1°C), but the fact that fluorine, unlike many other substances, remains highly reactive even at such low temperatures. Dewar and Moissan unwisely mixed liquid hydrogen with frozen fluorine: the result was a violent explosion. This, however, was only a beginning, and, as Dewar predicted, 'with hydrogen as a cooling agent we shall get from 13 to 15 degrees of the zero of absolute temperature, and . . . open up an entirely new field of scientific inquiry'.

Helium: the ultimate challenge

After hydrogen, there was still one great challenge, helium, which remained a gas at temperatures below the liquefaction points for all other gases. Helium had only been discovered on earth in 1895 – by Sir William Ramsay, one of the many people at odds with Dewar. Inevitably, supplies would be both limited and expensive – a formidable obstacle to research. None the less, Kamerlingh Onnes, working in his laboratory in Leiden, determined to surmount it and go on to achieve liquefaction.

Social life in the provincial Dutch city of Groningen, where Kamerlingh Onnes grew up, was highly stratified. University professors, mostly with very long tenure, defined the intellectual elite in such a way that outsiders would hardly feel at home. This excluded the Kamerlingh Onnes family, which was headed by a successful manufacturer of roofing-tiles (whom the professors would have regarded as unspeakably bourgeois). Even so, the Kamerlingh Onnes family greatly valued culture and refinement, which restricted them to a very narrow social circle – so much so that they much preferred their own company. To their son Heike, this was more of a help than a hindrance.

Studying physics at the university, Kamerlingh Onnes, always top of the class, won first prize for an essay describing alternative methods of measuring the density of gas vapours. This led to a fellowship at Heidelberg, where he worked with Bunsen and

Kirchhoff, followed by four years' research, not in Groningen, but in Leiden, the only Dutch university with an experimental physics laboratory. Following his doctoral thesis in 1879 (which was judged to be of exceptional merit) Kamerlingh Onnes went on, in 1882, to succeed to one of the university's chairs in physics. In his inaugural lecture, he stated the principle 'Through measurement to knowledge'[18] – words which he wished to see inscribed above every portal in the laboratory.

The programme of research announced by Kamerlingh Onnes was clearly indebted to van der Waals' equation of state, and the two men, one in Leiden and the other in Amsterdam, soon became firm friends and collaborators (and would both later become Nobel prizewinners). Van der Waals saw Kamerlingh Onnes as 'almost passionately driven to examine the merits of insights acquired on Dutch soil', which – by the implicit reference to the pioneering work of van Marum – helps explain Kamerlingh Onnes' interest in low-temperature research. (In appointing Kamerlingh Onnes as professor, Leiden, taking into account his unquestionable Dutch origin, had preferred him to Röntgen, who, although born in Prussia, had lived in Holland since he was three. Röntgen, after discovering X-rays, won the first Nobel prize for physics: at the end of the day Kamerlingh Onnes' work in physics may have been equally important.)

Although appointed professor in 1882, Kamerlingh Onnes was slow to get going, partly because he insisted on replicating the experiments of Cailletet and the two Poles: only in 1892 did he have in his laboratory a large-scale liquefaction plant good for temperatures down to –250°C. Much of the problem was funding, but here, in the end, Kamerlingh Onnes proved to be a master – so much so that at the beginning of the twentieth century Leiden had the world's best-equipped low-temperature laboratory.

In the 1890s, however, Kamerlingh Onnes faced a quite different problem. Holland has an unhappy history of explosions devastating its cities. (The last occasion was on 13 May 2000, when an exploding firework depot destroyed a large part of Entschede.) Two centuries earlier Leiden had suffered a similar fate when a ship exploded. In 1895, the City Council, learning of Kamerlingh Onnes' work with compressed hydrogen, ordered the suspension of the whole low-temperature operation.

Kamerlingh Onnes, supported by testimony from Dewar (whose laboratory had actually suffered any number of explosions, costing two of his assistants the loss of an eye), went to the Dutch courts, and in the end the Supreme Court ruled that his laboratory research could resume. Many years had been lost, however, so that it was Dewar who first liquefied hydrogen. This left Kamerlingh Onnes to accept the greatest challenge of all, the liquefaction of helium.

An immediate problem was finding a sufficiently abundant natural source: in England the only source was controlled by Ramsey, while only Dewar could produce liquid hydrogen. The stand-off between the two blocked advance in their respective laboratories, until in 1901 one of Ramsay's colleagues constructed a workable hydrogen liquefier. But Ramsay wanted liquid hydrogen for experiments leading to the isolation of inert gases besides argon and helium (which he had successfully isolated in the 1890s). Dewar, ever intent on liquefying helium, was still blocked, so that when Kamerlingh Onnes, in 1905, was able to tap an American source, he was ahead in the race.

By this time Dewar had won access to the English source, but had refused to share it with Kamerlingh Onnes: his pretext was that the supply was too limited. Kamerlingh Onnes, on his side, was able to crow, stating that 'the preparation of helium in large quantities became chiefly a matter of perseverance and care'.[19] He was on a roll: a confidant of the young Queen Wilhelmina, he enjoyed the support of everyone who counted in Holland. Physicists from all over the world wanted to work in the well-equipped Leiden laboratory. A letter from a certain Albert Einstein did not even merit a reply (although he and Kamerlingh Onnes would later become good friends).

The helium target was contained in a pressure chamber, surrounded by a glass jacket for containing liquid hydrogen. The key experiment started at 5.45 a.m. on 10 July 1908. By 4.20 p.m. the temperature in the apparatus (including the helium in the thermometer) had been reduced, according to plan, to −180°C. Helium was then let into the pressure chamber, and the glass jacket filled with liquid hydrogen. By steadily adding more hydrogen and increasing the pressure on the helium, the temperature differential became negative at 6.35 p.m.: the helium was now colder than the hydrogen, but it still was not liquid. At every stage, as the pressure was raised, some of the helium was released through a porous plug, so that what remained became colder. The process was repeated until, at around 7 p.m., Kamerlingh Onnes used his last bottle of liquid hydrogen. In despair at the possibility of failure, he looked through the glass bottom of the pressure chamber, and saw the outline of a liquid. He had succeeded after all: his only disappointment was that van der Waals, who had remained in Amsterdam, was not present to share his triumph.

Dewar was notified by telegram. True to character, he sent a reply claiming much of the credit:

> CONGRATULATIONS GLAD MY ANTICIPATIONS OF THE POSSIBILITY OF THE ACHIEVEMENT BY KNOWN METHODS CONFIRMED MY HELIUM WORK ARRESTED BY ILL HEALTH BUT HOPE TO CONTINUE LATER ON.

Kamerlingh Onnes allowed Dewar the credit for predicting, correctly, many properties of liquid helium – that it would be difficult to see, have low surface tension, and boil at 5°K.[20] (Dewar only proved wrong on density: helium was eight times lighter than water, and not five times as he had predicted).

The achievement, although remarkable, was just the beginning of scientific work with liquid helium: for one thing, Kamerlingh Onnes never succeeded in solidifying it, even though he did reduce the temperature to 1.04°K. He also got a production line going, good for 4 litres an hour, and in 1911 he developed a viable container for storing it – absolutely essential for further research.

Kamerlingh Onnes now had the means to work on a phenomenon first predicted in 1854, and already researched by the two Poles in the 1880s: this was the decline in the electrical resistance of metals at progressively lower temperatures. The temperatures reached with liquid helium promised an astonishing result: closed circuits at this temperature level could maintain an electric current indefinitely. This was superconductivity, finally achieved by Kamerlingh Onnes in 1911.[21]

The strange world on the threshold of absolute zero

This was the end of a long journey. For one thing, Kamerlingh Onnes had to contend with an argument put forward by Lord Kelvin (who had only died in 1907) that resistance, as it approached 0°K, would tend to infinity. But then, in 1906, the German physical chemist Walther Nernst (1864–1941) argued that the energy of an isolated system remained constant: only interaction outside the system, defined by both an input of work and the absorption of heat, could bring about a change in energy. This became the third law of thermodynamics. Applied to low-temperature physics, its result is that as temperature approaches 0°K entropy tends to zero, while at the same time the required conversion energy tends to infinity. If this were true – and Kamerlingh Onnes was persuaded that it was – then Lord Kelvin was mistaken, and electrical resistance would tend to zero.

After one or two false starts with other metals, notably platinum, Kamerlingh Onnes decided to work circuits made of mercury, which his laboratory could produce in a state of exceptional purity. The basic apparatus was designed to use helium to cool mercury in a U-shaped tube with wires running out at both ends connecting it to a galvanometer, which would measure its resistance.

As the temperature fell to around 20°K, the resistance declined in step with it – in mathematical terms, a linear relationship. Below this point, the decline continued, but in irregular fashion, to 4.19°K, at which point it fell abruptly to a point at which the galvanometer could not measure any resistance at all. This result was so astonishing that the experiment was repeated with improved apparatus. The result was always the same: at 4.19°K, the galvanometer reading consistently fell to zero. In 1913 Kamerlingh Onnes won a Nobel prize: in his acceptance speech he was unable to explain the phenomena he had observed, but he did suggest that 'they could possibly be connected with the quantum theory'.[22] As explained in Chapter 6, this theory, stated by Max Planck in 1900, was already revolutionizing the state of the art in physics.

A whole new world was opening up. Scientists had long predicted that ultra-low temperatures would be a special case. Back in 1887, Dewar, in a lecture at the Royal Institution, had predicted that 'molecular motion would . . . cease, and . . . the death of matter ensue'.[23]

Dewar, true to character, probably overstated the case, but in 1907 Einstein offered a quantum explanation for a phenomenon first observed by Cailletet in 1875. This related to the specific heat of metals: the basis was a standard figure for each metal, measuring the amount of heat required to raise the temperature of 1 gram by 1°C. As far back as 1819 this had been shown to vary with temperature according to Dulong and Petit's law, named after the two French physicists who had first stated it. This, a simple algebraic equation, proved to be no longer valid once experiments could be conducted at the low-temperature levels reached in the last quarter of the nineteenth century.

In particular, at about 20°K the specific heat of copper suddenly dropped by a factor of about 30. According to Einstein, the copper atoms, one by one, then cease vibrating, a quantum phenomenon that would explain the temperature drop. He was able to state a new equation, valid from the realm governed by Dulong and Petit down to 10°K. This showed for the first time that quantum theory could explain a phenomenon, at the same time confirming Nernst's recently stated third law of thermodynamics.

Finally Kamerlingh Onnes, as a result of experimenting with the specific heat of liquid helium, had to reckon with Einstein, and found him wanting. The year was 1924, and Kamerlingh Onnes, eighty years old, had left the actual laboratory work to a young American chemist, Leo Dana. What they discovered was a pronounced increase in the specific heat of liquid helium at 2.2°K. Already, in 1911, Kamerlingh Onnes had noted that its density peaked at the same temperature – so that in the lowest temperature range liquid helium actually expands. Graphs showing the changes of specific heat and density in this ultimate temperature range both have a characteristic shape similar to the Greek letter lambda. Consequently, such a diagram is known as a λ-curve, and the critical temperature, 2.2°K, as the λ-point.

Dana returned to America, and two years later Kamerlingh Onnes died. Three weeks later, his successor to the Leiden Chair, Willem Keesom, succeeded in solidifying helium, although only at a pressure of some 25 atmospheres – this was just the beginning of a new voyage of discovery. Keesom, realising that the fundamental properties of helium changed radically at the λ-point, proposed that two distinct phases, He I and He II, of liquid helium existed, respectively, above and below this point. What is more, he found that at a pressure below 25 atmospheres, liquid He II would extend as close to 0°K as it would ever be possible to reach.[24]

The discovery of the λ-point related directly to a theoretical construct, first suggested by Einstein in 1925, following up a theory of the Indian physicist, Satyendra Bose (1894–1974) – an expert on quantum physics. Einstein contended that as atoms, in a domain close to 0°K, approached standstill (the ultimate limit that can never be achieved), their wavefunctions merged, to form an entirely new form of matter. He II opened the way to this process, whose end-point – if it were ever reached – would be this new Bose–Einstein condensate (BEC). It took until 1995, seventy years later, to close the gap between the λ-point and the first drop of BEC – hardly surprising, seeing that the temperature at which this emerged was 0.00000000017°K.[25]

In this period, one development followed another with He II. The most spectacular was the discovery, by Peter Kapitsa (1894–1984) in 1935 of superfluidity. Kapitsa, born and brought up in Kronstadt, a Russian naval base near St. Petersburg, studied electrical engineering in the capital city's Polytechnic Institute. By the time he graduated, in 1918, the capital had moved to Moscow following the October Revolution of 1917. but Kapitsa stayed on as a lecturer in what had now become the Leningrad Polytechnic. There, a leading physicist, Abram Joffe, recognising his extraordinary gifts, had him invited to joint a Soviet scientific delegation to Western Europe. (The invitation was particularly welcome after Kapitsa had lost his son to scarlet fever and his wife and daughter to the flu epidemic that followed the First World War.) The delegation reached its destination, but without Kapitsa: Germany, France and Holland had all refused visas to this apparent youthful agitator – a completely false assessment. Kapitsa, left stranded in Estonia, had the same idea as Niels Bohr in 1909: he would go to England to work with Rutherford, who by this time had become director of the Cavendish Laboratory in Cambridge.

Rutherford, not surprisingly, first set Kapitsa (who apparently had no trouble getting a British visa) to work on radioactivity. This led him to a Clerk Maxwell Studentship in 1923 and a fellowship of Trinity College in 1925. He also remarried, and became the father of two more children. As Kapitsa's reputation grew, so did his freedom to write his own research agenda – which meant turning his attention to liquid helium. In 1934, C. J. Gorter and H. B. G Casimir, colleagues of Keesom in Leiden, suggested that helium existed in two states at very low temperatures. While in one state electrons continued to behave normally, the other began to assume the characteristics of the as yet unattained BEC – so that with zero entropy it was unable to transport heat.

The year 1934 was a critical turning-point for Kapitsa: this followed a visit to his mother in the Soviet Union. Another fellow of Trinity, the mathematician A. S. Besicovitch (1891–1970), had warned him against the visit, but Kapitsa would not listen. Besicovitch, who had escaped from the Soviet Union in 1925, proved right: Kapitsa was not allowed to return to Cambridge. Instead he was appointed director of the Moscow Institute for

Physical Problems, where, in 1946, he would be suspended – for eleven years – after refusing to work on nuclear weapons.

Although the Cavendish Laboratory allowed all Kapitsa's apparatus to be sold to the Moscow Institute, low-temperature research continued there, and with remarkable results. The apparatus was simple. An inverted glass bulb, containing a heating element, was placed in a bath of liquid helium. To begin with the levels of the helium, inside and outside the bulb, are the same, but when the heater is switched on, the rise in temperature inside the bulb causes the helium vapour to expand and the level of the liquid helium to fall. The result of the increase in pressure causes a difference in the two levels and this difference is a measure of the temperature difference.

This effect will cease, that is, the difference in levels will disappear, as a result of heat being conducted from the bulb into the bath, and the rate at which this occurs is then a measure of the conductivity of the liquid helium. All the differences were minuscule, so the apparatus had to be made with extreme precision if they were to be observed at all. With He II the results were astonishing: first, the rate of conductivity increased as the temperature difference decreased, and, second, at the lowest levels the conductivity proved to exceed that of He I by a factor measured in millions.

That was not all. At the very lowest temperatures, heating the He II inside the bulb caused its level to rise – the reverse effect to that described above. This could not have been caused by a difference in vapour pressure, because then the He II closest to the heater would actually have become colder – a result too paradoxical even in this extreme situation. This result was then confirmed with apparatus having an open-ended bulb, thereby ensuring that all the He II is subject to the same pressure. The effect observed was entirely new: He II was a liquid which flowed, when heated, in the direction of the heat source. This is known as the 'thermomechanical' effect.

Experimenters in Oxford demonstrated the reverse, the 'mechanocaloric' effect. In this case – without any heater – He II, flowing out of the bulb, became cooler in the process. Oxford and Cambridge together had shown, therefore, that with He II the flow of mass and the flow of heat occur in opposite directions: what is more, reversing one reverses the other.

In the course of all these experiments, another phenomenon was observed. Apparatus that at temperatures above 2.2°K appeared to be perfectly sealed leaked at lower temperatures, to the point that the apparatus failed completely. It was becoming plain that below the λ-point there was a pronounced drop in viscosity. This was the first appearance of superfluidity.[26] First measurements suggested that viscosity below the λ-point was about a tenth of that above it, but Kapitsa, using apparatus similar to that of the Oxford team, showed that the viscosity of the flow through the narrow channels could drop to as low as a millionth. The phenomenon was then confirmed in England by an experiment in which a small glass beaker was lowered into a bath of He II. In due course the levels of the liquid in the bath and the beaker stabilized at the same height: this could only occur by helium flowing up the sides of the beaker, and over its brim – an extraordinary result.[27]

At the same time, Lazlo Tisza, working at the Collège de France, proposed a two-fluid model of He II: one component is simply He I, while the other, superfluid, component increases in proportion as the temperature approaches 0°K, at which point there would be no He I at all. This explains the high level of viscosity produced by Kapitsa's apparatus, since only the superfluid will flow without friction through the narrow channels. This also explains the thermomechanical and mechanocaloric effects introduced above.

An explanation of the difference between He I and He II was published by Kapitsa in 1941 (by which time the war had put an end to low-temperature experiments in England), but he gave the credit to his colleague L. D. Landau (1908–68). The basis was a virtual particle called the *phonon*, which automatically implies a quantum effect. Phonons only appear as the result of the vibration of the helium atom, so that at absolute zero they would not be present. Helium is then conceived of as a superfluid matrix through which phonons travel. The matrix defines He II, while the phonons define He I.

After the Second World War, when the making of the atom bomb had led to new techniques in isotope separation, experiments were carried out with the light isotope ^{3}He (instead of the normal ^{4}He) of helium. The main problem was scarcity: only one in 10,000 helium atoms is ^{3}He. The difference was significant for the equations on which the Bose–Einstein theory was based, since these assumed the common ^{4}He isotope. For ^{3}He there was the alternative Fermi–Dirac[28] theory (introduced in Chapter 7), but it was not known how this would relate to the experimental results achieved with ^{4}He. One significant result was that helium, cooled below 0.8°K, was no longer homogeneous, but split into two separate phases: one, based on ^{4}He, being superfluid; and the other, based on ^{3}He, remaining in the normal state. In fact superfluid ^{3}He was only achieved in 1972 and then at the remarkably low temperature of 0.003°K. This stage is played out at the level of particle physics described in Chapter 10, and depends on the distinction between two sorts of elementary particles, bosons and fermions. This is a world that Kamerlingh Onnes never dreamt of.

To sum up, the liquefaction of helium creates a domain in which the character of key physical properties, such as viscosity, diamagnetism, and electrical and thermal conductivity, is radically changed. This can only be explained by subatomic phenomena occurring only at very low temperatures. Of these, one of the most remarkable is that of an electric current carried by Cooper pairs of electrons – named after the American Leon Cooper (1930–) who discovered them in 1956 – with equal but opposite momentum and spin, which means that they are free from the loss of momentum caused by the collision of single electrons; such an event, constantly occurring in the case of a normal current, explains electrical resistance. Without such resistance, an electric current will maintain itself indefinitely: this is superconductivity.

SQUIDs

Because Cooper pairs are not subject to normal electrical resistance, they can 'tunnel' across an insulating barrier from one superconductor to another. In a circuit containing such a barrier, the phenomenon only occurs with a current below a very low, but definite, critical threshold. Once above this threshold, tunnelling no longer occurs, and the current assumes the normal flow pattern of single electrons. This superconducting sandwich is known as the 'Josephson junction', after the British physicist Brian Josephson (1940–), who conceived it in 1962 while still a Cambridge research student. (He went on to win the Nobel prize for physics in 1973.) A single junction is minute: thousands, connected in series, can be printed on a silicon base the size of a playing card so as to constitute a useful potential difference. Their usefulness is immense: the change of state can be recorded with extreme accuracy, enabling very small currents to be measured. State-of-the-art technology maintains the instrument at the temperature of liquid helium, and it is now a standard component in a range of instruments known as SQUIDs – superconducting quantum interference devices.

The Josephson junction is a means for penetrating to the heart of the quantum world. Josephson himself discovered that a direct current voltage across the junction generated an alternating current frequency proportional to that voltage. This effect could then be used to determine one of the basic constants of physics, e/h (the electron's charge divided by Planck's constant), which is now the basis of an internationally recognised quantum standard of voltage.

Recent developments are even more breathtaking. Following the first production of BEC in 1995, a group at Harvard produced a temperature of 0.00000000005°K in a medium where the speed of light is reduced to just over 50 kilometres/hour: with this degree of cold to work with, it becomes possible to predict how matter would behave in the limiting case of zero entropy at the unattainable temperature of 0°K. *Ultima Thule* is in sight.

10

Big science

Particles and quanta

In Chapter 1, having stated that 'no-one can come to terms with the whole of big science', I then promised, perhaps unwisely, a final chapter telling what this topic has meant for the history of science in the years since the Second World War, when it came into its own. As for the war itself, Chapter 7 describes the galaxy of scientific talent assembled at Los Alamos, to work out the basic science needed for the development of the atomic bomb. One is left asking, what have all these people, to say nothing of younger generations, done since then? Some who were at Los Alamos are still alive: at least one of these, the physicist Hans Bethe, is unimpressed by more recent achievements. When well over eighty he noted that, with the development of quantum mechanics 'the understanding of atoms, molecules, the chemical bond and so on, that was all complete in 1928'.

This sweeps a lot of physics under the carpet, but this was also the view of Steven Weinberg, according to whom 'the development of quantum mechanics may be more revolutionary than anything before or after'. At all events, the 1920s, with Bohr, Heisenberg, Schrödinger, Kapitsa, Dirac, Pauli, Chadwick, Compton, and many others recounted earlier in this book, all active, was certainly a golden age in physics. Have we seen the like of these men in the last half-century?

What, for instance, is the contribution of particle physics, the *raison d'être* of much of the big science of the last half century? Once again, a leading American physicist, David Mermin, found the right put-down: 'All that particle physics has taught us about the central mystery is that quantum mechanics still works. Perfectly, as far as anybody can tell. What a let-down!'[1]

None the less today's domain of big science, for which Los Alamos can be taken as a prototype, is far from anything known in the 1920s. (Any outsider can see that the old Cavendish Laboratory at Cambridge and Los Alamos belong to different worlds.) The defining characteristic of big science is its scale. This can be measured according to any number of criteria, such as the number of people employed, the amount of money invested, the published reports of results, the dimensions, and above all the range and accuracy of the apparatus used. Compared with a base year, say 1942 (when Fermi's atomic pile, CP-1, went critical), the multiplication factor, for all relevant criteria, is a matter of thousands, if not millions.

So what, then, is the state of the game in particle physics at the beginning of the new century? With all that equipment something must have been achieved, to say nothing of the creative input of men such as Murray Gell-Mann, Richard Feynman and the Dutch Nobel prizewinner, Gerard 't Hooft, whose book *In Search of the Ultimate Building Blocks*[2] is the best attempt so far to explain everything to the layman.

The message is deceptively simple: the basic Rutherford–Bohr atom, extended by Chadwick's discovery of the neutron – all of which is described in Chapter 7 – is not the end, but the beginning of the story, as physicists such as P. A. M. Dirac were beginning to suspect even before the end of the quantum revolution of the 1920s. The ultimate building blocks are the twenty-odd particles, which can be classified according to the components listed in the table in Appendix B,[3] representing the *standard model*. These, confusingly, are not homogeneous; quarks, as can be seen from their charges, can occur only in combination, so their existence is a question of pure deduction from experimental observation.

The particles implicit in the standard model are extremely small and extremely short-lived. This means that every single one of them is bound to decay into other particles as a result of fundamental interactions, which occur constantly and with breathtaking rapidity. Beginning in about 1930, the problem facing experimental physics has been to design and construct apparatus that can produce observable events, confirming the properties of the fundamental interactions proposed by the theorists. The results, in practical terms, are described later in this final chapter. If the new apparatus works according to plan – which is a very big 'if' – then particle physics, and our understanding of the universe, will reach the end of the road. The reason is given by 't Hooft:[4]

> I would like to stress emphatically how extraordinary the standard model really is, even though it contains twenty numbers for which we do no know why they have the values they have, so that we also do not know how to calculate them from first principles. But once these numbers are given, we can, 'in principle', calculate any other physical phenomenon. All the properties of the fundamental particles, the hadrons, the atomic nuclei, atoms, molecules, substances, tissues, plants, animals, people, planets, solar systems, galaxies and perhaps even the entire universe, are direct consequences of the standard model... I should hasten to add that all these statements are of not much more than philosophical significance and that they mean quite little in practice. We are not at all able to deduce the properties of a cockroach using our standard model, and this will never change.

We can breathe again. Even so, it is possible, even in this book, to go a little further and look at the structure of contemporary particle physics. This is explained in Murray Gell-Mann's *The Quark and the Jaguar*[5] (whose title could have substituted 't Hooft's cockroach for the jaguar). Chapter 13 in this book, 'Quarks and All That: The Standard Model', outlines the basic structure, and clothes it with some flesh and blood. The model takes two forms: the first, quantum electrodynamics (or QED), although predicted by Dirac some forty years earlier, was developed in the 1960s, with Feynman playing the lead role; the second, quantum chromodynamics (or QCD), came into its own some ten years later, and was largely the brainchild of Gell-Mann himself. Both forms depend upon fundamental interactions, commonly represented in the form of Feynman diagrams, described by Gell-Mann[6] as 'funny little pictures... which give the illusion of understanding what is going on'. Figure 10.1 presents an example of a Feynman diagram, showing the basic electron–electron scattering interaction. The wavy line represents the elementary particle that is the agent of the interaction, in this case virtual photons. Such diagrams then represent the fundamental processes occurring in vast numbers on an infinitesimal timescale in all matter throughout the universe: things may slow down at temperatures close to absolute zero (as explained in Chapter 9), but in the world as we know it the pace never slackens. Even with the help of the giant

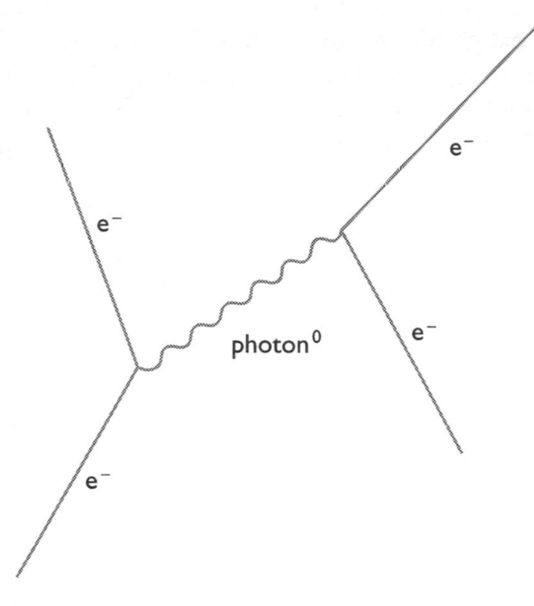

Figure 10.1 Feynman diagram showing two electrons exchanging a virtual photon, which gives rise to the electromagnetic force between them.

apparatus described later in this chapter, almost everything takes place beyond the limits of direct observation.

Modern physics is concerned with four fundamental interactions: gravitational, electromagnetic, strong and weak. Gravitation, critical for astronomy, because of its action between massive celestial bodies separated by vast distances, has yet to be successfully integrated into quantum theory (although quantum gravity may have been critical for the big bang). When it comes to gravity, general relativity – as first propounded by Einstein – is the name of the game, although the first general theory originated with Newton. Maxwell's equations (see Chapter 7) are the foundation of modern electromagnetic theory, but this now belongs to QED, and the basic Feynman interaction is electron–electron scattering, represented by exchanges of virtual photons, as illustrated in Figure 10.1. QED is particularly important for the opportunities it offers for experimental confirmation. Because this allows for unprecedented accuracy at the level of one part in 1,000,000,000, QED is the most accurate theory known to physics.

The strong interaction, operating at very short range inside the atomic nucleus, is the reason for the nucleus' great stability: this is the realm of QCD. The basic interaction for such experimental observation as the state of the art allows at CERN and other such places is an exchange of virtual mesons between hadrons, but then what are mesons and hadrons? Hadrons, quite simply, are subatomic particles that interact by the strong interaction, and mesons are a subclass of hadrons. Because they consist of quarks, they are not truly elementary. According to theory, mesons consist of a quark and an antiquark, so that – from the table in Appendix B – the charge is 0, while baryons, defining the alternative class of hadrons, consist of three quarks, with a combined charge of 1 (in the case of a proton) or 0 (in that of a neutron).

The result is that relevant Feynman diagrams substitute quarks for electrons and gluons for photons – but what then are gluons? They are simply quanta, named as such, because they 'glue' quarks together. This process is essential to the existence of protons and neutrons as the constituent elements of the atomic nucleus. The role of quarks in QCD is, however, much more complicated than that of electrons in QED. As is shown in Appendix B, there are three groups of quarks, I, II and III, each containing two quarks, one with a positive charge, 2/3, and the other with a negative charge, −1/3. Quarks can be one of six 'flavours': u (up), d (down), s (strange), c (charm), b (bottom) or t (top). They also possess 'colour' (red, green or blue), a property that, for the strong interaction, corresponds to charge for the electromagnetic interaction. Neither the flavours nor the colours (which account for the name quantum-*chromo*-dynamics) describe actual properties of quarks, which by their nature can be neither tasted nor seen. The notation was simply adopted as a convenient way of showing how quarks operated in different circumstances, or combined to form protons and neutrons: in this latter case, the three different fundamental colours of the component quarks produces a distinctive white nucleon. This may seem to be unnecessarily esoteric, but in fact the notation does in fact explain the different interactions, and their results, in which quarks are the basic particle.

Significantly, this is not all pure theory. Three American physicists, J. I. Friedman, H. W. Kendall and R. E. Taylor, working with the SLAC accelerator described in Chapter 7, discovered the first actual traces of quarks as early in 1967 – so that they were not merely figments of Gell-Mann's imagination. Experiments continuing over the succeeding years confirmed and extended the earlier results, and the three physicists finally won the Nobel prize in 1990. The basic operation was to bombard protons in the atomic nucleus with a stream of electrons at different energy levels. The quantum effect of every collision is then the exchange of a photon, whose wavelength is resolved by an electron microscope. At a certain stage, when the wavelength was sufficiently short, the results of experiment indicated that the photons were being absorbed within the proton. This could only happen as the result of interaction with elementary particles at a lower level (in a way reminiscent of that which led Rutherford to discover the atomic nucleus in 1911 – confirming what Gell-Mann had long been insisting on, that is, that the proton and the neutron are not elementary particles. The gluon is then required as a construct to explain what binds the component quarks together.

The strong interaction, which binds the component protons and neutrons together within the atomic nucleus, then resolves into interactions between quarks. Since, however, quarks can never be separated from the particles of which they are the constituent parts, it is impossible to work with them directly. This is one reason why the new accelerator being installed in the 26 kilometre tunnel at CERN is a Large *Hadron* Collider (LHC).

The Higgs boson

Some time in this first decade of the third millennium the Nobel prize for physics is almost certain to be awarded for the discovery of the Higgs boson. As far back as 1993, William Waldegrave, then the British Minister of Science, offered five bottles of champagne as prizes for the best one-page essay explaining why the Higgs boson was so important. The prizewinners were all top particle physicists, but none of them so far has found the Higgs boson.[7] What is this elusive particle, and why is it so important? And who, incidentally, was Higgs?

A fundamental question in particle physics is what gives particles mass? According to the simplest theoretical version of the standard model, the masses of all particles are zero. Since our very existence depends on each separate elementary particle having its own mass according to the class to which it belongs – as shown in the table in Appendix B – there must be some factor explaining reality as we know it. In the 1960s, the British physicist, Peter Higgs (1929–) suggested that the whole of space was permeated by the eponymous 'Higgs field' (comparable to the electromagnetic field), with the property that all particles, as they pass through it, acquired mass. The principle of wave–particle duality, fundamental to quantum physics, then requires a particle, the 'Higgs boson', to be the agent of interaction.

Theory suggests that the Higgs field can best be approached via the weak interaction, which relates to β-decay of the atomic nucleus (see Chapter 7), the generation of solar energy (Chapter 8) and superconductivity at low temperatures (see Chapter 9). Then, according to electroweak theory,[8] which gives a unified description of the electromagnetic and weak interactions, the Higgs boson, if it exists, must have properties related to the W boson[9] and Z boson, the elementary particles involved in the weak interaction. (The actual W and Z bosons were first observed experimentally in 1983.[10])

The discovery of the W and Z bosons was the result of the largest single project ever undertaken in particle physics: the vast power required would still not have been sufficient to reveal the Higgs boson. This would have been a task for the abandoned Texas supercollider (described below on page 353), but success must now wait upon the new LHC at CERN, although Fermilab's Tevatron might just get in first.

The computer revolution

Whatever Hans Bethe and David Mermin may say about the achievements of particle physics in the last half century, computers have increased the capacity to process data by a factor so large as to add an entirely new dimension to what can be achieved by experiment and observation in almost all branches of science – including those beyond the scope of this book. Yet the computer, as a working instrument available to science, did not exist in 1942, so that Fermi, as he watched CP-1 go critical, was the whole time busy with a slide-rule – today a decidedly obsolete instrument. When IBM first began to work on computers, its chairman, Thomas Watson, predicted that there would be perhaps five in the whole world.

The whole computer scene was transformed, first by the invention of the transistor in 1948,[11] and then by that of integrated circuit – the familiar 'chip' – in 1960.[12] This is still the basic state of the art, although the speed of operation, and the range of programs, has increased exponentially over the past forty-odd years. The result for practising scientists, in any field, is that they can work with volumes of data far exceeding anything previously possible. At the same time, state-of-the-art equipment, such as the most advanced particle counters, produces data at a commensurate rate.

What this means in another field, chemistry, can be seen from research carried out by two Americans, Herbert Hauptman and Jerome Karle, into the determination of molecular structures, both inorganic and organic. Chapter 6 shows how this is a classic topic in chemistry, as was the means adopted to explore it: X-ray crystallography. None the less Hauptman and Karle still won a Nobel prize in 1985, for devising systems of equations, based on the measured intensity of deflected X-rays, which could only be solved by state-of-the-art computer programs. The usefulness of the final results, to be read off from computer printouts, was based on a statistical procedure whose accuracy depended on a large number of repeated applications – impossible without a computer. In a parallel field, molecular biology, Aaron Klug, in England, had already won a Nobel prize, in 1982, for his use of an electron microscope for examining viruses, nucleic acids (including DNA) and chromosomes: although the basic equipment goes back to the 1930s, the transmission and scanning electron microscopes available at the end of the twentieth century greatly increased the range of possible observation.

Next to the computer, the laser, invented in 1960, is probably the most significant new instrument available to science – and also to medicine. The technology of the laser is not transparent, but the result, a narrow beam of 'collimated' light of a single wavelength – typically that of the red part of the spectrum – is easy to observe (although it should be noted that the laser is not confined to optical wavelengths). In quantum terms, the laser is essentially a beam of photons, but a comparable phenomenon, based on a beam of atoms, was produced at the American MIT laboratory in 1997. This development had to await the successful production of the Bose–Einstein condensate described in Chapter 9: any number of practical applications are expected.[13]

A remarkable use of lasers earned Ahmed Zewail – both an Egyptian and an American – a Nobel prize in 1999 in a field known as 'femtochemistry'. The focus is on the actual chemical reaction, whose duration is measured in femtoseconds, that is in multiples of 10^{-15} seconds (which is to a second as a second is to 32,000,000 years). Laser flashes of the order of 10–100 femtoseconds then allow the methods of spectroscopy (described in Chapter 6) to be applied on the timescale of the vibrations of the atoms in a molecule, which is the same as that of the chemical reaction. (This can be seen as a super-refined form of flash photography.)

The results were spectacular, because chemists could observe for the first time substances known as intermediates formed in the course of the transformation from the first to the final stage of the reactive process. Zewail's method, being sufficient for the fastest possible reactions, needs no further improvement. This is a fitting climax to research into the rate of chemical reactions, carried out at the end of the nineteenth century by Svante Arrhenius and J. H. van't Hoff (in 1901 the first winner of a Nobel prize for chemistry).

At the end of the day we must return to big science in the domain of particle physics, whatever the doomwatchers like Hans Bethe may be thinking. All that money spent on supercolliders, as at CERN, SLAC and Fermilab, must have had some underlying rationale, even if in 1993 this proved too much for the United States Congress (which in March of that year stopped funding the Texas SSC – 'superconducting supercollider' – but only after 16 out of 54 miles of tunnel had been built at a cost of $2 billion). Without the Texas SSC, Gell-Mann despairs of ever producing experimental results to bridge the 'supergap', which is the heart of the ultimate 'superstring' theory, whose purpose is to unify the whole of particle physics, the final apocalypse, inherent – if Bethe is right – in the quantum mechanics developed in the 1920s.

The underlying theory, which is purely mathematical, is the brainchild of the American physicist Edward Witten, a man spoken of with bated breath, whose citation record is almost beyond credibility. The theory is based on the premise, well-established among particle physicists, that to every instance of matter there is a corresponding instance of antimatter. Going back to the end of the 1920s, when Bose–Einstein statistics complemented Fermi–Dirac statistics (as related in Chapter 7), supersymmetry requires, in the fundamental boson–fermion dichotomy, that every boson has a partner that is a fermion, and vice versa. In a perfect realisation this would produce a state of 'supersymmetry', in which everything would cancel out, and the history of the universe, going back to the big bang, would have been a 'super-non-event'.

Gell-Mann's *supergap* is the result of the fact that this did not happen. In the event some particles were without partners – relatively speaking not very many, but still more than enough to create the cosmos as it is still being discovered by astronomers (as described in Chapter 8). Witten's superstring theory is designed to explain this, but the problem is that the sort of laboratory events that might confirm it require an energy level that could only have been provided by the Texas SSC. The first congress of the Clinton presidency plainly has much to answer for, although Gell-Mann hopes that CERN's LHC – only due to open in 2005 – might just do the trick,[14] along with revealing the Higgs boson.

The problem can be simply stated: Witten's superstring theory, by incorporating the standard model of particle physics, represented it in a form that was 'renormalisable' – in other words, it could be brought it down to earth to the point that certain predictable phenomena could be confirmed by experiment – provided the apparatus could operate at a sufficiently high energy level.

The discovery of the τ-lepton in 1995 meant that the last component of the standard model had been found experimentally. Subsequent experiments at CERN have confirmed that there are no more 'standard' families of quarks and leptons waiting to be discovered.

All this begs the question as to the way in which the particle events, which underlie the whole range of experiments, are observed. The cloud chamber described in Chapter 7 held the field until the bubble chamber was invented by Donald Glaser in the 1950s: charged particles (produced by an accelerator) pass through a chamber containing superheated liquid. The particle interactions then leave a trail of bubbles (such as are to be seen when water boils in a kettle), and these can be photographed.

The system worked so well that it led CERN to construct the giant 1000 tonne Gargamelle detector in 1970. This was used for experiments with neutrinos until 26 October 1978 when a large crack appeared in the chamber body. By this time Georges Charpak had developed a quite different, and much better, system for studying reactions between elementary particles. The bubble chamber had had its day, the Gargamelle was dismantled, and for more than twenty years the hall which had housed it – the highest building at CERN – was empty. The end of the 1990s witnessed the start of the process of converting it so as to house the ATLAS tile calorimeter for measuring the energy of particles emerging from LHC collisions. This should be complete in 2003, two years ahead of the LHC itself.

Charpak's counter works on the same principle as the ionisation chamber introduced in Chapter 7, but the anode consists of a large number of thin parallel wires contained between two cathode planes, as shown in Figure 10.2. Each separate wire can then register particle events, and state-of-the-art amplification then allows them

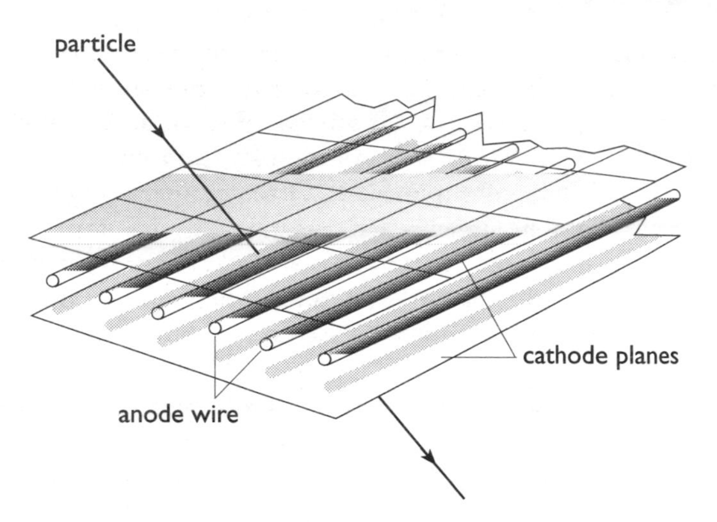

Figure 10.2 Charpak's particle detector.

to be recorded by computers – in some cases at a rate of several hundred thousand per second. Charpak's original discovery was in 1968, and from the 1970s it has led to such exotic, but still fundamental, results as the discovery at CERN of the charm quark in 1974 and the intermediate W and Z bosons in 1983.[15]. It has now led to a new 'drift chamber', where events can be recorded with a resolution of 60 micrometres.

With accelerators and colliders defining the apparatus used in particle physics, CERN can be taken as exemplary for their operation. This, in itself, is simple enough, and goes back to the early days, described in Chapter 7, when Rutherford was using radioactive elements to produce α-particles. Within a generation their range proved to be too limited for research requirements, which explains why Lawrence developed the cyclotron, where massive electric power would boost the energy of particles by a factor measured in millions. Unbelievably, his first cyclotron was only 10 centimetres in diameter. That of the tunnel at CERN is more than 800,000 times greater, and the electromagnetic coils surrounding the accelerator chamber contained inside the tunnel provide energy at a comparable increase in scale.

The results are to be seen in the vast energy of the charged particles accelerated within the chamber when they hit their target. The results of any experiment with the collider depend on two factors: the first is the particle chosen for acceleration; the second is the composition of the target with which it must eventually collide.

In addition to electrons and positrons, the particles accelerated by the original LEP collider, the LHC will operate not only with protons and antiprotons but also with ions, that is, the charged nuclei of atoms, such as oxygen, sulphur and lead – all of which are suited for particular kinds of experiment. The particles originate with linear accelerators, or 'linacs', with one for protons and ions, and another for the much lighter electrons and positrons. These then go, via booster circuits, to the proton synchrotron, continue through the super proton synchrotron, and on to the main LEP/LHC collider circuit.

As for targets, the field is wide open, and every experimenter will make his choice according to the results he is looking for. Whatever the experiment, these are likely to be particle collisions that can be registered and counted by particle detectors, such as that of Charpak described above.

The earthbound supercolliders still cannot compete with the vast energy of stars, as observed by modern astronomy, whose instrumentation, described in Chapter 9, also belongs to big science. The difference in scale is a matter of trillions. There is no better measure of big science than the fact that it works either with the smallest constituent particles of matter or with energy levels observed at the furthest reaches of the universe. The two cases can be taken to coalesce in the Never-Never Land of Witten's theoretical physics.

Quantum measurement

At the same time, and at a more mundane level, familiar units of measure, such as the second and the metre, are now established to a degree of accuracy – about one part in 10,000,000 – beyond the actual requirements of engineers and physicists.[16] This result, however, also belongs to big science. Its basis is quantum measurement, based on counting quantum events. This is easier said that done, since such events are microscopic in scale and occur at astronomical frequency levels. This is the problem that faced the Walrus and the Carpenter when they contemplated counting the grains of sand on the seashore.

None the less, in two remarkable cases, the problem has been solved. The first case is that of the Josephson junction, described in Chapter 9. The problem is that, in an operating range defined by the microwave radiation band (10–100 gigahertz), each junction corresponds to a voltage in the range 0.00002 to 0.0002 volts. On the other hand, since the microwave frequency that determines the actual voltage, helped by the caesium atomic clock,[17] can be regulated with an accuracy of one part in 1,000,000,000,000, the voltage of a Josephson junction can be set at the same level of accuracy. Modern technology can now produce a Josephson junction, as part of an integrated circuit, on a scale such that its length and breadth are about one-twentieth of a millimetre: miniaturisation then allows tens of thousands to be incorporated in series into a single circuit, to produce a useful aggregate voltage in the range 1–10 volts. The fact that the Josephson effect requires superconductivity does of course mean that any instrument based upon it must be kept very cold indeed: this is ensured by tank-loads of liquid helium that would have made Kamerlingh Onnes gasp with amazement.

The second case depends on a phenomenon first discovered by the American Edwin Hall in 1879. If a current I is passed through an appropriate semiconductor placed in a magnetic field at right angles to it, a potential difference V_H arises in a direction perpendicular to both the current and the magnetic field. This then defines the 'Hall resistance' R_H, in accordance with Ohm's law, so that $V_H = IR_H$. The Hall effect is produced with a semiconductor at a very low temperature, such as is required by the Josephson junction. (The Hall apparatus, a sort of chip, is also on a microscopic scale.) R_H then proves to be quantised, that is, it occurs only in multiples of 25812.8 ohms. This, according to quantum theory, is exactly equal to h/e^2, where h is Planck's constant, and e is the (negative) electric charge of a single electron. This has led to a new unit, the 'klitzing', to measure Hall resistance: this is defined as $h/4e^2$ – equivalent to 6453.2 ohms. Once again, because it is quantised, this is a unit that can be standardised within extreme narrow limits – of the order of one part in 10,000,000.

Since the Josephson effect measures e/h and the quantum Hall effect measures e/h^2, elementary algebra allows the two to combine to provide extremely accurate values for both e and h.[18] At the same time, the normal principles of the conversion of energy, outlined in Chapter 5, should eventually allow a standard unit of mass to be derived. This would then replace the prototype platinum kilogram kept at the International Bureau of Weights and Measures at Sèvres, just outside Paris. Since the metre is already defined in terms of c, the velocity of light in a vacuum, and the second in terms of the radiation frequency of the caesium atom,[19] this final step would mean that all basic physical constants were defined according to standard invariable phenomena, whose occurrence does not require any earthbound context. A second is a second, and a metre is a metre, in even the furthest reaches of the universe, and would remain so even if some catastrophe were to eliminate the planet on which the inventors of these standards once happened to live. Since 1 January 1990, the same has been true of units such as the volt and the ohm used in electricity. This followed an international agreement (which required a minute reduction in British laboratory standards – apparently unnoticed by the Euro-bashers). Only the kilogram (known in France as 'le grand K') remains domesticated, but at the level of the whole planet earth: for the time being, humankind still cannot do without Sèvres, but this will not last long.

One is left to ask whether such accuracy and the big science which is indispensable to it are necessary. The two are in fact dependent on each other, with feedback in both directions. Space exploration, at the present state of the art, would be impossible without the caesium clock and, closer to earth, the same is true of the global positioning system (GPS), which with the help of satellites is now the universal basis of navigation. Miniaturisation of computer components is now at a level measured in hundreds of atoms. The kilogram is now regarded as the ultimate stumbling-block, and the Watt balance, now under development at the US National Institute of Standards and Technology, is intended to define a standard weight in terms of counteracting electromagnetic forces, which are now, with the help of the Josephson junction and the quantum Hall effect, measurable in universal standard units. A cavernous underground vacuum chamber, immune to environmental influences, is being constructed to house the latest model of the Watt balance.

The dethronement of 'le grand K', first defined after all the trials in France of the 1790s described in Chapter 3, will be a landmark in history. It could happen at any time, although some resistance is to be expected from the French, who regard Sèvres as a national shrine. The event, when it comes, may be hardly worth a headline. After the new dawn has broken, science, which is nothing without measurement, will no longer be domesticated to this world, which, so far as we know, contains the only community capable of sustaining it. Looking back at the turn of the twenty-first century, future historians of science may well see the discovery of the Higgs boson, the final unravelling of the human genome and the dethronement of 'le grand K' as part of the same cosmic process.

Notes

Preface

1. G. Bell & Sons: the last edition appeared in 1965.
2. New York, Sygnet Science Library, 1964.
3. Cambridge University Press, 1997.
4. 't Hooft, op. cit., p.110.
5. This is an acronym for Centre Européen pour la Recherche Nucléaire; the work of this institution (now known as the European Organization for Nuclear Research) has gone far beyond the atomic nucleus, but the name CERN survives.
6. For much fuller treatment, see Peter Galison and Bruce Hebly (eds.), *Big Science: The Growth of Large Scale Research*, Stanford University Press, 1992, but of course the picture has changed considerably in the last ten years.
7. The bomb that destroyed Hiroshima was developed by a different process.
8. This may be an original use of the term, but post-modernism has long been a well-established theme in political and social science.
9. *Kunde* is a somewhat archaic word meaning much the same as *wetenschap*. The adjective, *kundig*, is still current, and connotes 'expertise'.
10. This point is well made in James Conant's forward to T. S. Kuhn, *The Copernican Revolution*, Harvard University Press, 1957.
11. See T. Crump, *The Anthropology of Numbers*, Cambridge University Press, 1991, pp. 45f.
12. Compare W. C. Dampier, op. cit., p. xiii: 'But at an early stage men almost invariably took a wrong path.'
13. In Holland's oldest university, Leiden, lectures were in Latin until 1839.
14. Quoted, J. and M. Gribbin, *Halley in 90 Minutes*, Constable, 1997, p. 65.

Chapter 1 – The mastery of fire

1. A. N. Whitehead, *Science and the Modern World*, Cambridge University Press, 1946, p. 144.
2. Paris, Plon, 1962.
3. See Philip Lieberman, *Uniquely Human: The Evolution of Speech, Thought and Selfless Behaviour*, Harvard University Press, 1991. For the communicative and cognitive potential of chimpanzees, see pp. 154f.
4. MIT Press, 1962.
5. Mouton, The Hague.

6. This statement is based on Section VII, 'Child Language Acquisition', in David Crystal (ed.), *The Cambridge Encyclopedia of Language*, 1987. The tendency of modern research is to discover linguistic competence at ever earlier stages in life.
7. 'Back to the Future: Trophy Arrays as Mental Maps in the Wopkaimin's Culture of Place', in Roy Willis (ed.), *Signifying Animals: Human Meaning in the Natural World*, Routledge, 1994, pp. 63–73. This whole book is relevant to the present stage of the argument.
8. The oldest zoo in Europe, in the gardens of the imperial palace of Schönbrunn in Vienna, dates from the eighteenth century.
9. Dorothy Sayers and Robert Eustace's *The Documents in the Case* is a detective novel, which turns on this question. Some mushrooms also contain psilocybin, a well-known hallucinogen.
10. Compared with the English-speaking world, the Slavonic world takes mushrooms very seriously and is much better informed about them.
11. Walter Ong's *Orality and Literacy: The Technologizing of the Word*, Methuen, 1982, p. 50.
12. Ibid., p. 72.
13. Gerard 't Hooft, *In Search of the Ultimate Building Blocks*, Cambridge University Press, 1997, p. 5.
14. On the very first day of this research four random informants produced between them a list of 650 plant names. D. E. Breedlove and R. M. Laughlin, *The Flowering of Man: A Tzotzil Botany of Zinacantan*, Smithsonian Institution Press, 1993, vol. I, p. 1.
15. Ibid., p. ix.
16. 'The Descent of Man and Selection in Relation to Sex', in P. H. Barrett and R. B. Freeman (eds.), *The Works of Charles Darwin*, vols. 21–23, William Pickering, London, 1989, p. 49.
17. *Eléments de chimie*, Leiden, 1752, vol. I, p. 144.
18. J. Goudsblom, *Fire and Civilization*, Penguin 1994, p. 162. This book provides much of the material for the present section.
19. Who thus became the first man in history known to have discovered an element: phosphorus was obtained by reducing human urine by boiling. The white residue that then appeared was given the name 'bearer of light' because it glowed in the dark. Little was recorded of Brand's work, and phosphorus was rediscovered in 1680 by Robert Hooke, whose work is well documented.
20. It is described for Zinacantan in Breedlove and Laughlin, op. cit., vol. II, pp. 531–2.
21. Harper Colophon Books, 1975, trans. from *Le cru et le cuit*, Librairie Plon, 1964. The early origins of cooking are discussed in R. W. Wrangham et al., The Raw and the Stolen, Cooking and the Ecology of Human Origins, *Current Anthropology*, vol. 40, 1999, pp. 567–94.
22. *Problemata*, III, 43.
23. Silicon is a metalloid, which means it has certain metallic properties: this explains why it is so useful in semiconductors and hence gives it name to Silicon Valley.
24. Some, such as mild steel, are alloyed with non-metals, in this case carbon.
25. Richard Harwood, *Chemistry*, Cambridge University Press, 1998, p. 65.
26. Silicon dioxide, SiO_2.
27. A. L. Oppenheim, *Ancient Mesopotamia: A Portrait of a Dead Civilization*, University of Chicago Press, 1977, pp. 323f.
28. Calcium oxide, CaO.
29. Sodium carbonate, $NaCO_3$.

30. Written Chinese, contrary to popular belief, does not have this property: see John de Francis, *Visible Speech: The Diverse Oneness of Writing Systems*, University of Hawaii Press, 1989. The words in quotes are to be found on pp. 5 and 21.
31. This remains true, even though specific spellings may distinguish the meanings of homophones, e.g. 'their' and 'there' in English.
32. Michael Ventris, who first proved this by deciphering inscribed clay tablets from Mycenaean sites in Greece and Crete, was an architect; his remarkable story is told by John Chadwick, *The Decipherment of Linear B*, Cambridge University Press, 1959.
33. For greater detail, see Thomas Crump, *The Anthropology of Numbers*, Cambridge University Press, 1991, p. 45.
34. This definition is an adaptation of that in Crump, ibid., p. 73.
35. The significance of resistance in the theory of electric currents follows the work of Georg Ohm (1787–1854), whose name is now that of the unit in which it is measured.
36. This is the sidereal month: for different ways of defining a month or a year, see W. M. Smart, *A Textbook of Spherical Geometry*, 6th edn., Cambridge University Press, 1977.
37. Why this is so is explained in T. S. Kuhn, *The Copernican Revolution*, Harvard University Press, 1957, pp. 266.
38. The Islamic lunar calendar, by failing to do this, begins some eleven days earlier, in every successive solar year. This is critical for the timing of the ninth month, prescribed for the fast of Ramadan, which at present is in the middle of winter.
39. See the diagrams in T. S. Kuhn, op. cit., pp. 9 and 12.
40. Which first appeared in Egypt in the fifteenth century BC, but it was only introduced into Greece by Anaximander (611–547 BC) some thousand years later.
41. The division of the hour into 60 minutes (*partes minutae*) and the minute into 60 seconds (*partes minutae secundae*) is part of the legacy of the Babylonian system of counting on a base of sixty.
42. Karl Menninger, *Number Words and Number Symbols: A Cultural History of Numbers*, MIT Press, 1969, p. 162. This book well describes the linguistic as well as metrical chaos of the premodern world.
43. The inch comes from the *uncius*, a lexical form denoting a twelfth, in this case, of a *pes*, or foot. This in turn defined a 'pace', and a mile was then a thousand (*mille*) paces.
44. There was an intermediate unit, the *solidus*, defined by 12 *denarii*, so that there were 20 *solidi* in a *libra*. This was the origin of the familiar l.s.d. of the old British currency.
45. C. A. Ronan and J. Needham, *The Shorter Science and Civilization in China*, Cambridge University Press, 1981, vol. II, p. 137.
46. See, for example, E. Ohnuki-Tierney, *Illness and Culture in Contemporary Japan: An Anthropological View*, Cambridge University Press, 1984.
47. The record is analysed in detail in T. Crump, *Solar Eclipse*, Constable, 1999, pp. 81f.
48. The words quoted come from the entry in the *Larousse Dictionary of Scientists*, Larousse, 1994.
49. Kuhn, T. S., op. cit., p. 128. Plato's scientific thought is mainly to be found in his *Timaeus*.
50. Ibid., p. 129.
51. Aristotle, *Physics* (trans. Robin Waterfield), Oxford University Press, 1999.

52. This explains the word 'planet', which comes from the Greek for 'wanderer': in the case of the sun, its movement across the firmament of stars can only be observed just after sunset or before sunrise. During the hours of daylight, the light of the sun makes it impossible to see the stars, although the moon is occasionally visible.
53. Aristotle, op. cit., IV, 5.
54. Ibid., VIII, 7.
55. Ibid., VIII, 4b24.
56. Ibid., Introduction by Robin Waterfield, p. xvi.
57. Ibid., p. xxxi.
58. Ibid., p. xlii.
59. Claudius Ptolemaeus.
60. Claudius Galenus.
61. *c.* 460–377 BC. The Hippocratic oath is still fundamental in Western medical ethics.
62. For a full explanation, see M. Hoskin (ed.), *The Cambridge Illustrated History of Astronomy*, Cambridge University Press, 1997, p. 41.
63. For a full explanation, see Hoskin, ibid., p. 44.
64. G. E. R. Lloyd (ed.) *Hippocratic Writings*, Penguin 1978, Introduction, p. 51.
65. See Kuhn, op. cit., p. 104.
66. B. Russell, *A History of Western Philosophy*, Allen & Unwin, 1946, p. 551.

Chapter 2 – The rebirth of science

1. Kuhn, T. S., *The Copernican Revolution*, Harvard University Press, 1957, pp.101–2.
2. See J. A. Weisheipl (ed.), *Albertus Magnus and the Sciences*, Toronto Pontifical Institute of Medieval Studies, 1980, pp. 9–10.
3. Meryl Jancey, *Mappa Mundi: The Map of the World at Hereford Cathedral*. Hereford Cathedral Enterprises, 1995, pp. 27–8.
4. Kuhn, op. cit., p. 112.
5. Bertrand Russell, *A History of Western Philosophy*, Allen & Unwin, 1946, p. 485, tells us that Aquinas 'is not engaged in an inquiry the result of which it is impossible to know in advance . . . he already knows the truth, it is declared in the Catholic faith'.
6. See Crump, *The Anthropology of Numbers*, Cambridge University Press, 1991, pp. 45–6.
7. The famous Gutenberg Bible was printed some five years later.
8. This is now in the Science Museum in London. The yard, before 1497 known as the 'ulna', was defined as equal to three feet by a statute of 1305, according to which the foot was also equal to twelve inches, and the inch, to three grains of barley.
9. Kuhn, T. S., *The Copernican Revolution*, Harvard University Press, 1957, p. 136.
10. Ibid., p. 140.
11. Ibid., p. 148.
12. Ibid., p. 155.
13. Ibid., p. 170.
14. Ibid., p. 126.
15. Tycho's nova we now know to be a supernova, a quite different phenomenon, as explained in Chapter 8.
16. This meant 5400 separate divisions, each of about a fiftieth of an inch.

17. Apollonius (*c.* 270–220 BC) was the founder of the geometry of conic sections, of which the circle is a special case. As an astronomer he considered eccentric orbits for planets, but for some 1800 years the idea was hardly followed up.
18. As noted in the book of Exodus (xiii: 21): 'And the Lord went before them . . . by night in a pillar of fire, to give them light.'
19. E. G. Ruestow, *The Microscope in the Dutch Republic: The Shaping of Discovery*, Cambridge University Press, 1996, p. 6.
20. Remarkably, *Science* (6 April 2001) reported that physicists at the University of California, San Diego, had succeeded in producing a transparent medium with a negative refractive index, so that both the incident and the refracted light are on the same side of the normal. The medium consisted of a lattice of glass plates, each some 0.25 millimetres thick, with copper strips etched on one side, and copper circles on the other. Refraction was produced with microwaves (wavelength 3 centimetres) and a refractive index of −2.7. In 1968 the Russian physicist V. G. Vesalago had proved that such an effect was not incompatible with Maxwell's electromagnetic theory (introduced in Chapter 4): the actual realisation promises a number of practical uses, but so far the effect has not been achieved with normal light.
21. The law was discovered by the little-known English mathematician Thomas Harriot (1560–1621), some twenty years before Snel. It was also later discovered independently by Descartes, who was much better known: see Bernard Williams, *Descartes: The Project of Pure Inquiry*, Pelican Books, 1978, p. 18. At all events, it is known to physics as Snel's law.
22. Originally known as Gerhard van Lauchen.
23. Probably, although not quite certainly, on 15 February.
24. Mathematics was essential to medicine in the preparations of the astrological charts used to prescribe correct treatments of illness. See I. Grattan-Guinness, *The Fontana History of the Mathematical Sciences*, 1997, p. 177.
25. Actually a German, Christoph Clau (1538–1612).
26. Enrico Caetani.
27. Michael Sharratt, *Galileo: Decisive Innovator*, Cambridge University Press, 1994, p. 70.
28. This happened in 1630 when Francesco Fontana produced telescopes with convex eyepieces in place of Galileo's plano-convex eyepieces.
29. This must be attributed to Christiaan Huygens (1629–95) in 1656, twenty-four years after the death of Galileo.
30. Thomas S. Kuhn, *The Copernican Revolution: Planetary Astronomy in the Development of Western Thought*, Harvard University Press, 1957, p. 220.
31. Ibid., p. 225.
32. Sharratt, op. cit., p. 107.
33. The term was actually coined by the Roman Church in the nineteenth century.
34. Aristotle (supported by Plato and Socrates) is portrayed as the greatest of the numerous figures from the ancient world mentioned by Dante: see *Inferno* IV: 131–4 and *Purgatorio* III: 43.
35. Now best known for its frescos by Fra Angelico.
36. Sharratt, op. cit., p. 114–15.
37. Ibid., p. 127.
38. Ibid., p. 129.
39. Ibid., p.130.
40. Stillman Drake (1970), trans. from the original *Dialogo* '. . . dove nei congressi di quatro giornate si discorre sopra i due massimi sistemi del mondo tolemaico e copernicano'.
41. Sharratt, op. cit., p. 161.

42. Ibid., p. 178.
43. E. G. Ruestow, *The Microscope in the Dutch Republic: The Shaping of Discovery*, Cambridge University Press, 1996. p. 7. This book provides much of the material for the present section.
44. Ibid., p. 178.
45. Ibid., p. 184. Circulation itself was discovered by William Harvey (1578–1657).
46. Nicolaas Hartsoeker's claim to prior discovery in 1674 is very dubious: ibid., p. 23.
47. The most accessible study is Simon Schama, *The Embarrassment of Riches: An Interpretation of Dutch Culture in the Golden Age*, Collins, 1987.
48. Ruestow, op. cit., p. 156.
49. The English botanist Nehemiah Grew (1641–1712), who, with the help of the microscope, produced the first complete account of plant anatomy, did much of his work in Leiden.
50. See his *De Mulierum Organis Generationi Inservientibus Tractatus Novus*.
51. See his *De Virorum Organis Generationi Inservientibus*.
52. Ruestow, op. cit., p. 105.
53. Ibid., p. 139.
54. Ibid., p. 116.
55. Ibid., p. 119.
56. Ibid., p. 63.
57. T. Craanen (ed.) *Tractatus Physico-Medicus de Homine*, Leiden, Petrum van der Aa, 1689, p. 234.
58. Ruestow, op. cit., p. 92.
59. Ibid., p. 76.
60. Ibid., p. 79.
61. Ibid., p. 116.
62. Ibid., p. 201.
63. Ibid., pp. 211–12.
64. Ibid., p. 212.
65. H. Collins and T. Pinch, *The Golem: What You Should Know about Science*, Cambridge University Press, 2nd edn., 1998, Chap. 4.
66. Ibid., p. 223.
67. See Thomas Crump, *Man and his Kind*, Darton, Longman & Todd, 1973, p. 15.
68. Published in his *Exercitationes de Generatione Animalium*. London, typis Du-Gardianis, 1651.
69. Ruestow, op. cit., p. 232.
70. *Philosophical Experiments and Observations of the Late Doctor Robert Hooke*, London, Frank Cass, reprint, 1967, pp. 261, 268.
71. See particularly two papers published in the *Philosophical Transactions of the Royal Society*: 'Concerning the Various Figures of the Salts Contained in the Several Substances' (vol. XV, 1685, p. 1073) and 'Concerning the Figures of the Salts of Crystals' (vol. XXIV, 1705, pp. 1906–17).
72. Ruestow, op. cit., p. 284.
73. Ibid., p. 322. The Italian, Giovanni Amici (1786–1863), was, in 1812, the first to produce an achromatic microscope.
74. For medicine see Roy Porter, *The Greatest Benefit to Mankind*, Fontana, 1999, p. 609.
75. R. S. Westfall, *The Life of Isaac Newton*, Cambridge University Press, 1993, p. 20, citing Newton's contemporary, John Strype. Westfall provides much of the biographical material used in this chapter.

76. Quoted, T. S. Kuhn, Introduction to Newton's Optical Papers in I. B. Cohen (ed.), *Isaac Newton's Papers & Letters on Natural Philosophy and related documents*, Harvard University Press, 1958, p.28.
77. Ibid. p. 25. This is a variant of what had already been said by Aristotle, referring to Plato: *Nicomachaean Ethics*, Book I, 1096a16.
78. Bernard Williams, *Descartes: The Project of Pure Inquiry*, Pelican Books, 1978, p. 24.
79. I. Gratton-Guinness, *The Fontana History of the Mathematical Sciences*, Fontana Press, 1997, pp. 223–5.
80. As reflected in his best known principle: *Cogito, ergo sum* (I think, therefore I am).
81. This description of Cartesian theory is based on that in Michael Hoskin (ed.), *The Cambridge Illustrated History of Astronomy*, 1997, pp. 135–7.
82. Westfall, op. cit. p. 28.
83. Kuhn, op. cit., p. 39.
84. Newton's own explanation of this figure is as follows: 'In the annexed design of this Experiment, *ABC* expresseth the prism set endwise to sight, close by the hole *F* of the window *EG*. Its vertical angle *ACB* may conveniently be about 60 degrees: *MN* designeth the Lens. Its breadth $2\frac{1}{2}$ or 3 inches. *SF* one of the streight lines, in which difform Rays may be conceived to flow successively from the Sun, *FP*, and *FR* two of those rays unequally refracted, which the Lens makes to converge towards *Q*, and after decussation to diverge again. And *HI* the paper, is divers distances, on which the colours are projected: which in *Q* constitute Whiteness, but Red and Yellow in *R*, *r* and *t*, and Blue and Purple in *P*, *p*, and *w*.'
85. See Oldenburg's letter to Newton, Westfall, op. cit., p. 82.
86. Ibid., p. 83.
87. Kuhn, op. cit., p.28.
88. Ibid., p. 30.
89. See Bella Bathurst, *The Lighthouse Stevensons*, Flamingo, 2000, pp. 137–44.

Chapter 3 – Science, technology and communication

1. G. L'E. Turner, *Scientific Instruments 1500–1900: An Introduction*. Philip Wilson, 1998, p. 42.
2. The whole question is discussed in T. Crump, *The Anthropology of Numbers*, Cambridge University Press, 1991, pp. 23.
3. Fourth Estate, 1995.
4. Quoted, Sobel, op. cit., p. 52.
5. In 1936, the German airship, *Hindenburg*, whose lift was provided by hydrogen, was consumed by fire on its arrival in the United States. This was the end of the use of hydrogen. Helium, an inert gas, is now used, but being heavier than hydrogen it does not provide the same lift and it is also much more expensive.
6. One example, reported in *The Times*, 27 April 2000, is an experiment called 'Boomerang', carried out by an international team working with a balloon circling the South Pole: tiny temperature variations measured in background radiation provided important evidence of what the universe was like some 300,000 years after the big bang, at a time when it was 1000 times smaller and hotter than it is now.
7. In the course of the eighteenth century, three French expeditions, by timing a pendulum of constant length in Lapland, Peru and South Africa, had by 1752 determined conclusively that the earth is flattened at the poles.

8. The arithmetic is straightforward: if the degrees in the two latitudes are L and L', and the distance separating them along the meridian is d, then the 90° length of the quarter-meridian is $d \times 90/(L - L')$.
9. D. Guedj, *Le mètre du monde*, Editions du Seuil, 2000, p. 92.
10. Volts measure the potential of the external electric supply, amps the current in the circuit of any appliance, and watts the energy of the appliance. Amps are commonly specified in fuses, where they indicate the maximum current possible before the fuse blows. This is the result of the heat generated by the current melting the fuse wire.
11. The London Science Museum has an exhibit consisting of a glass canister containing both a propeller and a jet engine, each revolving in its own horizontal plane around a central vertical axis. The air is then slowly evacuated, to the point that the propeller engine no longer revolves, while the speed of the jet engine is greatly increased.
12. J. Needham, *The Shorter Science and Civilisation in China*, (abridged by C. A. Ronan), Cambridge University Press, 1978, vol. 1, p. 52.
13. E. Stuhlinger and F.I. Ordway, *Wernher von Braun*, Krieger Publishing Company, 1994, p. 226.
14. The actual records are inconsistent, but see the tables in F. I. Ordway and M. R. Sharpe, *The Rocket Team*, Thomas Y. Crowell, 1979, p. 406.
15. E. Stuhlinger and F.I. Ordway, op. cit., p. 61.
16. In 1960 this was renamed the George C. Marshall Space Flight Center in honour of Gen. George C. Marshall.
17. Cited in Stuhlinger and Ordway, op. cit., p. 113.
18. As early as 1923, Hermann Oberth, in his *The Rocket into Interplanetary Space*, had seen the satellite as ideal for astronomical telescopes.
19. With an orbital period of 113.2 minutes it remained in orbit for more than twelve years.
20. Stuhlinger and Ordway, op. cit., p. 137. With observations designed by the American physicist, James van Allen (1914–), *Explorer I* had revealed an unexpectedly high level of radiation in the upper atmosphere. This was the result of the earth's magnetic field trapping high-speed charged particles from outer space in two doughnut-shaped zones – the 'van Allen belts'.
21. The first liquid propellant rocket was launched by the American Robert Goddard (1882–1945), in Worcester, Massachusetts, on 16 March 1926. Although he was an underfunded loner, who achieved relatively little, the Goddard Space Flight Center, outside Washington DC, is named after him.
22. The acronym is for the National Aeronautics and Space Agency, established by President Eisenhower in 1958.
23. Quoted in Stuhlinger and Ordway, op. cit., p. 251.

Chapter 4 – Discovering electricity

1. Quoted in I. B. Cohen, *Benjamin Franklin, Scientist and Statesman*, Charles Scribner's Sons, 1975, p. 49.
2. In 1759, Franz Aepinus, a contemporary and admirer of Franklin, published a book containing a revised version of the theory, which did account for negative repulsion. At this stage, however, Franklin appears to have lost interest. Significantly, Joseph Priestley, who discovered oxygen, correctly argued for an inverse square law governing the interaction between charged bodies.
3. Cohen, op. cit., p. 62.
4. Ibid., p. 64.

5. In terms of abundance on or close to the earth's surface, gold measures at five parts in 100,000,000, and silver at one part in 1,000,000. The occurrence of both precious metals is far from uniform. Deep-mining of gold is only economic at better than one part in 10,000, so it is not surprising that workable seams are so few.
6. Lead measures two parts, tin, three parts and zinc, ten parts in 100,000.
7. Pyrites derives from the Greek *pyr*, meaning 'fire', indicating the appropriate means of refinement.
8. This is because the oxygen content is reduced; at the other electrode it increases, the process known as oxidation. The combined process is known as 'redox'.
9. This was put to practical use in the South Foreland lighthouse in 1858: see B. Bowers, *A History of Electric Light & Power*, Science Museum 1982, p. 21.
10. D. Gooding and F. A. J. L. James (eds.), *Faraday Rediscovered*, Macmillan 1985, Introduction, p. 1.
11. Quoted, G. N. Cantor, 'Reading the Book of Nature: The Relation between Faraday's Religion and Science', in D. Gooding and F. A. J. L. James, op. cit., pp. 69–81.
12. For further details, see the entry in the *Oxford Dictionary of the Christian Church*, Oxford University Press, 1999.
13. Quoted, S. Forgan, 'Faraday – From Servant to Savant: The Institutional Context', in D. Gooding and F. A. J. L. James, op. cit., p. 61.
14. A photograph from 1856 shows the Prince Consort presiding, with his two older sons – one the future King Edward VII – in the audience.
15. D. M. Knight, 'Davy and Faraday: Fathers and Sons', in D. Gooding and F. A. J. L. James, op. cit., p. 46.
16. Faraday's own chemistry lecture notes associate electricity and magnetism as attractive forces as early as 1816: see R. D. Tweney, 'Faraday's Discovery of Induction: A Cognitive Approach', in D. Gooding and F. A. J. L. James, op. cit., p. 192.
17. L. P. Williams, 'Faraday and Ampère: A Critical Dialogue', in D. Gooding and F. A. J. L. James, op. cit., p. 90.
18. Ibid., p. 91.
19. Ibid., p. 96.
20. Quoted, B. Bowers, *A History of Electric Light & Power*, Science Museum 1982, p. 16.
21. Quoted, ibid., p. 96.
22. The name derives from the fact that a permanent magnet produces the field containing the moving coil. The earliest magnetos produced no more than a series of electric pulses. None the less, until the 1930s, they still provided the ignition system – no more than a series of sparks – for internal combustion engines.
23. The material on Wheatstone comes mainly from B. Bowers, 'Faraday, Wheatstone and Electrical Engineering', in D. Gooding and F. A. J. L. James, op. cit., pp. 163–174.
24. N. J. Nersessian, 'Faraday's Field Concept', in D. Gooding and F. A. J. L. James, op. cit., pp. 175–187.
25. Quoted in Tweney, op. cit., p. 193.
26. Bowers, op. cit., covers the field admirably.

Chapter 5 – Energy

1. The man himself was extraordinary. He was actually an American, Benjamin Thompson, born in 1753. At the age of thirty-one, after backing the wrong

side in the American revolution, he found employment with the king of Bavaria, to whom he rendered various services, such as introducing the potato into the kingdom, and supervising the royal arsenal. For these he was raised to the nobility of the Holy Roman Empire, taking his title from his hometown in Massachusetts.

2. *Oxford Dictionary of Quotations*, 4th edn., 1992, p. 494 (which also gives the Latin original).
3. *Dialogue Concerning Two New Sciences*, 1638, cited in I. B. Cohen, *The Birth of a New Physics*, Heinemann, 1960, p. 163.
4. I. B. Cohen, op. cit., p. 160.
5. R. S. Westfall, *The Life of Isaac Newton*, Cambridge University Press, 1993, p. 159.
6. I. B. Cohen, op. cit., p. 155.
7. Westfall, op. cit., p. 171. The proof, in its final form, is Proposition LXXI in Book I of Newton's *Principia*.
8. These figures are derived from W. M. Smart, *Textbook on Spherical Trigonometry*, 6th edn., Cambridge University Press, 1977, Appendices A and B.
9. After whom the Cavendish Laboratory at Cambridge (founded in 1872) was named.
10. Quoted, Westfall, op. cit., p. 177.
11. Quoted, ibid., p. 190.
12. T. Crump, *Solar Eclipse*, Constable, 1999, pp. 75–81.
13. *Preliminary Discourse of the Study of Natural Philosophy*, 1830, pp. 79–80.
14. B. Russell, *Introduction to Mathematical Philosophy*, George Allen & Unwin, 1920, p. 194.
15. B. Russell, *A History of Western Philosophy*, George Allen & Unwin, 1946, pp. 218, 225.
16. Ibid., p. 222.
17. Ibid., p. 566.
18. Crump, op. cit., pp. 133–35. Although H. Collins and T. Pinch, *The Golem: What You Should Know about Science*, 2nd. edn., Cambridge University Press, 1998, Chap. 5, questions the validity of these observations, more recent, and much more accurate observations, have confirmed Einstein's theory.
19. See the catalogue of the exhibition, *Les photographies et le ciel*, Musée d'Orsay, Paris, 2000.
20. G. H. Hardy, in his *A Mathematician's Apology*, Cambridge University Press, 1940, pp. 32–35, gives two examples, one attributed to Euclid, the other to Pythagoras.
21. M. Fisch, *William Whewell: Philosopher of Science*, Oxford, Clarendon Press, 1991, p. 34.
22. B. Russell, *A History of Western Philosophy*, George Allen & Unwin, 1946, pp. 605.
23. See A. R. Hall, *Philosophers at War: The Quarrel between Newton and Leibniz*, Cambridge University Press, 1980.
24. I. B. Cohen, *Introduction to Newton's Principia*, Cambridge University Press, 1971, p. 291.
25. Reproduced in Hall, op. cit., pp. 263–4.
26. B. Russell, *History of Western Philosophy*, George Allen & Unwin, 1946, p. 182. Russell added that 'in logic, this is still true at the present day'.
27. See G. L'E. Turner, *Scientific Instruments 1500–1900: An Introduction*, University of California Press, 1998, p. 8.
28. Michael Sharratt, *Galileo: Decisive Innovator*, Cambridge University Press, 1994, p. 207.

29. Quoted from 'De Motu', R. S. Westfall, *The Life of Isaac Newton*, Cambridge University Press, 1993, p. 168.
30. Michael Sharratt, op. cit., pp. 51, 234.
31. Ibid., p. 47.
32. The story is told in detail in Herbert Butterfield, *Origins of Modern Science, 1300–1800*, G. Bell & Sons, 1965, 10–11.
33. Ibid., p. 6.
34. Ibid., p. 11.
35. Sharratt, op. cit., p. 57.
36. Ibid., p. 78.
37. Bernard Williams, *Descartes: The Project of Pure Inquiry*, Pelican Books, 1978, p. 255, referring to Newton's *Principia*, II: 37, 39.
38. Alan Gabbey, Huygens and Mechanics, in H. J. M Bos et al. (eds.), *Studies on Christiaan Huygens*, Lisse, Swets & Zeitlinger, 1980, p. 168.
39. Ibid., p. 169.
40. Ibid., p. 171.
41. The only uncertainty here is its pronunciation: does it rhyme with cool, cowl or coal? The balance of opinion now favours the first of these alternatives. See D. S. L. Cardwell, *James Joule: A Biography*, Manchester University Press, 1989.
42. Quoted, ibid., p. 15.
43. Ibid., p. 17.
44. The exact figure is 4.1868×10^7 ergs per calorie, although Joule did not work with the metric system.
45. Quoted Cardwell, op. cit., p. 76. The difference in temperature at Niagara would only be 0.5°F.
46. Given that absolute zero is now known to be −459.67°F (−273.15°C), Joule's result is remarkably accurate.
47. *Uber die Erhaltung der Kraft*, Ostwald's klassiker, 1847.
48. Cardwell, op. cit., p. 93.
49. An interesting example of this is to be found at Kinlochfine, on the west coast of Scotland. This is the site of a deep-water harbour, a hydroelectric power station, and a vast electrolytic plant for the refinement of ore. The sole purpose of this complex is to produce aluminium for use in industry. Electrolysis is the only practical means for refining the ore containing the third most abundant element in the earth's surface: it requires, however, a massive supply of electric power. The location of Kinlochfine, surrounded by mountains at the end of a sea loch, in an area of high rainfall, makes it ideal for this process. The costs of distributing the finished product from a relatively remote site are small in relation to the savings realised by its location, as described above. The process, it should be noted, realises the potential energy of the water used to drive the electric turbines so as to produce electricity, which in turn powers the electrochemical process of electrolysis. All this was foreshadowed by Joule's experiments.

Chapter 6 – Chemistry

1. A. Donovan, *Antoine Lavoisier*, Cambridge University Press, 1993, p. 83.
2. The chemical formula is $CaSO_4.2H_2O$. There are in fact five varieties, of which only one, gypsite, is suitable for this purpose.
3. Quoted, Donovan, op. cit., p. 47.
4. See ibid., p. 144.
5. A. Lavoisier, *Oeuvres* (6 vols.). Imprimerie Impériale Paris, 1862–93, II, p. 7.
6. Donovan, op. cit., p. 95.

7. Ibid., p. 134.
8. The chemical formula is simple: NH_3.
9. Donovan, op. cit., p. 147.
10. Lavoisier, op. cit., II, p. 226.
11. Ibid., pp. 228–232
12. Cited, W. H. Brock, *The Fontana History of Chemistry*, Fontana, 1992, p. 123.
13. E. C. Bentley, *Clerihews Complete*, Werner Laurie, 1951.
14. In the early nineteenth century more than half the world's coal was produced in Britain.
15. D. Knight, *Humphry Davy: Science and Power*, Blackwell, 1992, p. 40.
16. Carbon occurs somewhere between zinc and aluminium on the reactivity scale.
17. J. Davy, *Memoirs of the Life of Sir Humphry Davy*, Longman, vol. 1, p. 136.
18. Knight, op. cit., p. 62.
19. This meant that the anode was silver, the cathode zinc, so that, following the definition of the two electrodes, the current flowed from the former to the latter.
20. Quoted, Knight, op. cit., pp. 63–4.
21. Quoted, ibid., p. 66.
22. Quoted, B. Dibner, *Alessandro Volta and the Electric Battery*, Franklin Watts, New York, 1964, p. 95.
23. Ions are atoms that have either gained or lost electrons – in the right circumstances, an extremely common physical event, with far-reaching consequences, particularly for electric conduction. Heat, as one way of creating ions (a continuous process in the sun's atmosphere), produces the phenomenon of thermal ionisation.
24. H. B. Jones, *Faraday*, vol. 1, p. 210.
25. Quoted, Knight, op. cit, p. 98.
26. H. Davy, 'Some Experiments on the Combustion of the Diamond', *Phil. Trans. of the Royal Society*, vol. 105, 1815, pp. 99–100.
27. A. Donovan, *Antoine Lavoisier*, Cambridge University Press, 1993, p. 97.
28. H. Davy, 'Some Experiments on a Solid Compound of Iodine and Oxygene', *Phil. Trans. of the Royal Society*, vol. 105, 1815, pp. 203–19.
29. Quoted, A. Thackray, *John Dalton: Critical Assessment of His Life and Science*, Harvard University Press, 1972; this book provides most of the biographical material in the present chapter.
30. Quoted, ibid., p. 12.
31. Manchester, 1793.
32. NH_3.
33. H_2O_2: this compound readily decomposes into water (H_2O) and oxygen (O_2) according to the formula $2H_2O_2 \rightarrow 2H_2O + O_2$.
34. OH.
35. W. H. Brock, *The Fontana History of Chemistry*, Harper Collins, 1992, p. 135.
36. H. C. Urey, 'Dalton's Influence on Chemistry', in D. S. L. Caldwell (ed.), *John Dalton and the Progress of Science*, Manchester University Press, 1968, p. 332.
37. W. H. Brock, 'Dalton versus Prout: The Problem of Prout's Hypothesis', in Caldwell, op. cit., p. 249.
38. Thackray, op. cit., p. 62.
39. A. Donovan, *Antoine Lavoisier*, Cambridge University Press, 1993, p. 157.
40. But see H. G. Söderbaum, *J. J. Berzelius; Autobiographical Notes*, Baltimore, 1934.
41. Sir Harold Hartley, quoted, C. A. Russell, 'Berzelius and the Development of Atomic Theory', in Caldwell, op. cit., pp. 259–60.
42. *Annals of Philosophy*, iii (1814), p. 51.

43. Chlorophyll is a natural pigment, whose molecules belong to the category of porphyrins, in which nitrogen atoms are typically coordinated to metal ions: in the case of chlorophyll these are magnesium, an element essential to all known biological species.
44. R. Harwood, *Chemistry*, Cambridge University Press, 1998, p. 195.
45. Since, with the discovery of the periodic table, these were found to belong to the class of halogens, the salts are now known as halides.
46. 'Arbeiten is schön, aber Erwerben ist ekelhaft', quoted, K. Danzer, *Robert W. Bunsen und Gustav R. Kirchhoff: Die Begründer der Spektralanalyse*, 1972, Leipzig, Teubner. Much of my material comes from this book (which although published in the German Democratic Republic is tainted only marginally by Marxist orthodoxy).
47. This remained the cheapest source of electric current until Werner von Siemens' (1816–92) invention of the self-acting dynamo in 1867.
48. Sodium chloride (NaCl).
49. Entweder ein Unsinn oder eine ganz große Sache.
50. D. Dewhurst and M. Hoskin, 'The Message of Starlight: the Rise of Astrophysics', in M. Hoskin (ed.), *The Cambridge Illustrated History of Astronomy*, 1997, p. 264.
51. Quoted, K. Danzer, op. cit., p. 48.
52. Chemical analysis through spectral observations.
53. As early as 1835 Charles Wheatstone (1802–75) used a prism to analyse the spectrum of a spark produced with a mercury electrode, going on to other metals – zinc, cadmium, bismuth, lead and tin in the molten state. He concluded that 'the appearances are so different that by this mode of examination, the metals may be readily distinguished from each other'. Contrary to Ångstrom's observations, Wheatstone's results did not differ according to the medium containing the spark. See B. Bowers, 'Faraday, Wheatstone and Electrical Engineering', in D. Gooding and F. A. J. L. James (eds.), *Faraday Rediscovered*, Macmillan, 1985, p. 167.
54. K. Danzer, op. cit., p. 50.
55. Ibid., p. 51.
56. Ibid., p. 52: the thirteen elements are cadmium, barium, magnesium, chrome, nickel, copper, zinc, strontium, cadmium, cobalt, manganese, aluminium and titanium. It is now known that at least sixty-six elements are present in the sun.
57. Penguin, 2000.
58. The Trans-Siberian Railway only opened in 1905.
59. O. N. Pisarzhevsky, *Dimitry Ivanovich Mendeleyev: His Life and Work*, Foreign Languages Publishing House, Moscow, 1954, p. 15.
60. Among Mendeleyev's friends was Alexander Borodin (1833–87), then a talented young scientist, now remembered as a composer.
61. Quoted, Pisarzhevsky, op. cit., p. 29.
62. See H. Collins and T. Pinch, *The Golem: What You Should Know about Science*, Cambridge University Press, 2nd edn., 1998, chap. 4.
63. Pisarzhevsky, op. cit., p. 35.
64. Ibid., p. 38.
65. Ibid., p. 41.
66. P. Atkins, *The Periodic Kingdom*, Weidenfeld and Nicolson, 1995, p. 86.
67. This is the final, and now standard, form: originally Mendeleyev listed the elements vertically, and not horizontally.
68. The correct value is 69.723.
69. The correct values are Te, 127.6; I, 126.9; Co, 58.9 and Ni, 58.7.

70. Some clarification is required: Mendeleyev used the Roman I to VIII to number his columns, but from row 4 the numbers divided into two sets, IA to VIIIA and IB to VIIIB, to accommodate the so-called transition elements, mostly hard, strong and dense metals such as iron and copper. There were two additional columns VIII, defined by cobalt and nickel in row 4, making 18 columns in all. These are now designated as groups, numbered from 1 to 18, while the rows are periods 1 to 7. Even so, there is an additional block in periods 6 and 7, containing in period 6 the so-called lanthanides, or 'rare earths', and in period 7 the actinides (all of which are radioactive).
71. Quoted, W. H. Block, *The Fontana History of Chemistry*, Fontana Press, 1992, p. 208.
72. In a process particularly characteristic of mammals, one function of the liver is to convert ammonia (NH_3) into the much less toxic urea ($CO(NH_2)_2$), which is then excreted in solution.
73. Ibid., p. 201.
74. Quoted, ibid., p. 211.
75. Ibid., p. 212.
76. There are also secondary and tertiary alcohols, each with their own 'ladder'.
77. Quoted, ibid., p. 251.
78. In his *Lehrbuch der Chemie* (1866–68), quoted, Brock, op. cit., p. 252.
79. The actual term 'valency' only came into general use after 1865.
80. Van't Hoff also coined the term 'stereochemistry' in 1890.
81. Quoted, Brock, op. cit., p. 256.
82. Quoted, ibid., p. 265.
83. For the reasons, see ibid., p. 208.
84. Quoted, J. G. Burke, *Origins of the Science of Crystals*, University of California Press, 1966, p. 43.
85. *On the Heavens*, Book III, chapter 8, 306b–307a.
86. Also known as 'rhombic'.
87. Burke, op. cit., p. 69.
88. This development of the chemistry of crystals in described in Burke, op. cit., pp. 24–29.
89. Ibid., p. 42.
90. Ibid., p. 78.
91. E. Nagel, *The Structure of Science*, New York, 1961, p. 31.
92. S. Toulmin, *The Philosophy of Science*, New York, 1960, p. 53.
93. Burke, op. cit., p. 59.
94. This paragraph may oversimplify the diversity of possible cases by excluding the possibility that the optical axis is not a crystal axis: see Burke, op. cit., p. 143.
95. Ibid., p. 160.
96. R. H. Stuewer, *The Compton Effect: Turning Point in Physics*, Science History Publications, 1975, p. 68.
97. The name 'Silicon Valley' tells it all.
98. 'Crystallography', *Dictionary of Science and Technology*, San Diego Academic Press, 1992, p. 559.
99. In ionic bonding, such as in common salt, NaCl, there is an exchange of electrons so that the sodium (Na) becomes positively and the chlorine (Cl) negatively ionised: this explains why salt is sometimes represented as Na^+Cl^-.
100. *Linus Pauling in His Own Words*, Touchstone, 1995, p. 86.
101. Until 1968 only four crystalline forms of carbon, two types of graphite and two of diamond, were known. Two more were discovered in 1968 and 1972, but in 1985 Harold Kroto in England and Robert Curl and Richard Smalley

in Texas, working with vaporised carbon condensed in a stream of helium to a temperature close to absolute zero, produced crystalline clusters with sixty and seventy carbon atoms. In 1990, the synthesis of macroscopic quantities of C_{60}, achieved by using an arc between two graphite electrodes to vaporise carbon, confirmed a molecular structure of perfect regularity, defining a polyhedron with twelve pentagonal and twenty hexagonal faces. Because this was the structure of the architect Buckminster-Fuller's geodesic dome (and also of the standard European football), it became the prototype for a whole new class of carbon molecules known as 'fullerenes', whose remarkable properties are still being explored. Kroto, Curl and Smalley were Nobel Laureates for chemistry in 1996.

102. This states the case a little too simply because mica and asbestos are both generic terms, covering many different instances.

Chapter 7 – The new age of physics

1. H. Bondi, *Assumption and Myth in Physical Theory*, Cambridge University Press, 1965, p. 5.
2. For radio astronomy, see Chapter 9.
3. A. Pais, *Inward Bound*, Clarendon Press, 1986, pp. 479.
4. Op. cit., p. 478.
5. This result, dating from 1905, is well explained in op. cit., pp. 87–8.
6. The European location of Mallinckrodt Medical, the leading producer of radioisotopes for use in medicine, is part of the Dutch Euratom reactor complex.
7. There was a possible hidden agenda here. Callaghan's energy policies had been condemned by Tony Benn, a former cabinet colleague. Winning JET was a round to Callaghan.
8. www.jet.efda.org/pages/history-of-jet.html.
9. As defined on page 229.
10. Quoted, A. Pais, *Niels Bohr's Times, in Physics, Philosophy and Polity*, Oxford, 1991, pp. 105–6.
11. Compare Wheatstone's figure, given in Chapter 5.
12. See A. Pais, *Inward Bound*, Clarendon Press, 1986, p. 86. Thomson's figures, 6.8×10^{-10} electrostatic units for the charge and 3×10^{-26} grams for the mass of the electron, although much too high, were of the right order of magnitude. The mass of the electron is now calculated at $9.1093897 \times 10^{-28}$ grams.
13. The exact figure, 1/1836, was found by R. A. Millikan (1869–1953) in 1912: his method, based on observing the motion of charged oil-drops, was a refinement of Thomson's own work.
14. Quoted, Dieter Hoffman, 'Heinrich Hertz and the Berlin School of Physics', in David Baird et al. (eds.), *Heinrich Hertz, Classical Physicist, Modern Philosopher*, Kluwer, 1981, pp. 1–8.
15. Quoted, John H. Bryant, 'Heinrich Hertz's Experiments and Experimental Apparatus: His Discovery of Radio-waves and the Delineations of their Properties', in David Baird et al., op. cit., pp. 39–58.
16. These are known as Coulomb forces after Charles Auguste Coulomb (1736–1806), who discovered them.
17. Bryant, op. cit., p. 52.
18. Quoted, Susan Quinn, *Marie Curie: A Life*, Simon & Schuster, 1995, p. 41.
19. Ibid., p. 142.
20. Quoted, ibid., p. 151.

21. A. Pais, *Inward Bound: Matter and Forces in the Physical World*, Clarendon Press, 1986, p. 113.
22. He was in fact the examiners' second choice, but their first choice, J. C. Maclaurin, had just got married and chose to take up an appointment in New Zealand.
23. Becquerel had just noted a similar effect with uranium and uranium-X.
24. Quoted, David Wilson, *Rutherford: Simple Genius*, Hodder & Stoughton, 1983, p. 165; this book provides much of the material for the present section.
25. Quoted, ibid., p. 291.
26. Rutherford, quoted, ibid., p. 292: to make sure of this point, Rutherford had taken the trouble to attend a standard university course on statistics.
27. E. Rutherford, 'The Scattering of Alpha- and Beta-rays and the Structure of the Atom', submitted to Manchester Literary and Philosophical Society, 7 March 1911.
28. E. Rutherford, *Radioactive Substances and Their Radiations*, Cambridge University Press, 1913.
29. The precise measure, now known as the 'relative atomic mass', is defined as the ratio of the average mass per atom of the element in its natural state to 1/12 the mass of a carbon-12 atom.
30. In formal terms: ${}^{4}He + {}^{14}N = {}^{1}H + {}^{17}O$.
31. From the Duchy of Mecklenburg, now part of Germany.
32. The state religion in Denmark.
33. Quoted, Abraham Pais, *Niels Bohr's Times, in Physics, Philosophy and Polity*, Clarendon Press, 1991, p. 121.
34. Balmer calculated $R = 3.29163 \times 10^{15}$ per second: the correct value is now known to be $R = 3.28984186 \times 10^{15}$ per second, a difference of less than one part in a thousand.
35. See Abraham Pais, op. cit., p. 144.
36. This was not actually observed until the 1980s: ibid., p.152.
37. Quoted, ibid., p. 192.
38. Quoted, ibid., p. 193.
39. Methuen, 1998.
40. Quoted, Pais, op. cit., p. 202.
41. See, for example, Richard Harwood, *Chemistry*, Cambridge University Press, 1998, pp. 60-62.
42. Quoted, Pais, op. cit., p. 209.
43. See, for example, Martin Gardner, *The Night is Large*, Penguin, 1997, p. 36.
44. Since, without the fourth quantum number, the number of k states in any shell is $2k - 1$, the number of all such states in shell n is the sum of the first n odd numbers: this by elementary algebra is n^2, so that, taking in account the fourth quantum number, the total number of possible states is $2n^2$.
45. There is actually a small glitch in this series: two 3*d*-electrons are added in the step from nickel (28) to copper (29) with one electron lost to the outer shell: this is then added back in the following step, from copper (29) to zinc (30).
46. Quoted, Pais, op. cit., p. 216.
47. Quoted, Pais, op. cit., p. 264.
48. Thomas Powers, *Heisenberg's War: The Secret History of the German Bomb*, Penguin, 1994, pp. 36–7.
49. Ibid., p. 110.
50. Ibid., p. 112.
51. Ibid., p. 118.
52. Quoted, Pais, op. cit., p. 486.
53. Quoted, ibid., p. 496.

54. Quoted, ibid., p. 501.
55. Quoted, ibid., p. 502.
56. Quoted, ibid., p. 491.
57. Quoted, A. Brown, *The Neutron and the Bomb*, Oxford University Press, 1997, p. 5.
58. Kapitsa's letter to his mother in Russia (quoted, ibid., pp. 76–7), describing the wedding is a gem. Although, unfortunately, too long to quote in full, it contains one great sentence: 'There are some further ceremonies and the clergyman exhorts the man and wife about the purpose of marriage as if they themselves had no idea.'
59. Quoted, ibid., p. 137.
60. For the use of chlorine in detecting solar neutrinos, see page 231.
61. Brown, op. cit., p. 52.
62. This term was later coined by Rutherford: in the Bakerian lecture he referred to the hydrogen nucleus. Brown, op. cit., p. 53.
63. J. L. Glasson, a Cavendish research student, quoted, ibid., p. 54.
64. Rutherford's work with scintillation counters is described on page 205.
65. This is simple arithmetic: two α-particles have a combined atomic mass of 8, which with the addition of a neutron becomes 9, the atomic mass of beryllium.
66. The analogy here is with the Compton effect (see page 275), which is the result of X-rays (like γ-rays, electromagnetic) colliding with electrons, but there is a vast difference in mass, by a factor of nearly 2000, between electrons and protons.
67. Already in the 1930s the oscilloscope was standard laboratory equipment. The basic unit (as with television) is a cathode ray tube, which acts as a screen on which any series of events, capable of being represented by an electric current, can be observed and recorded photographically.
68. Paraffins are saturated hydrocarbons, so that the molecule consists only of hydrogen and carbon atoms (with the general formula C_nH_{2n+2}). Paraffin is a residue produced by the refinement of petroleum, the most common form in which hydrocarbons occur in nature.
69. Quoted, C. P. Snow, *The Physicists*, Papermac, 1982, p. 85.
70. Quoted, Brown, op. cit., p. 108.
71. Quoted, H. Childs, *An American Genius: The Life of Ernest Orlando Lawrence*, Dutton, 1968, p. 58.
72. The term was coined by Lawrence at a very early stage, but it was only officially recognised in 1936, when the new Radiation Laboratory was opened at Berkeley.
73. Childs, op. cit., p. 193.
74. Volt-protons measure the particle energy of the protons, which is not the same as (although related to) the voltage potential of the electric field of the cyclotron.
75. These conferences, of which the first was held in 1912, were set up and endowed by Ernest Solvay (1838–1922), who became a wealthy industrialist as a result of finding a much improved method of manufacturing soda.
76. Childs, op. cit., p. 206. Rutherford would later become more enthusiastic: ibid., p. 255. In 1937, a month after Rutherford's death, the Royal Society awarded Lawrence the Hughes Medal 'for the most important instrument of physical research since the C. T. R. Wilson expansion chamber': ibid., p. 278.
77. There was a price to pay: the new laboratory would be Crocker Radiation Laboratory, named after William H. Crocker, the Regent who had arranged the necessary finance.

78. Since the neutron has no charge, a synchrotron can only provide a neutron beam indirectly: this is achieved by the proton bombardment of a target that emits neutrons (as in Chadwick's original experiment described on page 250).
79. This produced a brilliant limerick from a colleague, Lee du Bridge:
    A handsome young man with blue eyes
    Built an atom-machine of great size,
    When asked why he did it
    He blushed and admitted,
    'I was wise to the size of the prize'.
80. P. de Latil, *Enrico Fermi: The Man and his Theories* [translated from French], Souvenir Press, 1965, p. 98. Where not otherwise attributed, this book is the source of material in the present chapter.
81. Excerpts, intelligible to the non-specialist, from many of Fermi's papers relating to neutron bombardment are to be found in de Latil, op. cit., pp. 147–162.
82. The radon was itself a decay product of radium.
83. Paraffin is always a compound containing twice as many hydrogen as carbon atoms, i.e. C_nH_{2n}.
84. This is somewhat confusing, because the two heavy isotopes of hydrogen, ^2H and ^3H, each have their own names, respectively deuterium and tritium. The latter, with a half-life something over 12 years, does not occur in nature, but is now manufactured on a large scale.
85. Of the order of 200 megaelectronvolts.
86. This term applied to any element occurring in more than one form, so that, for example, ozone is an allotrope of oxygen. Carbon has three allotropes: diamond, graphite and fullerite (which was only identified in 1985).
87. *Adventures of a Mathematician*, Charles Scribner's Sons, 1976, p. 5.
88. The background to this letter is to be found in Richard Rhodes, *The Making of the Atomic Bomb*, Penguin, 1986, pp. 305–11.
89. See ibid., pp. 282–88 for a simple account of Bohr's reasoning. The actual paper, 'Resonance in Uranium and Thorium Disintegrations and the Phenomenon of Nuclear Fission', is to be found in *Physical Review*, vol. 56, 1939.
90. This was named after an English nanny, Maud Ray, who had once worked for Niels Bohr in Copenhagen: how this came about is a long involved story, for which see R. Rhodes, op. cit., pp. 340–1.
91. Quoted, ibid., p. 369.
92. In geometric terms, a flattened rotational ellipsoid.
93. Quoted in Robert Jungk, *Brighter than a Thousand Suns: A Personal History of the Atomic Scientists*, Harcourt, Brace, 1958, p. 113.
94. Named after the Russian physicist, Pavel Cherenkov (1904–). The blue light is the result of the water being bombarded by γ-rays produced in the course of the reaction.
95. Planck was so highly esteemed in Germany that after the Second World War the Kaiser Wilhelm institutes for scientific research were renamed after him. This was no more than his due: he had resigned his chair in 1935 out of protest against the Nazis, and in 1944 his son, Erwin, was executed for his part in the plot to assassinate Hitler.
96. A. Pais, *Inward Bound*, Clarendon Press, 1986, p. 136.
97. Named after Lord Rayleigh (1842–1919) and James Jeans (1877–1946), who together developed this formula to describe the distribution of energy of enclosed low-frequency radiation.

98. After a colloquium presenting the new quantum theory, a colleague, Peter Debye (1884–1966), noted: 'We did not know whether the quanta were fundamentally new or not.' Quoted, Pais, op. cit., p. 134.
99. Ibid.
100. See Pais, op. cit., p. 135, for the mathematical formula.
101. B. R. Wheaton, *The Tiger and the Shark: Empirical Roots of Wave-Particle Dualism*, Cambridge University Press, 1991, p. 108.
102. Ibid., p. 106.
103. Ibid., p. 109.
104. This is described in detail in T. Crump, *Solar Eclipse*, Constable 1999, p. 133–8.
105. The Bose–Einstein combination comes up again in Chapter 5.
106. Wheaton, op. cit., p. 220.
107. Ibid., p. 229.
108. The critical threshold is $\hbar/mc$, where m is the rest-mass of the particle and c is the speed of light: this turns out to be 2.4263×10^{-12}. For the Compton effect, where the particle is an electron, the value is 3.8616×10^{-13}.
109. For this simple explanation, see Vincent Icke, *Wasknijpers*, in NRC Handelsblad, 17 March 2001. Icke is a leading Dutch astronomer and populariser of science; his title means 'clothes-pegs'.
110. R. H. Stuewer, *The Compton Effect: Turning Point in Physics*, Science History Publications, New York, 1975, p. 204.
111. A. Compton, The Scattering of X-rays, *Journal of the Franklin Institute*, 198 (1924), 61–2.
112. A. Compton, 'A Quantum Theory of the Scattering of X-rays by Light Elements', *Phys. Rev.*, 21, 1923, 484.
113. Stuewer, op. cit., p. 273.
114. Quoted in Wheaton, op. cit., p. 286.
115. M. de Broglie, *Les rayons x*, Paris, 1922, p. 19.
116. This explains the well-known limerick:

    There was a young lady named Bright,
    Who travelled much faster than light,
    She departed one day,
    In a relative way,
    And arrived the previous night.

117. The mathematics, reproduced in Wheaton, op. cit., pp. 290–2, although not difficult, is still too involved for the present text.
118. L. de Broglie, *Thèses: Recherches sur la théorie de quanta*, Masson, Paris, 1963, p. 53.
119. Wheaton, op. cit., p. 294.
120. This phenomenon was first observed, serependitiously, by Clinton Davisson (1881–1958) and Lester Germer (1896–1971) in America on 3 March 1927. They shared the 1937 Nobel prize for physics with G. P. Thomson.
121. The Russian physicist, Igor Tamm, once said of Dirac, 'He seems to talk only with children, and they have to be under ten.'
122. Quoted in A. Pais, op. cit., p. 288.
123. The biographical information comes from H. Kragh, *Dirac: A Scientific Biography*, Cambridge University Press, 1990.
124. Dirac's 'The Quantum Theory of the Electron' was received by the *Proceedings of the Royal Society* on 2 January 1928.
125. Fermi was first to work out the Fermi–Dirac statistics, but Dirac produced his contribution, unaware of Fermi's work. For this he apologised, but the statistic, in its present form, incorporates the contributions of both sides.

Chapter 8 – Astronomy

1. Quoted, Bernard Lovell, *Times Literary Supplement*, 13 July 2001.
2. Recent research shows that Leonard Digges (1520–59) invented a reflecting telescope more than a hundred years before Newton; he may also have invented a refracting telescope. See John Gribbin, *Companion to the Cosmos*, Orion Books, 1996, p. 143.
3. Or vice versa in Ptolemaic astronomy.
4. See Table 1 in T. Crump, *Solar Eclipse*, Constable, 1999, p. 49.
5. Quoted, Peter Lancaster-Brown, *Halley and His Comet*, Blandford Press, 1985, p. 69.
6. Quoted, ibid., p. 52. Saturn would be the furthest known planet until almost the end of the eighteenth century.
7. Quoted, ibid., p. 70.
8. Totality, for the last English eclipse, on 11 August 1999, was confined to parts of Devon and Cornwall.
9. For further discussion, see T. Crump, *Solar Eclipse*, Constable, 1999, pp. 117-18.
10. Quoted (without source), Lancaster-Brown, op. cit., p. 80.
11. In the nineteenth century, the Marquis de Laplace pointed out that the error would have been reduced to 13 days if Clairaut had known the true mass of Saturn: ibid., p. 88. The influence of Uranus and Neptune, unknown in 1759, would also account for the error.
12. An extract is to be found in J. C. Beaglehole, *The Life of Captain James Cook*, Stanford University Press, 1974, pp. 100–101.
13. At least one popular book has already appeared: Eli Major, *June 8, 2004 – Venus in Transit*, Princeton University Press, 2000.
14. Quoted in M. A. Hoskin, *William Herschel and the Construction of the Heavens*, Oldbourne, 1963, p. 20.
15. Ibid., p. 23.
16. As in the classic Greek couplet: *ἀστερας ἐἰσαθρεις ἀστηρ ἐμος. ἐἰθε γενοιμην οὐρανος ὡς σε πολλοις ὀμμασιν βλεπω.* (You look at the stars, my star. Would that I were heaven so that I could see you with many eyes.)
17. Quoted in K. Ferguson, *Measuring the Universe: The Historical Quest to Quantify Space*, Headline Books, 1999, p. 152.
18. See A. J. Berry, *Henry Cavendish: His Life and Scientific Work*, Hutchinson, 1960, p. 15.
19. Ibid., p. 161.
20. Ibid., p. 164.
21. Somewhat ironically, Cavendish's result would have been 5.45 – a less accurate figure – but for an arithmetical error discovered by Francis Baily in 1841.
22. On 1 May 2000, Prof. Jens Grundlach of the University of Washington, Seattle, reported to the American Physical Society the most accurate measure so far of the earth's mass: this is 5.972×10^{24} kilograms: significantly the professor's apparatus was based on the same principles as that of Cavendish.
23. *Cours de philosophie positive*, Paris, 1835, 11, 8.
24. His name was adopted for the units measuring the wavelength of electromagnetic waves.
25. See Q. Bajac et al. *Dans le champ des étoiles: Les photographes et le ciel 1850–2000*, catalogue of an exhibition at the Musée d'Orsay, Paris, 14 June to 24 September 2000.
26. Fizeau was also the first ever to devise apparatus for measuring the velocity of light on earth, although his results were not as accurate as those calculated astronomically.

27. M. Hoskin (ed.) *The Cambridge Illustrated History of Astronomy*, Cambridge University Press, 1997, p. 196.
28. Sirius, at 8.7 light-years, is the seventh closest star to the sun.
29. The R stands for Russell, the H for Ejnar Hertzsprung (1873–1967), who in 1912 had plotted a similar but much more limited diagram for the Hyades star cluster, then at the extreme outer range of stellar parallax.
30. The first observation was in 1650, when the Jesuit astronomer, Johannes Riccioli (1598–1671), with the help of a telescope observed that ζ Ursae Majoris was a double star: it was more than a century later before it was suggested that the two components were in close proximity.
31. In 1922 the Russian mathematician Alexander Friedmann discovered solutions to Einstein's equations governing general relativity which required an expanding universe.
32. Norton, 1998, p. 12.
33. Hoskin, op. cit., p. 344.
34. And independently discovered by Carl von Weizsäcker (1912–).
35. Beryllium-9 is the only natural isotope.
36. The reason why is to be found in T. Crump, *Solar Eclipse*, Constable, 1999, pp. 64–72.
37. *Times Literary Supplement*, 13 July 2001, p. 4.
38. Penzias and Wilson, but not Gamow, were awarded Nobel prizes in 1965.
39. In addition to Hoyle, Geoffrey and Margaret Burbage, and Willy Fowler, the 1983 Nobel Laureate.
40. Elements beyond plutonium (94) do not occur naturally; technetium and promethium (atomic numbers 43 and 61) are also radioactive.
41. H. Friedman, *The Astronomer's Universe*, Norton, 1998, p. 165.
42. A quasar some 12 billion light-years away in the constellation Sextans is the most distant object ever observed by astronomers. Its discovery, reported early in 2000, occurred in the course of the Sloan Digital Sky Survey, using a telescope at Apache Point, New Mexico.
43. Prof. Roger Tayler FRS, University of Sussex.
44. For these phenomena, see Friedman, op. cit., pp. 166–73, 185–6, 221–7.
45. Twinkling, which makes stars seem larger, is an atmospheric effect of the earth.
46. See N. Calder, *The Manic Sun*, Pilkington Press, 1997, p. 27.
47. Not ten, because Fe I is not ionised.
48. Freeman Dyson, quoted in T. Crump, *Solar Eclipse*, Constable, 1999, p. 134. This book provides most of the present material relating to eclipses.
49. H. Collins and T. Pinch, *The Golem: What You Should Know about Science*, Cambridge University Press, 1998, p. 50.
50. See page 302.
51. There are actually three types of neutrino, but only one, the electron-neutrino, is relevant here. The question of their mass in discussed in G. 't Hooft, *In Search of the Ultimate Building Blocks*, Cambridge University Press, 1997, pp. 130–3.
52. The Japanese also have a neutrino detector, the Kamiokande, located at the bottom of a mine, but its results support those of the other detectors.
53. At the Fermilab in July 2000, the τ-neutrino became the last fundamental particle to be directly observed.
54. When I visited Los Alamos in October 1995, the main thrust of the experiments being carried out with the linear accelerator of the Meson Physics Facility was to investigate the possible mass of the neutrino.
55. This is an acronym for *Astronomy with a Neutrino Telescope and Abyss environmental RESearch.*

Chapter 9 – Physics: ground zero

1. Quoted, T. Shachtman, *Absolute Zero and the Conquest of Cold*, Houghton Mifflin, 1999, p. 125.
2. Quoted, A. Pais, *Inward Bound*, Clarendon Press, 1896, p. 137.
3. *Science* (10 December 1999) reported the presence of bacteria in the fresh-water Lake Vostok in Antarctica, which is covered by 4000 metres of polar ice. At the other extreme, deep-sea hydrothermal vents, with temperatures up to 350°C, according to research carried out by the Bridge project of the University of Leeds, provide support for microbial communities which thrive in the temperature range 85–105°C, so-called 'hyperthermophilic niche'.
4. One of these, the Joint European Taurus (JET), located at Culham, just south of Oxford, is described in Chapter 7.
5. There is actually a place called Thule in the north of Greenland, notorious in the 1960s for the crash of an American aircraft carrying nuclear bombs. There is also a rare metallic element thulium (atomic number 69), discovered in 1879, but with no known practical use.
6. The word 'refrigerator' was coined around 1800 by an American inventor, Thomas Moore, to describe a well-insulated double-walled cedar tub, with snow packed in between the two walls. This he used to take butter from his farm to market, where buyers willingly paid a premium.
7. This was discovered around 1860 by an Irish physician, Thomas Andrews, experimenting with carbon dioxide.
8. The solid form, 'dry ice', is a common commercial refrigerant (also used for stage smoke effects). It was first produced when Charles Thilorier, in 1834, attained a temperature of −110°C for liquid hydrogen. His apparatus incorporated a thermometer based on helium. This made use of the gas equation, by which, if volume is kept constant (as in the thermometer), temperature is proportional to pressure: Thilorier's instrument measured pressure, and so, derivatively, temperature.
9. *Reflexions sur la puissance motrice du feu* was the French title.
10. The figures have now been refined to 273.15°K and 373.15°K.
11. Donald Cardwell, quoted, T. Shachtman, op. cit., p. 107.
12. See G. 't Hooft, *In Search of the Ultimate Building Blocks*, Cambridge University Press, 1997, p. 5.
13. Poland was then partitioned between Russia, Prussia and Austria.
14. Shachtman, op. cit., p. 125.
15. Tadeusz Estreicher, quoted, ibid., p. 139.
16. Quoted, ibid., p. 145.
17. Quoted, ibid., p. 135.
18. In Dutch, 'door meten tot weten' – a much more telling statement.
19. Shachtman, op. cit., p. 175.
20. The actual figure is 4.4°K.
21. In the course of time, since 1911, superconducting materials have been found at steadily higher temperatures. In 2001 *Nature* (1 March) reported that a Japanese team led by Jun Akimutsu had found that magnesium diboride (MgB_2) became superconducting at 39°K.
22. Shachtman, op. cit., p. 194.
23. Ibid., p. 141.
24. The facts relating to He II come from K. Mendelssohn, *The Quest for Absolute Zero*, Taylor and Francis, 1977, chap. 11.
25. This was soon reduced to 0.000000000002°K – more than a million times colder than interstellar space.

26. It is uncertain who first used this term: Shachtman, op. cit., p. 210, attributes it to Gorter and Casimir in 1934, while Mendelssohn, op. cit., p. 251, notes its first appearance in an article by Kapitsa in *Nature* in 1938. Kapitsa is followed here, because of the contribution of his experiments to establishing the phenomenon.
27. Mendelssohn, op. cit., p. 254: the author himself participated in this experiment.
28. After E. Fermi (1901–54) and P. A. M. Dirac (1902–84).

Chapter 10 – Big science

1. The three quotations above all come from J. Horgan, *The End of Science*, Little, Brown, 1996, pp. 78–9.
2. Cambridge University Press, 1997.
3. As appears on page 351, the Higgs particle has yet to be discovered.
4. Ibid., p. 117.
5. Abacus, 1994.
6. Ibid., pp. 178–9.
7. For this information, I am indebted (12 July 2001) to Prof. Tom Kibble of Imperial College, one of the five prizewinners, and a key-player in the Higgs field for more than thirty years.
8. This is also known as the GWS model, after the American physicists Sheldon Glashow, Steven Weinberg and Abdus Salam, who proposed it at the end of the 1960s and became Nobel Laureates in 1979.
9. The origins of this term go back to the Bose–Einstein statistics introduced in Chapter 7.
10. By Carlo Rubbia and Simon van der Meer, both working at CERN. They became Nobel Laureates in 1984.
11. At the Bell Telephone Laboratory.
12. At Texas Instruments (TI), by Jack Kilby. TI's $150,000,000 silicon manufacturing research centre is named after him.
13. Development is now centred at the US National Institute of Standards and Technology at Gaithersburg, MD.
14. See M. Gell-Mann, op. cit., p. 204.
15. Charpak finally won a Nobel prize in 1992.
16. B. Kibble and T. Hartland, 'A Quantum Leap to Better Standards', *New Scientist*, 5 May 1990, p. 48.
17. On 29 March 2001, a group of American scientists and one German reported an all-optical atomic clock based on the 1.064 petahertz transition of a single trapped positive ion of ^{199}Hg, the third most common isotope of mercury. This, with the help of a mode-locked femtosecond laser, produces an output consisting of pulses at a rate of 1 gigahertz. The actual event governing the operation of the clock occurs 7,000,000,000,000,000 times per second. This, the equivalent of the number of seconds since the big bang, provides the measure of the new clock's accuracy – no more than 1 second out since the beginning of the universe. (See *Science* online, 12 July 2001).
18. $e = 1.60217635 \times 10^{-19}$ coulombs and $h = 6.62606821 \times 10^{-34}$ joule-seconds.
19. The second is the time taken by a caesium-113 atom to vibrate 9,192,631,770 times, and a metre is the distance light travels in a vacuum in 1/299,792,458th of a second.

Bibliography

Atkins, P. *The Periodic Kingdom*, Weidenfeld and Nicolson, 1995.

Baird, D., et al. (eds.), *Heinrich Hertz, Classical Physicist, Modern Philosopher*, Kluwer, 1981.

Bajac, Q., et al., *Dans le champ des étoiles: Les photographes et le ciel 1850–2000*, Musée d'Orsay, Paris, 2000.

Beaglehole, J. C., *The Life of Captain James Cook*, Stanford University Press, 1974.

Bentley, E. C., *Clerihews Complete*, Werner Laurie, 1951.

Berry, A. J., *Henry Cavendish: His Life and Scientific Work*, Hutchinson, 1960.

Biot, J.-B., *Traité de physique*, Paris, 1802.

Bondi, H., *Assumption and Myth in Physical Theory*, Cambridge University Press, 1965.

Bowers, B., *A History of Electric Light & Power*, Science Museum 1982.

Bowers, B., 'Faraday, Wheatstone and Electrical Engineering', in D. Gooding and F. A. J. L. James (1985), pp.163–174.

Boyle, R., *The Sceptical Chymist*, London, 1661.

Breedlove, D. E., and Laughlin, R. M., *The Flowering of Man: A Tzotzil Botany of Zinacantan*, Smithsonian Institution Press, 1993.

Brock, W. H., *The Fontana History of Chemistry*, Harper Collins, 1992.

Brock, W. H., 'Dalton versus Prout: The Problem of Prout's Hypothesis', in Caldwell (1968).

Broglie, L. de, *Thèses: Recherches sur la théorie de quanta*, Masson, Paris, 1963.

Broglie, M. de, *Les rayons x*, Paris, 1922.

Brown, A., *The Neutron and the Bomb*, Oxford University Press, 1997.

Bryant, J. H., 'Heinrich Hertz's Experiments and Experimental Apparatus: His Discovery of Radio-waves and the Delineations of their Properties', in David Baird et al., op. cit., pp. 39–58.

Buffon, Comte de, *Histoire naturelle*, Paris, 44 vols., 1749–67.

Burke, J. G., *Origins of the Science of Crystals*, University of California Press, 1966.

Butterfield, H., *Origins of Modern Science, 1300–1800*, G. Bell & Sons, 1965.

Calder, N., *The Manic Sun*, Pilkington Press, 1997.

Caldwell, D. S. L. (ed.), *John Dalton and the Progress of Science*, Manchester University Press, 1968.

Cantor, G. N., 'Reading the Book of Nature: the Relation between Faraday's Religion and Science', in D. Gooding and F. A. J. L. James (1985), pp. 69–81.

Carnot, S., *Reflections on the Motive Power of Fire*, London, 1824.

Chadwick, J., *The Decipherment of Linear B*, Cambridge University Press, 1959.
Childs, H., *An American Genius: The Life of Ernest Orlando Lawrence*. E.P. Dutton, 1968.
Chomsky, N., *Syntactic Structures*, Mouton.
Cohen, I.B., *Benjamin Franklin, Scientist and Statesman*, Charles Scribner's Sons, 1975.
Collins, H., and Pinch, T., *The Golem: What You Should Know about Science*, Cambridge University Press, 2nd edn., 1998.
Comte, A. *Cours de philosophie positive*, Paris, 1835.
Copernicus, N., *De Revolutionibus Orbium Coelestium*, 1542.
Craanen, T. (ed.), *Tractatus Physico-Medicus de Homine*, Leiden, Petrum van der Aa, 1689.
Crump, T., *The Anthropology of Numbers*, Cambridge University Press, 1991.
Crump, T., *Man and his Kind*, Darton, Longman & Todd, 1973.
Crump, T., *Solar Eclipse*, Constable, 1999.
Crystal, D. (ed.), *The Cambridge Encyclopedia of Language*, Cambridge University Press, 1987.
Dalton, J., *Meteorological Observations and Essays*, Manchester, 1793.
Dampier, W. C., *A History of Science and its Relations with Philosophy and Religion*, Cambridge University Press, 1942.
Danzer, K., *Robert W. Bunsen und Gustav R. Kirchhoff: Die Begraünder der Spektralanalyse*, B. G. Teubner, Leipzig, 1972).
Darwin, C., 'The Descent of Man and Selection in relation to Sex', in P.H. Barrett and R.B. Freeman (eds.), *The Works of Charles Darwin*, vols. 21–3, William Pickering, London, 1989.
Davy, H., 'Some Experiments on the Combustion of the Diamond', *Philosophical Transactions of the Royal Society*, vol. 105, 1815, pp. 99–100.
Davy, H., 'Some Experiments on a Solid Compound of Iodine and Oxygene', *Philosophical Transactions of the Royal Society*, vol. 105, 1815, pp. 203–19.
Davy, J, *Memoirs of the Life of Sir Humphry Davy*, Longman.
Dewhurst, D., and Hoskin, M., 'The Message of Starlight: The Rise of Astrophysics', in Hoskin (1997), pp. 256–343.
Dibner, B., *Alessandro Volta and the Electric Battery*, Franklin Watts, New York, 1964.
Dictionary of Science and Technology, San Diego Academic Press, 1992.
Donovan, A., *Antoine Lavoisier*, Cambridge University Press, 1993.
Ferguson, K., *Measuring the Universe: The Historical Quest to Quantify Space*, Headline Books, 1999.
Forgan, S., 'Faraday – From Servant to Savant: The Institutional Context', in D. Gooding and F.A.J.L. James (1985), pp. 52–67.
Francis, J. de, *Visible Speech: The Diverse Oneness of Writing Systems*, University of Hawaii Press, 1989.
Franklin, B., *Poor Richard: An Almanack*, 1727.
Franklin, B., *Experiments and Observations on Electricity, Made at Philadelphia in America*, 1751.
Friedman, H., *The Astronomer's Universe*, W.W. Norton, 1998.
Gabbey, A., 'Huygens and Mechanics', in H.J.M. Bos et al. (eds.), *Studies on Christiaan Huygens*, Lisse, Swets & Zeitlinger, 1980, pp. 166–199.
Galilei, G., *The Assayer*, Florence, 1623.
Galilei, G., *Dialogues Concerning Two New Sciences*, appeared in 1638.
Galison, P., and Hebly, B. (eds.), *Big Science: The Growth of Large Scale Research*, Stanford University Press 1992.
Gardner, M., *The Night is Large*, Penguin, 1997.
Gell-Mann, M., *The Quark and the Jaguar*, Abacus, 1994.

Gilbert, W., *On the Magnet*, 1600.
Gooding, D., and James, F. A. J. L. (eds.), *Faraday Rediscovered*, Macmillan 1985.
Goudsblom, J., *Fire and Civilization*, Penguin, 1994.
de Graaf, R., *De Virorum Organis Generationi Inservientibus*, Amsterdam, 1669.
de Graaf, R., *De Mulierum Organis Generationi Inservientibus Tractatus Novus*, Amsterdam, 1673.
Grattan-Guinness, I., *The Fontana History of the Mathematical Sciences*, Fontana, 1997.
Gribbin, J., *Companion to the Cosmos*, Orion Books, 1996.
Gribbin, J., and Gribbin, M., *Halley in 90 Minutes*, Constable, 1997.
Guedj, D., *Le mètre du monde*, Editions du Seuil, 2000.
Hall, A. R., *A Brief History of Science*, New York, 1964.
Halley, E., *Catalogus Stellarum Australium*, London, 1679.
Halley, E., *A Synopsis of the Astronomy of Comets*, London, 1705.
Hardy, G. H., *A Mathematician's Apology*, Cambridge University Press, 1940.
Harvey, W., *Exercitationes de Generatione Animalium*, typis Du-Gardianis, London, 1651.
Harwood, R., *Chemistry*, Cambridge University Press, 1998.
Haüy, R. J., *Traité de mineralogie*, Paris, 1806.
Helmholtz, H. von, *Uber die Erhaltung der Kraft*, Ostwald's klassiker, 1847.
Hoffman, D., 'Heinrich Hertz and the Berlin School of Physics', in David Baird et al., op. cit., 1981, pp. 1–8.
't Hooft, G., *In Search of the Ultimate Building Blocks*, Cambridge University Press, 1997.
Hooke, R., *Philosophical Experiments and Observations of the Late Doctor Robert Hooke*, Frank Cass, London, 1967 (reprint).
Horgan, J., *The End of Science*, Little, Brown, 1996.
Hoskin, M. A., *William Herschel and the Construction of the Heavens*, Oldbourne, 1963.
Hoskin, M. A. (ed.), *The Cambridge Illustrated History of Astronomy*, Cambridge University Press, 1997.
Huygens, C., 'De Iis Quae Liquido Supernatant', 1650, *Oeuvres Complètes*, 1908.
Huygens, C., 'De Motu Corporum ex Percussione', *Opuscula Postuma*, 1703.
Huygens, C., *Treatise on Light*, 1690.
Icke, V., *Wasknijpers*, NRC Handelsblad, 17 March 2001.
Jancey, M., *Mappa Mundi: The Map of the World*, Hereford Cathedral Enterprises, 1995.
Jungk, R., *Brighter than a Thousand Suns: A Personal History of the Atomic Scientists*, Harcourt, Brace, 1958.
Kekulé, F. A., *Lehrbuch der Chemie*, 1868.
Kepler, J., *Cosmographic Mystery*, 1596.
Kibble, B., and Hartland, T., A Quantum Leap to Better Standards, New Scientist, 5 May 1990, pp. 48–51.
Knight, D. M., 'Davy and Faraday: Fathers and Sons', in D. Gooding and F. A. J. L. James (1985), pp. 33–49.
Knight, D. M., *Humphry Davy: Science and Power*, Blackwell, 1992.
Kragh, H., *Dirac: A Scientific Biography*, Cambridge University Press, 1990.
Kuhn, T. S., *The Copernican Revolution*, Harvard University Press, 1957.
Latil, P. de, *Enrico Fermi: The Man and his Theories* [translated from French], Souvenir Press, 1965.
Lavoisier, A., *Eléments de chimie*, Paris, 1789.

Leeuwenhoek, A. van, 'Concerning the Various Figures of the Salts Contained in the Several Substances', *Philosophical Transactions of the Royal Society*, vol. 15, 1685, p. 1073.

Leeuwenhoek, A. van, 'Concerning the Figures of the Salts of Crystals', *Philosophical Transactions of the Royal Society*, vol. 24, 1705, pp. 1906–17.

Lévi-Strauss, C. *Le cru et le cuit*, Librairie Plon, 1964.

Lieberman, P., *Uniquely Human: The Evolution of Speech, Thought and Selfless Behaviour*, Harvard University Press, 1991.

Lloyd, G. E. R. (ed.) *Hippocratic Writings*, Penguin, 1978.

Major, E., *June 8, 2004 – Venus in Transit*, Princeton University Press, 2000.

Marinacci, M., *Linus Pauling in His Own Words*, Touchstone 1995.

Maxwell, J. C., *Treatise on Electricity and Magnetism*, 1873.

Mendelssohn, K., *The Quest for Absolute Zero*, Taylor and Francis, 1977.

Menninger, K., *Number Words and Number Symbols: A Cultural History of Numbers*, MIT Press, 1969.

Nagel, E., *The Structure of Science*, New York, 1961.

Nersessian, N. J., 'Faraday's Field Concept', in D. Gooding and F. A. J. L. James (1985), pp. 175–187.

Newton, I., *Philosophiae Naturalis Principia Mathematica*, London, 1687.

Oberth, H., *The Rocket into Interplanetary Space*, 1923.

Ohnuki-Tierney, E., *Illness and Culture in Contemporary Japan: An Anthropological View*, Cambridge University Press, 1984.

Ong, W., *Orality and Literacy: The Technologizing of the Word*, Methuen, 1982.

Oppenheim, A. L., *Ancient Mesopotamia: A Portrait of a Dead Civilization*. University of Chicago Press, 1977.

Ordway, F. I., and Sharpe, M. R., *The Rocket Team*, Thomas Y. Crowell, 1979.

Pais, A. *Inward Bound*, Clarendon Press, 1986.

Pais, A., *Niels Bohr's Times , in Physics, Philosophy and Polity*, Oxford, 1991.

Pisarzhevsky, O. N., *Dimitry Ivanovich Mendeleyev: His Life and Work*, Foreign Languages Publishing House, Moscow, 1954.

Powers, T., *Heisenberg's War: The Secret History of the German Bomb*, Penguin, 1994.

Quinn, S., *Marie Curie: A Life*, Simon & Schuster, 1995.

Rhodes, R., *The Making of the Atomic Bomb*, Penguin, 1986.

Ronan, C. A., and Needham, J., *The Shorter Science and Civilization in China*, Cambridge University Press, 1981.

Ruestow, E. G., *The Microscope in the Dutch Republic: The Shaping of Discovery*, Cambridge University Press, 1996.

Rumford, Count, 'An Experimental Inquiry concerning the Source of Heat Excited by Friction', *Proceedings of the Royal Society*, 1798.

Russell, B., *History of Western Philosophy*, George Allen & Unwin, 1946.

Russell, C. A., 'Berzelius and the Development of Atomic Theory', in D. S. L. Caldwell (1968), pp. 259–60.

Rutherford, E. *The scattering of Alpha- and Beta-rays and the Structure of the Atom*, Manchester Literary and Philosophical Society, 1911.

Rutherford, E., *Radioactive Substances and Their Radiations*, Cambridge University Press, 1913.

Sayers, D., and Eustace, R., *The Documents in the Case*, Gollancz, 1934.

Schama, S. *The Embarrassment of Riches: An Interpretation of Dutch Culture in the Golden Age*, Collins, 1987.

Shachtman, T., *Absolute Zero and the Conquest of Cold*, Houghton Mifflin, 1999.

Sharratt, M., *Galileo: Decisive Innovator*, Cambridge University Press, 1994.

Smart, W. M., *A Textbook of Spherical Geometry*, 6th edn., Cambridge University Press, 1977.

Snow, C. P., *The Physicists*, Papermac, 1982.
Sobel, D., *Longitude: The True Story of a Lone Genius Who Solved the Greatest Scientific Problem of His Time*, Fourth Estate, 1995.
Söderbaum, H. G., *J. J. Berzelius; Autobiographical Notes*. Baltimore, 1934.
Stuewer, R. H., *The Compton Effect: Turning Point in Physics*, Science History Publications, 1975.
Stuhlinger, E., and Ordway, F. I., *Wernher von Braun*, Krieger Publishing Company, 1994.
Swammerdam, J., *Historia Insectorum Generalis*, Amsterdam, 1669.
Thackray, A., *John Dalton: Critical Assessment of His Life and Science*, Harvard University Press, 1972.
Toulmin, S., *The Philosophy of Science*, New York, 1960.
Turner, G. L'E., *Scientific Instruments 1500–1900: An Introduction*. University of California Press, 1998.
Ulam, S., *Adventures of a Mathematician*, Charles Scribner's Sons, 1976.
Urey, H. C., 'Dalton's Influence on Chemistry', in D. S. L. Caldwell (1968).
Vygotsky, L. S., *Thought and Language*, MIT Press, 1962.
Watts, I., *Improvement of the Mind*, 1809.
Weisheipl, J. A. (ed.), *Albertus Magnus and the Sciences*, Toronto Pontifical Institute of Medieval Studies, 1980.
Westfall, R. S., *The Life of Isaac Newton*, Cambridge University Press, 1993.
Wheaton, B. R., *The Tiger and the Shark: Empirical Roots of Wave-Particle Dualism*, Cambridge University Press, 1991.
Whewell, W., *History and Philosophy of the Inductive Sciences*, Cambridge, 1837–60.
Whitehead, A. N., *Science and the Modern World*, Cambridge University Press, 1946.
Williams, B., *Descartes: The Project of Pure Inquiry*, Pelican Books, 1978.
Williams, L. P., 'Faraday and Ampère: A Critical Dialogue', in D. Gooding and F. A. J. L. James (1985), pp. 83–104.
Willis, R. (ed.), *Signifying Animals: Human Meaning in the Natural World*, Routledge, 1994.
Wilson, D., *Rutherford: Simple Genius*, Hodder & Stoughton, 1983.
Wrangham, R. W., et al., 'The Raw and the Stolen, Cooking and the Ecology of Human Origins', *Current Anthropology*, vol. 40, 1999, pp. 567–594.

Glossary

Abacus an elementary calculator, of great antiquity and still widely used in the Far East, operated by sliding beads along parallel wires.

Aberration in optics, a distortion of the image formed by a lens or mirror, either because of its curvature, or because of differences in wavelength of the light transmitted. From the nineteenth century, both these types, *spherical* and *chromatic*, have largely been corrected in optical instruments.

Absolute zero at this temperature, 0°K (kelvin), the lowest theoretically attainable, atoms and molecules are at zero-point energy (making the gaseous state impossible). This is the basis for the standard temperature scale in which 0°C is equivalent to 273.15°K.

Académie des Sciences in France, the official institution for protecting and advancing the interests of science. Its members are all elected for their exceptional scientific achievements.

Accelerator in high-energy physics, any apparatus designed to increase the momentum of fundamental particles for experimental purposes.

Air the basic constituent of the earth's atmosphere, known, since the end of the eighteenth century, to consist mainly of oxygen and nitrogen.

Andromeda nebula a nebula in the constellation Andromeda, in which many key astrophysical phenomena have been observed in the past century.

Anode see *electrode*.

Apollo moon programme launched by President Kennedy, undertaken by *NASA*, using for the first time spacecraft to explore the moon. The programme ended with six landed missions, the first, *Apollo 11*, in July 1969, and the last, *Apollo 17*, in December 1972.

Atomic nucleus the positively charged central core of the atom, consisting of protons and neutrons, discovered by Ernest Rutherford (1871–1937) in 1911.

Atomic number the number of protons in the *atomic nucleus*, which defines every separate element in the *periodic table*.

Atomic pile a large aggregate of *radioactive* elements, embedded in a moderator, designed so as to sustain a continued nuclear *reaction* – as in a nuclear power station.

Atomic weight commonly known as 'relative atomic mass', because it relates the mass per atom of the naturally occurring form of an element to 1/12 that of a carbon-12 atom (taken as the standard of measurement).

Avogadro's hypothesis at the same temperature and pressure, equal volumes of all gases contain an equal number of molecules.

Balmer's formula an elementary algebraic formula for calcuating the light frequencies corresponding to the lines in the visible hydrogen spectrum.

Base metals common metals, such as lead and iron, which deteriorate on exposure to air, moisture and heat.

Benzene the archetypical aromatic compound, C_6H_6, whose molecule has a ring structure (see page 189), fundamental in organic chemistry.

Big bang an explanation of the origin of the universe, at a finite moment in the past, as a result of the explosion of a state of matter of enormous density and temperature.

Binary stars any pair of stars revolving around a common centre of mass, an astronomical phenomenon observed, not necessarily by telescope, in several different forms.

Black-body radiation electromagnetic radiation emitted by a black body, defined by the property that it absorbs and emits almost all the radiation falling on it. Although a perfect black body can only exist as a theoretical concept, the phenomenon, first observed at the end of the nineteenth century, was critical for the formulation of *quantum physics* by Max Planck (1858–1947) and others in the early twentieth century.

Bonding in chemistry the strong attractive force, taking a number of different forms, which holds atoms together in a molecule or crystal.

Bose–Einstein condensate the ultimate state of matter at extremely low temperatures (around 2×10^{-7} °K), at which stage thousands of atoms of a single element become a single superatom, with remarkable physical properties.

Bose–Einstein statistics describes a system of particles obeying the rules of quantum rather than classical mechanics, in which any number can occupy a given quantum state. Such particles are known as *bosons*, in contrast to *fermions*, governed by Fermi–Dirac statistics.

Boson see *Bose–Einstein* statistics.

Boyle's law a particular case of the *gas laws*, governing the relation between mass and volume at a constant temperature.

Bragg equation defines algebraically the angle of incidence at which the intensity of X-rays reflected from a crystal is at a maximum (see page 200).

Bubble chamber based on liquid, typically hydrogen, maintained by pressure slightly above its boiling point. Ionisation radiation can then be observed, because the passage of ionised particles reduces the pressure, leading to the formation of a trail of bubbles.

Caesium atomic clock whose extreme accuracy, based upon the frequency equivalent to the energy difference between two states of the caesium-133 atom, now defines the second in SI units.

Calculus in its original seventeenth-century forms related small increments of a variable x both to change, and the rate of change, in a function $f(x)$ based upon it, and in doing so became indispensable to the development of scientific theory.

Carbon the defining element of all organic compounds, and as such fundamental to life.

Carbon dioxide a colourless, odourless gas, CO_2, present in the earth's atmosphere, and the essential source of carbon for plants, which absorb it by photosynthesis; the supply is continuously replenished by respiration by living organisms, and combustion of fossil fuels, the latter now being responsible for the greenhouse effect.

Cartesian systems describe and allow for the analysis of geometrical forms and theory in algebraic terms, often with the assistance of *calculus*.

Cathode see *electrode*.

Cathode rays streams of electrons emitted from the *cathode* within a vacuum chamber containing both a cathode and an *anode* – a phenomenon providing the experimental basis of much of modern physics.

Cavendish Laboratory since 1873 the physics laboratory of Cambridge University, long renowned for many key discoveries.

Celestial dynamics the classic formulation, associated with Isaac Newton (1642–1727), of the motion of celestial objects in gravitational fields.

Cepheids a wide range of stars, with luminosity varying over a period of 1–50 days, in such a way that the observed brightness is a measure of their distance from the earth.

Ceramics originally produced by the baking of clay, together with other inorganic material, in an irreversible process leading to chemically unreactive end products of extreme stability and a wide range of uses.

CERN originally the Conseil Européen pour la Recherche Nucléaire, now, as the European Laboratory for Particle Physics, a world leader in this field. See also *LHC*.

Chain reaction defined by a series of steps, in which each is the result of the one preceding it, sometimes with a multiplier effect leading to a chemical or nuclear explosion.

Cherenkov radiation a phenomenon, characterised by the emission of blue light from a transparent medium, such as water, as a result of passage of atomic particles at a velocity greater than that of light in that medium.

Cloud chamber used in the classic early days of particle physics, contains supersaturated vapour, which, by condensing in drops on ions, allows the passage of ionising radiation to be observed.

Combustion a chemical reaction between oxygen and some other substance producing heat and light.

Complementarity the use of two different but complementary concepts (e.g. particles vs. waves) to explain *quantum* phenomena.

Compton effect the loss of energy of X- or γ-ray *photons* when they are scattered by free *electrons* (see page 275).

Conservation the principle that the magnitude of a physical property of a system, such as energy, mass or charge, remains unchanged while its distribution within the system is altered.

Copernican astronomy proposed, for the first time, the sun, in place of the earth, as the centre of the solar system. See also *Ptolemaic Astronomy*.

Cosmic rays high-energy charged particles entering the earth's atmosphere, either from the sun or from outer space.

Cosmology the study of the origin and evolution of the universe at large.

Crab nebula a glowing cloud of gas and dust in the constellation Taurus, a remnant of a *supernova* explosion observed in China in the year 1054 (see page 312).

Cryometer any instrument for measuring extremely low temperatures.

Crystal a solid having a regular internal arrangement, based on polyhedra, of atoms, ions or molecules.

Cyclotron an *accelerator* in which charged particles, introduced at the centre, are accelerated along an outward spiral path.

Dalton's atomic theory first proposed a standard atom, which could neither be created nor destroyed, as the fundamental unit defining any element.

Domestication the historical process by which humankind has come to control the number, distribution and selective development of specific plants and animals, primarily for consumption as food.

Doppler effect the observed change in the frequency of a wave as a result of the motion of its source in relation to the observer.

Electric arc historically the first means of using an electric current for lighting, is essentially an incandescent discharge caused by the current between two *electrodes*, in which the *ionisation* of the gap separating them maintains a conductive medium allowing the effect to continue indefinitely.

Electrode a conductor that emits (*anode*) or collects (*cathode*) electrons in a cell, vacuum tube, semiconductor, etc.

Electrolysis the process by which the passage of an electric current through a conducting liquid (the electrolyte), causes the concentration of positive ions at the *anode*, and negative ions at the *cathode*.

Electromagnetism the fundamental phenomenon inherent in the relation between electricity and magnetism.

Electron the elementary negatively charged particle, present in all atoms, grouped in electron shells surrounding the nucleus.

Electron microcope using a beam of *electrons* instead of a beam of light, produces much larger images of the object observed than the classic optical *microscope* (whose resolution is limited by the comparatively long wavelengths of the visible *spectrum*).

Element in chemistry a substance that, because it cannot decompose into simpler substances, is defined by a single atom.

Elementary particles the fundamental constituents of all matter throughout the universe – whose presence (to be distinguished from that of atoms) was first revealed by the discovery of the *electron* in 1897.

Euclidean geometry originated in Euclid's *Elements* about 300 BC, with fundamental axioms relating to plane figures, typically comprising straight lines and circles.

Euratom the European Atomic Energy Commission, belonging to the European Union and devoted to promoting the peaceful use of atomic energy.

Femtochemistry the study of chemical *reactions*, typically the breaking and formation of individual chemical bonds in compounds, which occur within a timespan measured in femtoseconds (10^{-15} seconds).

Fermi–Dirac statistics describes a system of particles obeying the rules of *quantum* rather than classical mechanics, in which only one particle can occupy a given quantum state. Such particles are known as *fermions*, in contrast to *bosons*, governed by *Bose–Einstein* statistics.

Feynman diagrams a space-time diagram illustrating fundamental *quantum* interactions in particle physics.

Field theory governs the forces acting on one body as the result of the presence of another, as in the classical case of Newtonian gravitation. See also *fundamental interactions*.

Fire a continuous form of *combustion* producing, typically, both heat and light.

Fission in nuclear physics the process by which a heavy *nucleus* splits into two parts in a process leading to the emission of two or three *neutrons*, accompanied by a release of energy.

Fraunhofer lines dark lines in the *spectrum* of sunlight caused by the absorption, by elements in the sun's outer surface, of corresponding wavelengths of the radiation emitted from its interior. Their discovery was critical to the development of *spectroscopy*.

Fundamental interactions the four basic interactions between separate bodies – gravitational, electromagnetic, strong and weak – which together account for all forces observed to occur in the universe.

Fusion physics relates to nuclear reactions, of a kind only possible in *plasma* at temperatures of the order of 10^8 °K, in which two lighter atoms combine to form a heavier one, emitting at the same time a subatomic particle accompanied by a release of energy greater than that of comparable chemical reactions by a factor of 10^6.

Gas laws govern the relation between temperature, pressure and volume in ideal gases; see also *Boyle's law*.

Geiger counter devised in 1908 as one of the earliest means of detecting and measuring *ionising* radiation.

Globular cluster a densely packed ball of stars (sometimes counted in millions), occurring in many galaxies including our own.

Gluon a chargeless particle, with no rest mass, visualised as being exchanged between *quarks*, in maintaining the strong interaction between them.

Gravitational constant in Newton's law of gravitation, the multiplier necessary to give the correct quantity for the attraction between two bodies.

Gravity the particular case of the gravitational force (the weakest of the four *fundamental interactions*) operating an any object with mass within the earth's gravitational field.

Hadron the class of subatomic particles (including *protons* and *neutrons*) related to the strong interaction (and believed to consist of *quarks*).

Half-life the time taken for *radioactive* decay to transform half the radioactive particles present in any aggregate.

Hall effect the production, by a current flowing through a strong transverse magnetic field, of a difference in electric potential at right angles to both current and the field; see page 357.

Heisenberg's uncertainty principle discovered in 1927, according to which there is a quantum limit, $h/4\pi$, to the accuracy to which both the position and the momentum of an elementary particle can be determined. See also *Planck's constant*.

Helium a colourless, inert gas, with no known compounds and the lowest boiling point of all elements.

Hippocratic oath the fundamental ethical code governing the behaviour of doctors since ancient times.

Homo sapiens sapiens the last stage in the evolution of humankind, occurring some 100–120 thousand years ago, and characterised by the use of language.

H–R diagram or Hertzprung–Russell diagram, in astronomy a kind of graph on which the temperature of each star is plotted against its absolute *magnitude*.

Hubble space telescope operating from a satellite orbiting the earth (built jointly by *NASA* and the European Space Agency) and producing superb images from above the earth's atmosphere.

Hydrocarbon any one of a vast range of chemical compounds containing only *hydrogen* and *carbon*.

Hydrogen the first and lightest element in the *periodic table*, and the only one with no neutrons in the nucleus of its most common *isotope*, is a highly reactive, colourless, odourless gas, present in water and all organic compounds

Incandescence light emitted as a result of raising any substance (such as a lamp filament) to a sufficiently high temperature.

Induction in physics, the change in the state of a body as the result of its being placed in a field, typically *electromagnetic*.

Inertia the inherent property of matter causing it to resist any change in its motion.

Interference the interaction of wave motions so as to produce a resultant wave whose form characterises the original waves; in optics, a phenomenon strongly supporting the wave theory of light.

Inverse square law requires that the strength of physical forces (e.g. gravity) operating over a given distance is inversely proportional to the square of that distance.

Ion an atom whose shell has lost or gained one or more electrons: the former case is positive *ionisation*, the latter, negative. The phenomenon, which is extremely common in many different contexts, is fundamental in many different branches of science, from astronomy to neurology.

Ionisation see *ion*.

Isotope defines an atom of a given element element according to the number of *neutrons* in the nucleus, so allowing any element to have a number of different isotopes (some with special properties, such as *radioactivity*).

JET (Joint European Torus) essentially the *tokamak* at the EU laboratory at Culham, near Oxford.

Josephson junction a device applying the electrical properties of matter at the ultra-low temperatures to measuring electric currents at the level of single electron pairs (see page 343).

Kepler's laws stated at the beginning of the seventeenth century, relate the dimensions of time and length in the elliptical orbits of planets in the solar system (see page 43).

Large Electron Positron (LEP) collider the original particle *accelerator* in the 26 kilometre long circular tunnel at *CERN*, now being replaced by the Large Hadron Collider (LHC), due to become operational in 2005.

Lens a curved, ground, polished piece of transparent material (typically glass) used as a key component in optical instruments.

Lepton any one of a class of elementary particles, including *electrons* and *neutrinos*, subject to both the electromagnetic and the weak *fundamental interactions*.

Leyden jar a glass jar, with a layer of metal foil on both the inside and outside, developed as a capacitor, with the power to accumulate an electric charge, in the eighteenth century.

Light scientifically defined by the range of electromagnetic frequencies (the visible spectrum) to which the eye is sensitive.

Los Alamos originally the site, in New Mexico (USA), of the scientific research of the Manhattan Project, but now operating as one of the *US National Laboratories*.

Magellanic Clouds two galaxies, only visible from the southern hemisphere, relatively close to our galaxy, and including groups of stars all at roughly the same distance from us.

Magnitude a measure of the luminosity of stars. Apparent magnitude is based upon their actual appearance, but fails to take into account the key factor of distance, which is critical in determining absolute magnitude.

Main sequence is the class of stars (including the sun), represented by a band in the *H–R diagram*, which shine as the result of fusion reaction of *hydrogen* and *helium*.

Manhattan Project joint British–American operation that in the Second World War led to the first ever development of atomic weapons.

Mauna Kea an extinct volcano, the highest mountain in Hawaii, chosen because of extremely favourable atmospheric conditions as the location for a remarkable number of different types of observatory.

Maxwell's equations four equations, fundamental in classic electrodynamics since first stated by James Clerk Maxwell in the 1860s, which describe mathematically the changes in an *electromagnetic* field over time.

Mechanical equivalent of heat the ratio between equivalent units of mechanical and thermal energy, fundamental to the conversion of energy as investigated experimentally during the nineteenth century. See also *thermodynamics*.

Messier catalogue compiled by Charles Messier in the late eighteenth century, and the first historical attempt to list and code faint astronomical objects systematically.

Metrology the science of measurement, in particular for scientific purposes requiring extreme accuracy.

Microorganisms a wide heterogeneous class, including bacteria, protozoa, etc., that can only be observed with the help of a microscope (whose invention in the seventeenth century led to the first discovery).

Microscope any device for obtaining a magnified image, beyond the range of normal vision, of small objects.

Molecular biology focused on large molecules, particularly proteins, DNA and RNA, characteristic of living organisms.

Molecule the basic unit in chemistry, typically comprising separate elements, defined as the smallest part of a compound that can take part in a chemical reaction.

NASA National Aeronautics and Space Administration, in the USA an independent government agency, with headquarters in Washington, with eight field centres across the country, each specialising in different aspects of air/space travel and exploration.

Natural philosophy until well into the nineteenth century, the generic term for all the exact sciences, except for optics and pure mathematics.

Nebula from the Latin for 'cloud', refers to any patch of light in the sky, not visible in greater detail, but now known to represent different phenomena, such as aggregates of stars or clouds of gas.

Neutrinos fundamental uncharged particles, capable only of relating to other particles by the weak interaction, emitted in vast quantities by the sun and stars (see page 321), and able to penetrate massive solid aggregates with almost zero interaction, so making detection extremely difficult.

Neutron the fundamental uncharged particle comprising the atomic nucleus, but in isolation subject to β-decay into a proton and an electron.

Newtonian dynamics based on *Newton's laws of motion* (formulated on the implicit assumption – substantially true at low energy levels – that a body's mass does not vary with its speed).

Newton's laws of motion three in number (see page 121), are the foundation of *Newtonian dynamics*.

Nitrogen a colourless gaseous element, constituting some three-quarters of the earth's atmosphere, and now known to be an essential constituent of proteins and nucleic acids in living organisms.

Noble gases the gaseous elements helium, neon, argon, krypton, xenon and radon, belonging to group 18 in the *periodic table*, all representing the termination of a period by closing the electron shell. The result is a capacity to form compounds so restricted that the first instances were only discovered in 1962.

Notation the use of special signs, or the idiosyncratic use of common signs, such as letters in chemical compounds, to designate entities with a special significance within the related branch of science.

Nova a short-lived bright visible object, or 'new' star, the result of an explosion of a faint star, possibly never observed before that event.

Nucleus the positively charged central core of the atom, consisting of *protons* and (except for *hydrogen*) *neutrons* bound together by the *strong interaction*.

Optics the classic science of light, particularly as propounded by Isaac Newton.

Oxidation originally a chemical reaction with oxygen, as opposed to one involving the loss of oxygen, or reduction. Now it defines a number, positive or negative, defined by the number of electrons over which an element has gained or lost control, as a result of it forming part of a compound. Since the process involves the transfer of electrons between the outer shells, it occurs only in *reactions* involving *ions* – which are, however, extremely common.

Oxygen a colourless, odourless, gaseous element, making up nearly half the earth's crust, and more than a quarter of its atmosphere, indispensable for all living organisms which acquire their energy by oxidizing nutrients to *carbon dioxide* and water.

Parallax in astronomy the apparent displacement of the position of a celestial object as a result of a change in the position of the observer, e.g. as the result of the motion of the earth within the solar system. This was the basis of the earliest methods used successfully to estimate the distance of stars.

Particle see *elementary particles*.

Particle accelerator see *accelerator*.

Pauli's exclusion principle in *quantum* mechanics excludes the possibility of there being two systems of particles (e.g. *electrons* in an atom), with an identical set of *quantum numbers* (see page 236).

Periodic table since about 1870, systematically arranges the elements, in rows and columns, according to the unit increase in *atomic number*, taking into account common properties (such as degree of reactivity) to determine the division into rows and columns. See Appendix A.

Phlogiston a hypothetical substance which, during the seventeenth and eighteenth centuries, was believed to be contained in all combustible susbtances, to be released in the process of combustion.

Phosphorus a common, highly reactive, combustible element, discovered in the seventeenth century, and essential for the support of life.

Photon (from the Greek *phōs*, 'light') the *quantum* of *electromagnetic* radiation, travelling at the speed of light with a rest mass of zero, and essential for explaining *inter alia* the photoelectric effect and non-wave phenomena.

Photosynthesis the chemical process, dependent upon sunlight, by which plants synthesise organic compounds from *carbon dioxide* and water.

Piezoelectricity the result of mechanical stress creating a difference in electric potential between opposing faces of a *crystal*; the reverse effect follows the creation of electric potential and, with an alternating electric field in phase with the natural elastic frequency of the *crystal*, provides the basis for such instruments as the quartz clock.

Pile a structure created by the aggregation of separate reacting elements, to increase the energy supplied, as with the electric current coming from the *voltaic pile*.

Place-value system any system of numeration in which the number represented depends not only on the numerals, 0, 1, 2, etc., but also on their place, representing a given power of the base number (e.g. 10 in Arabic numerals) in the form presented, so that e.g. $27 = 2 \times 10 + 7$.

Planck's constant h, the fundamental irreducible constant of *quantum physics*, defined so that $E = h\nu$, where E is a *quantum* of energy and ν is its frequency.

Plasma in high-energy physics, the state of matter, generally at extremely high temperatures, characterised by a high level of *ionisation*, abundant free *electrons*, leading to near impossibility of forming chemical bonds. The greater part of the material universe consists of plasma.

Polarisation confines the vibrations of a transverse wave to one direction. In optics, the phenomenon – occurring in a number of different ways – in which light waves are only transmitted in one plane.

Proton the particle within every atomic *nucleus* which accounts for its positive charge.

Proton–proton reaction the means by which energy is generated by nuclear fusion reactions, involving *hydrogen* and *helium*, within main sequence stars with mass comparable to or less than that of the sun.

Positron the elementary positively charged particle corresponding to the negatively charged *electron*, but only detected in the 1930s.

Prism a block of glass or other transparent material, with flat sides, used for deflecting light rays or dispersing them to different parts of the visible *spectrum*.

Proper motion the apparent motion of stars across the sky in relation to the whole firmament, best observed with powerful telescopes over long periods of time, and explained mainly by the closeness of any star observed to the observer.

Protozoa mainly microscopic elementary one-celled organisms, with asexual reproduction, abundant in marine, fresh water and moist terrestrial habitats.

Ptolemaic astronomy the ultimate form of the classic astronomical theory locating the earth at the centre of the solar system. See also *Copernican astronomy*.

Pulsar rapidly rotating neutron star, transmitting regularly timed pulses of *electromagnetic* waves at radio frequencies, capable of being detected by radio telescopes.

Pyrometer any instrument for measuring extremely high temperatures.

Quantum the minimum amount, proved to exist by Planck, Einstein, and others, by which certain physical properties, such as energy and momentum, can change within a given system.

Quantum chromodynamics describes the *strong interaction* in terms of *quarks* and antiquarks and the exchange of *gluons* (see page 349).

Quantum electrodynamics explains the interaction between electromagnetic radiation and charged matters in terms of *quantum* theory.

Quantum numbers characterise an atom in terms of *quantum* physical properties of the *electrons* in the different shells (see page 236).

Quantum physics is the study physical phenomena at the level of *quanta*, where the fact that relevant properties do not vary continuously, but in *quantum* leaps, requires a quite different theoretical approach.

Quark together with the antiquark, the basic component in the structure of *hadrons*, and subject to the *strong interaction* occurring by means of *gluon* exchange.

Quasar the highly energetic core of an active galaxy, first discovered by *radio astronomy*.

Radical in chemistry the stable part of a substance that retains its identity through a series of *reactions* even though a compound (see page 187).

Radioactivity the spontaneous disintegration of certain 'radioactive' atomic nuclei by the emission of α- or β-particles, or γ-radiation.

Radio astronomy the study of the universe by monitoring radiofrequency *electromagnetic* waves.

Reaction in chemistry, a change, spontaneous or deliberate, in one or more elements or compounds such that a new compound is produced.

Reactive series an ordered list of generally metallic elements based on their propensity to react with other substances, and indicating therefore their relative stability.

Reactor in nuclear physics a device in which a *chain reaction* based on atomic fission is sustained and controlled in order to produce new energy, radio-*isotopes* or new nuclides.

Reflection the return of all or part of a beam of particles or waves (typically of light) as a result of encountering the interface between two media.

Refraction the change in direction of a beam of particles or waves (typically of light) as it passes from one medium into another. Where the latter is a crystal, the beam may split into two components, each with distinctive optical properties such as *polarisation*: this is double refraction.

Relativity theories propounded by Albert Einstein (1879–1955) to account for discrepancies in Newtonian dynamics occurring with very-high-speed relative motion. The special theory is focused on light, the general theory on gravitation.

Resonance an oscillation of an atomic, acoustic, electric, mechanical, or other system at its natural frequency of vibration, enabling a large output to be produced by a comparatively small input.

Rocket a projectile, with a suitable aerodynamic profile, driven by the rapid *combustion* of specially designed fuel in a closed chamber, with an outlet opposite to the direction of motion.

Royal Institution a learned scientific society, founded in London in 1799, whose laboratories became Britain's first research centre.

Royal Society a scientific institution, in continuous existence since its foundation in 1660, whose members, known as 'Fellows', are elected on the basis of their original contribution to the advancement of science.

Rutherford–Bohr atom the standard model of the atom, including its *electron* shells, as established by Ernest Rutherford (1871–1937) and Niels Bohr (1885–1962) in the years following the discovery of the atomic *nucleus* in 1911 (see page 229).

Scintillation counter counts atomic events occurring in the laboratory by recording the flashes of light emitted by an excited atom reverting to its ground state after having been excited by a subatomic particle.

Semiconductor a crystalline solid, with conductivity substantially higher than that of an isulator and lower than that of a conductor, and electrical properties that ensure that according to input the flow of current is either enhanced or inhibited – a property extremely useful in computers and amplification systems.

SI units (Système International d'Unités), with the metric system as starting point, internationally recognised for all scientific purposes, and consisting of seven base units and two supplementary, with eighteen more derived from them.

Silicon the second most abundant element in the earth's crust, typically a component of sand and derivatively, glass, but now used extensively as the basis for *semiconductors*.

Snel's law defines, on the basis of elementary trigonometry, the relation between the angle at which light is incident on the interface between two refracting media and that of the same light after refraction.

Solar physics relates to distinctive physical phenomena both inside and on the surface of the sun, and their outside impact, particularly upon the earth.

Spectroscopy the use and theory of methods for producing and analysing *spectra*, typically of light, for discovering and controlling the distinctive properties of matter giving rise to particular spectral phenomena.

Spectrum originally the breakdown of rays of light according to different wavelengths of which they are composed, and by extension to all electromagnetic waves, and by further extension to any property, such as the charge to mass ratio of different ions capable of being ordered on a numerical scale, as is achieved by a mass spectrograph.

Standard model in particle physics, the accepted model of the interaction of elementary particles, classified according to the components listed in the table in Appendix B.

States of matter in classical physics and chemistry, solid, liquid and gas, but now, rock, ice, gas and *plasma*.

Sunspots black spots on the sun, first observed through Galileo's telescope in 1610, and still significant in *solar physics*.

Supercollider the ultimate *accelerator*, such as will be represented by the LHC at *CERN*, for producing the highest possible energy levels in charged particles.

Supernova a star exploding at 10^{10} times normal energy levels as a result of using up all its available nuclear fuel, so that with the light emitted it dominates the whole galaxy for a short period of time.

Superstring the ultimate unified theory of *fundamental interactions*, characterised by a length scale of 10^{-35} metres, and a corresponding energy scale of 10^{19} gigaelectronvolts, both beyond the range of any possible observation.

Telescope any device for obtaining a magnified image, beyond the range of normal vision, of distant objects. Originally an optical instrument, telescopes for astronomical use have now been developed to operate at wavelengths far outside the visible *spectrum*.

Thermocouple an electrical device, based on two wires of different metals, that can be used as a thermometer by measuring the electric potential arising when the two ends, joined together, are at different temperatures.

Thermodynamics the general study of the transfer of heat linked to the conversion of energy and its availability to do work; see also the *mechanical equivalent of heat*.

Tokamak a toroidal shaped thermonuclear reactor, containing a plasma of heavy *hydrogen isotopes*, isolated from the internal walls by a strong magnetic field, which also provides the energy for the fusion process designed to create new energy. Efficient reactors of this kind, defined by a sustained energy output greater than the input, have yet to be built, although *JET* is hard at work on this problem.

Torsion balance an instrument for measuring very weak forces by recording very slight displacements of a bar free to rotate about a central axis in a horizontal field. Towards the end of the eighteenth century, this was the first effective instrument for measuring G, the gravitational constant, and hence estimating the mass of the earth.

Transition elements the three main series, belonging the to the d-block in columns 4 to 12 in the *periodic table*, are typical metals, whose chemical properties are defined by unfilled d-orbitals, but this does not define the whole class, which extends also to the f-block; see page 372, note 70.

Trigonometry the mathematical resource, essential in almost every branch of science, defined by functions, varying with the angle between two straight lines; these functions are the six possible ratios between two different sides of a right-angled triangle – hence the name. In fact two, sine and cosine, are sufficient, for the remaining four can be defined in terms of them. The basic relationship is then $\sin^2 \theta + \cos^2 \theta = 1$, which is a trigonometrical statement of Pythagoras' theorem.

United States National Laboratories engaged in fundamental research in different areas, often in collaboration with university science departments, with federal government finance, with Argonne, Lawrence Livermore, Los Alamos, Oak Ridge and Sandia as well-known examples.

United States Space Flight Centers two of the eight Field Centers of *NASA*, Goddard (in Maryland), devoted to astronomy and earth sciences, and Marshall (in Alabama) to launch vehicles and space science.

Valency the combining power of an atom to form molecules, generally compound, expressed as a whole number according to the number of bonds capable of being formed by outer shell *electrons* (see page 188).

van Allen belts sources of intense radiation surrounding the earth, first discovered by James van Allen (1914–) using radiation detectors on Explorer satellites, and the result of high-energy charged particles trapped by the earth's magnetic field.

Voltaic cell the original, devised by Alessandro Volta (1745–1827), as the first means of producing and maintaining an electric potential by chemical means, revolutionised scientific research by providing a viable source of continuous electric current.

Voltaic pile an early form of battery, consisting of *voltaic cells* linked in series.

Wave theories fundamental in many branches of science, deal with periodic (and therefore regular) disturbances in a medium or space that propagate in such a way that the transit of any given point occurs at an observable frequency ν, with a corresponding length l separating successive points at the same phase, the two combining as νl to determine the wave's velocity.

X-ray crystallography a standard technique – particularly useful in the life sciences – using *X-ray diffraction*, for determining the microstructure of crystals or molecules.

X-ray diffraction the result of diffraction by a *crystal* in which the distance separating atoms is comparable with that of the wavelength of *X-rays* (see page 199).

X-rays short-wave electromagnetic radiation in the 10^{-11} to 10^{-9} metre band produced by the bombardment of atoms by high-quantum-energy particles, with the property, useful in medicine and industry, of passing through many forms of matter.

Appendix

A. The Periodic Table of the Elements

1	2	3	4	5	6	7	8	9	10	11	12	13	14	15	16	17	18
				1 H													2 He
3 Li	4 Be											5 B	6 C	7 N	8 O	9 F	10 Ne
11 Na	12 Mg											13 Al	14 S	15 P	16 S	17 Cl	18 Ar
19 K	20 Ca	21 Sc	22 Ti	23 V	24 Cr	25 Mn	26 Fe	27 Co	28 Ni	29 Cu	30 Zn	31 Ga	32 Ga	33 As	34 Se	35 Br	36 Kr
37 Rb	38 Sr	39 Y	40 Zr	41 Nb	42 Mo	43 Tc	44 Ru	45 Rh	46 Pd	47 Ag	48 Cd	49 In	50 Sn	51 Sb	52 Te	53 I	54 Xe
55 Cs	56 Ba	57* La	72 Hf	73 Ta	74 W	75 Re	76 Os	77 Ir	78 Pt	79 Au	80 Hg	81 Ti	82 Pb	83 Bi	84 Po	85 At	86 Rn
87 Fr	88 Ra	89† Ac															

s–block (groups 1–2); Transition elements (groups 3–12); *d*–block (groups 4–12); *p*–block (groups 13–18)

*Lanthanides	57 La	58 Ce	59 Pr	60 Nd	61 Pm	62 Sm	63 Eu	64 Gd	65 Tb	66 Dy	67 Ho	68 Er	69 Tm	70 Yb	71 Lu
†Actinides	89 Ac	90 Th	91 Pa	92 U	93 Np	94 Pu	95 Am	96 Cm	97 Bk	98 Cf	99 Es	100 Fm	101 Md	102 No	103 Lr

f–block (58–71, 90–103)

B. Table of Elementary Particles

Name	Symbol	Mass (MeV)	Charge	N_c	
Spin 1, gauge photons:					
Photons	γ	0	0	1	$U(1)$
Vector bosons for the weak force	Z^0	91,188	0	1	$SU(2)$
	W^+	80,280	+	1	
	W^-	80,280	−	1	
Gluon	A_s	0	0	8	$SU(3)$
Spin 0, Higgs:					
	H^0	>60,000	0	1	
Spin $\frac{1}{2}$, quarks:					
I up	u	5	$\frac{2}{3}$	3	
down	d	10	$-\frac{1}{3}$	3	
II charm	c	1600	$\frac{2}{3}$		
strange	s	180	$-\frac{1}{3}$	3	
III top	t	180,000	$\frac{2}{3}$	3	
bottom	b	4500	$-\frac{1}{3}$	3	
Spin $\frac{1}{2}$, leptons:					
I e-neutrino	ν_e	≈ 0	0	1	
electron	e	0.510999	—	1	
II μ-neutrino	ν_μ	≈ 0	0	1	
muon	μ	105.6584	—	1	
III τ-neutrino	ν_τ	≈ 0	0	1	
tau	τ	1771	—	1	
Spin 2, graviton:					
	g	0	0	1	

Index

Page numbers in bold refer to information contained in the glossary.

B

C

H

I

J

K

Q

R